PHYTOPATHOLOGY IN PLANTS

Research Progress in Botany

PHYTOPATHOLOGY IN PLANTS

Philip Stewart, PhD

*Head, Multinational Plant Breeding Program; Author;
Member, US Rosaceae Genomics, Genetics and
Breeding Executive Committee; North Central Regional Association
of State Agricultural Experiment Station Directors, U.S.A.*

Sabine Globig

*Associate Professor of Biology, Hazard Community
and Technical College, Kentucky, U.S.A.*

Apple Academic Press

TORONTO   NEW JERSEY

Research Progress in Botany Series

Phytopathology in Plants

© Copyright 2011*
Apple Academic Press Inc.

First Published in the Canada, 2011
Apple Academic Press Inc.
3333 Mistwell Crescent
Oakville, ON L6L 0A2
Tel. : (888) 241-2035
Fax: (866) 222-9549
E-mail: info@appleacademicpress.com
www.appleacademicpress.com

First issued in paperback 2021

ISBN 13: 978-1-77463-246-8 (pbk)
ISBN 13: 978-1-926692-80-7 (hbk)

Philip Stewart, PhD
Sabine Globig

Cover Design: Psqua

Library and Archives Canada Cataloguing in Publication Data
CIP Data on file with the Library and Archives Canada

CONTENTS

ACKNOWLEDGMENTS AND HOW TO CITE

The chapters in this book were previously published in various places and in various formats. By bringing these chapters together in one place, we offer the reader a comprehensive perspective on recent investigations into this important field.

We wish to thank the authors who made their research available for this book, whether by granting permission individually or by releasing their research as open source articles or under a license that permits free use provided that attribution is made. When citing information contained within this book, please do the authors the courtesy of attributing them by name, referring back to their original articles, using the citations provided at the end of each chapter.

INTRODUCTION

This volume includes the latest research into the diseases that affect plants. Phytopathology, the scientific study of plant diseases caused by pathogens (infectious diseases) and environmental conditions (physiological factors), is an ever-growing field, with research constantly bringing to light new information. Organisms that cause infectious disease include fungi, oomycetes, bacteria, viruses, viroids, virus-like organisms, phytoplasmas, protozoa, nematodes, and parasitic plants. Chapters within this book bring to light the most recent studies of pathogen identification, disease etiology, disease cycles, economic impact, plant disease epidemiology, plant disease resistance, how plant diseases affect humans and animals, pathosystem genetics, and management of plant diseases. The information provided here will allow readers to stay current with this field's ongoing research and ever-developing knowledge base.

— **Philip Stewart, PhD**

Expanding the Paradigms of Plant Pathogen Life History and Evolution of Parasitic Fitness Beyond Agricultural Boundaries

Cindy E. Morris, Marc Bardin, Linda L. Kinke, Benoit Moury,
Philippe C. Nicot and David C. Sands

Introduction

How do pathogens, whether they parasitize plants or animals, acquire virulence to new hosts and resistance to the arms we deploy to control disease? The significance of these questions for microbiology and for society at large can be illustrated by the recent worldwide efforts to track and limit the emergence of human transmissible strains of swine and avian influenza virus and of multidrug-resistant lines of human pathogenic bacteria, and to restrain the spread of Ug99, a strain of stem rust of wheat. Recent research in medical epidemiology has elucidated the impact of pathogen ecology in environmental reservoirs on the evolution of novel or enhanced pathogen virulence. In contrast, the evolution of virulence in plant pathogens has been investigated from a predominantly agro-centric perspective, and has focused overwhelmingly on evolutionary forces related to interactions with the primary plant host. Here, we argue that current concepts from the field of medical epidemiology regarding mechanisms that lead to acquisition of novel virulence, biocide resistance, and enhanced pathogenic fitness can serve as an important foundation for novel hypotheses about the evolution of plant pathogens. We present numerous examples of virulence traits in plant pathogenic microorganisms that also have a function in their survival and growth in nonagricultural and nonplant habitats. Based on this evidence, we make an appeal to expand concepts of the life history of plant pathogens and the drivers of pathogen evolution beyond the current agro-centric perspective.

Paradigms of Evolution of Virulence in Human "Environmental Pathogens"

The classification of diseases in terms of their epidemiology is a useful starting point for a comparison of plant and human pathogens [1]. In medical epidemiology, anthroponoses are diseases transmitted among humans that have no other known reservoirs for multiplication. Typhoid fever, smallpox, and certain venereal diseases are examples. Zoonoses, such as rabies, lyme disease, severe acute respiratory syndrome (SARS), and avian and swine influenzas, are transmitted to humans from living animals. Sapronoses are diseases transmitted to humans from environmental reservoirs where the pathogen thrives saprophytically. These habitats include soil, water, and decaying plant and animal matter. Examples include Legionnaire's disease, cholera, aspergillosis, and the emerging epidemics of melioidosis (Burkholderia pseudomallei). Human pathogens with saprophytic phases or residing in environmental reservoirs are also referred to as "environmental pathogens" [2]–[6].

Studies of virulence factors of human pathogens in environmental reservoirs have begun to reveal the importance of alternate hosts, of dual-use virulence factors, and in general of how environmental habitats can select for traits that confer enhanced fitness as human pathogens. For example, interactions with microbial eukaryotes seem to have led to the acquisition of traits useful for pathogenicity to mammalian cells. Numerous environmental pathogens, including Cryptococcus neoformans, Legionella spp., Chlamydophila pneumoniae, Mycobacterium avium, Listeria monocytogenes, Pseudomonas aeruginosa, and Francisella tularensis, might have acquired virulence traits via their resistance to predation by amoebae. This resistance, associated with the ability to grow inside the amoebae—which are essentially alternate hosts—has likely led to the selection of traits conferring survival in macrophages [7]. Resistance to macrophages involves the capacity of the bacteria to resist or debilitate the macrophage's phagosomes and to multiply in the cytoplasm. Many of the traits essential for virulence to humans likewise seem to play roles in adaption to the environments where the organisms are saprophytes (Table 1). These traits have dual roles in environmental and parasitic fitness and are thus referred to as "dual-use traits". Melanins, siderophores, and the capacity to form biofilms are among the frequently cited examples. C. neoformans provides one of the richest examples of dual-use traits. This fungus, frequently found in soils that contain high levels of bird guano and in association with certain plants, causes meningoencephalitis. A nonexhaustive list of its dual-use traits includes capsule formation and production of melanin, laccase, phospholipase, proteases, and ureases [8]. In the environment these traits contribute to survival and in human hosts they contribute to the capacity of C. neoformans to avoid host resistance mechanisms and to attack host tissue. Microbial efflux pumps have also evolved dual uses. These transport systems are used for managing toxic compounds in the environment of the microorganism and can have a broad spectra of activity leading to multidrug resistance among environmental microorganisms [9]. Human activities resulting in the disposal of a wide range of chemical products into the environment, including household cleaners that contain the broad spectrum antimicrobial triclosan, may be inadvertently exacerbating the abundance of multidrug-resistant bacteria [10].

Table 1. Examples of putative dual-use traits related to pathogenic and environmental fitness of human pathogens.

Organism	Trait or Gene	Role in Pathogenic Fitness	Role in Environmental Fitness	Reference
Vibrio cholera	Toxin co-regulated pilus	Virulence factor in humans	Biofilm formation on chitin	[59,60]
Legionella pneumophila	Eukaryotic-like proteins that mimic cellular functions of eukaryotic proteins; type II and type IV secretion systems, surface proteins involved in attachment, secreted effectors	Virulence factors in macrophages	Parasitism and multiplication in protozoa	[61]
Burkholderia cenocepacia	Quorum-sensing regulatory system	Regulation of virulence factors implicated in "cepacia syndrome"	Regulation of factors involved in nematode killing	[62]
Yersinia pestis	Extracellular polysaccharide production linked to the action of heme storage gene (hms) products	Transmission to the human host and protection from the action of leukocytes	Colonization of flea esophagus via biofilm formation	[63]
Cryptococcus neoformans, Alternaria fumigatus	Melanins	Protects microbial cells against phagocytosis	Protection against oxidation	[24]
Alternaria flavus, Histoplasma capsulatum, Aspergillus fumigatus, A. nidulans and numerous bacteria	Siderophores	Virulence factor in humans	Sequestering iron in the environment	[21–23]
Pseudomonas aeruginosa and Stenotrophomonas maltophilia	Efflux pumps	Intrinsic multidrug resistance	Exclusion of lipophilic toxic compounds from cells	[10,64,65]
Acinetobacter baummannii	Efflux pumps, genetic promiscuity, exopolysaccharides and biofilm formation, siderophore-like compounds	Multidrug resistance, attachment, stimulation of host inflammation, virulence factor in humans	Exclusion of toxic compounds from cells, resistance to desiccation, sequestering of iron	[66]

Virulence of environmental pathogens has been described as a set of cards, or a diverse set of attributes acquired as a function of the life history of a pathogen and its adaptation to different environments [3],[8]. It is becoming increasingly clear that evolutionary forces outside the context of human–pathogen interactions are responsible for the acquisition and maintenance of some virulence factors [11]. Genomics and phylogenetics are revealing the evolutionary link between, for example, commensal strains of Escherichia coli and modern pathogens such as enterohaemorrhagic strains of this species (such as O157). The mechanisms proposed to explain how these commensals have become pathogens are grounded in their ecology and life histories, culminating in the notion of ecological evolution ("eco-evo") [11]. The eco-evo approach to understanding the emergence of pathogens gives credence, from the perspective of genomics, to evolutionary and adaptive scenarios that are surmised from a thorough understanding of the ecology and life history of pathogens.

Links Between Plant Pathogenicity, Adaptation to Biotic and Chemical Stress, and Key Vital Functions

At present, epidemiological classifications of plant diseases are based on the interaction of the pathogen and the host (biotrophic or necrotrophic, obligate or facultative), on the number of cycles of propagule production (mono- and polycyclic diseases), on the importance of latency in symptom expression, and on the role of vectors, but there is no formalized equivalent of "sapronoses". Nevertheless, numerous plant pathogens are present in diverse nonagricultural habitats or survive saprophytically in agricultural contexts. These include a range of bacteria, fungi, and stable viruses (a nonexhaustive list of examples is presented in Table 2). A striking characteristic of many of the virulence factors of these plant pathogens is that they are linked to—or are in themselves—traits critical to adaptation to the nonplant environment, as will be illustrated below. This provides a compelling reason to adopt a holistic view of the life history and evolution of plant pathogens, to move beyond the traditional borders of agriculture and the presumed "primary" plant host. Adaptation to biotic and abiotic stresses, within or outside of agricultural habitats, likely plays as important a role in the evolution of parasitic fitness of plant pathogens as it does for human pathogens.

As illustrated above, traits that confer fitness in response to biotic and abiotic environmental stress can have dual-use as virulence factors in human pathogens. Toxins and toxin transport systems (including efflux pumps, in particular) are among the common adaptations for antagonizing and defending against the co-inhabitants of a habitat. In plant pathogens, the transport systems for toxins and antimicrobials can have broad spectrum activity, leading to resistance to agricultural fungicides and also contributing to virulence [12]. Genes coding for wide spectrum efflux pumps are present in the chromosomes of all living organisms [9]. The efflux pump BcAtrB of Botrytis cinerea confers resistance to antimicrobials produced by soil and plant microflora (2,4-diacetylphloroglucinol and phenazine antibiotics) [13],[14] and also to the fungicide fenpiclonil and the plant defensive phytoalexin resveratrol [15]. The transporter ABC1 from Magnaporthe grisea protects the fungus against azole fungicides and the rice phytoalexin sakuranetin [12]. Numerous plant pathogenic bacteria, including Erwinia amylovora,

Table 2. Examples of plant pathogens reported to thrive in nonagricultural habitats or to survive saprophytically in agricultural contexts in the absence of host plants.

Species	Nonagricultural Habitats or Substrates Where Microbe Has Been Detected	Putative Factors Conducive to Survival	References
Bacteria			
Burkholderia cepacea	Ubiquitous in soils and waters and associated habitats	Unusually large genome harboring genes for a multitude of traits related to ecological fitness including the capacity to use a large spectrum of carbon sources	[67]
Dickyea spp. including D. chrysanthemi and Pectobacterium carotovorum (formerly Erwinia chrysanthemi and E. carotovora)	Oceanic aerosols, soils, alpine rivers, and other surface water, snow	Capacity of pectolytic bacteria to obtain nutrients from rotting plant material and to use a wide range of carbon sources; cell surface properties than foster condensation of water vapor; growth and survival as a facultative anaerobe	[68–71]
Pantoea agglomerans	Fecal matter, soil, surface waters	This bacterium is generally an opportunistic plant pathogen that is normally a fit saprophyte	[72,73]
Pseudomonas syringae	Clouds, snow rain, epilithic biofilms, wild alpine plants (substrates linked to the water cycle)	Biofilm formation; production of toxins and siderophores; survival of freezing	[74,75]
Rhodococcus fascians	Soil, ice, polar seawater, lesions on animals, rinds of cheese	Sexual promiscuity favoring acquisition of diverse plasmid-borne traits; capacity to shift metabolic pathways as a function of food base	[76–78]
Streptomyces spp.	Ubiquitous in soil and water	Production of a diverse array of degradative enzymes critical to saprophytic lifestyle; capacity to produce a wide range of antibiotics important in species interactions; resistant to many antibiotics	[29]
Fungi			
Alternaria spp.	Most Alternaria species are common saprophytes; found in soil or decaying plant tissues and atmospheric aerosols	Derive energy as a result of cellulytic activity. Production of toxic secondary metabolites. Production of melanin protecting against environmental stress or unfavorable conditions (extreme temperatures, UV radiation and compounds secreted by microbial antagonists).	[79,80]
Aspergillus spp.	Marine and terrestrial habitats, soil; associated with insects, humans, and other animals	Production of toxins including aflatoxins; production of siderophores and degradative enzymes (pectinases, proteases)	[81–84]
Cladosporium spp.	Soil; atmospheric aerosols	Carbohydrate-binding protein modules (LysM effectors). No other suppositions found in the literature.	[18,79,81,83]
Fusarium spp.	Soil; extreme saline soil habitats; marine and fluvial habitats	Production of defense-related metabolites (antibiotics, trichotecenes, mycotoxins…) and of siderophores; vigor in competitive use of foods, ability to colonize a wide range of substrates	[81,83,85–89]
Leptosphaeria maculans	Can survive as a saprobe for many years on debris	Maintains numerous genes required for saprophytic life (for nutrient acquisition, competition with soil microflora), necrotrophic parasitism via toxins and degradative enzymes	[90]
Mucorales: Mucor spp., Rhizopus spp.	Soil and a variety of organic substrates; marine habitats including insect cadavers	Production of siderophores (by Rhizopus)	[81–83,91]
Pythium spp. (nonobligate parasitic oomycetes)	Soil and water	No suppositions found in the literature	[92]
Penicillium spp.	Soil, sediment-rich subglacial ice; atmospheric aerosols	Production of toxins and siderophores	[79,81–83,93]
Viruses			
Tomato mosaic virus	Clouds, glacial ice, soil of pristine forests	Overall stability of tobamoviruses	[94–96]

Dickeya spp. (formerly the multiple biovars of E. chrysanthemi), and Agrobacterium tumefaciens, also produce efflux pumps that are involved in their resistance to plant antimicrobials (reviewed by Martinez et al. [9]). Toxins themselves can have a broad spectrum of action. For example, mycotoxins, well known for their human and animal toxicity, have broad spectrum activity and are thought to have evolved as a defense against predators (nematodes) and antagonists (other microorganisms) [16]. One family of these, the trichothecenes, contributes significantly to the virulence of many Gibberella (Fusarium) species [17].

Adaptation to biotic stress also implicates systems for the detection or inhibition of arms of aggression used by co-inhabitants. Recent work on fungi suggests that systems to detect enzymes that degrade fungal cell walls are also deployed as virulence factors. Lysin motifs (LysMs) are carbohydrate-binding protein modules that have been found in mammalian and plant pathogenic fungi as well as in saprophytes [18]. Bolton et al. [19] demonstrated that the LysM protein Ecp6 acts as a virulence factor in the plant pathogenic fungus Cladosporium fulvum. As virulence factors they may suppress host defenses by sequestrating chitin oligosaccharides that are known to act as elicitors of plant defense responses [19] and also as activators of host immune responses in mammals [20]. de Jonge and Thomma [18] suggest that these proteins may also have a role in the protection of saprophytic fungi against chitinase-secreting competitor microbes or mycoparasites.

Protection against abiotic stress can involve molecules that have also become virulence factors. Siderophores [21]–[23] and various pigments including melanins [24] are virulence factors in some human pathogens. Siderophores contribute to resistance to oxidative stress and sequestering iron when it is rare in the environment. In the plant pathogens Alternaria brassicicola, Cochliobolus spp., Fusarium graminearum [25], and M. grisea [26], siderophores or their precursors are virulence factors. Melanins offer protection from extreme temperatures, UV radiation, and antimicrobials. In the plant pathogens M. grisea and Colletotrichum spp., melanins are also virulence factors via their essential role in the formation of tissue-penetration structures such as appressoria [17]. In many cases, toxins and siderophores are produced by nonribosomal peptide synthase or polyketide synthase pathways. These pathways, widely distributed in the microbial world, are highly adaptable and have given rise to a wide range of compounds with a plethora of activities, including many of pharmaceutical importance [27]. HC-toxin of Cochlobilus carbonum, victorin in C. victoriae, and T-toxin in C. heterostrophus are products of these pathways [28]. The key virulence factor of Streptomyces spp., thaxtomin [29], and the multitude of host-specific and nonspecific toxins in Pseudomonas syringae pathovars [30] are also produced by these pathways.

The capacity to detect changes in conditions of the abiotic environment has also become part of the virulence factors of some plant pathogens. For example, to detect changes in environmental conditions, organisms exploit two-component histidine kinase complexes. These are key elements of the machinery for signal sensing, allowing bacteria, yeasts, fungi, and plants to adapt to changing environments. In the plant pathogen B. cinerea, one of its multiple histidine kinases, BOS1, not only mediates osmosensitivity and resistance to fungicides, but is also essential for formation of macroconidia and expression of virulence [31].

Recognition and understanding of the full complexity of the life history of plant pathogens will enhance our capacity to evaluate the diversity and intensity of environmental stresses that microorganisms face and will contribute novel hypotheses concerning the role of environmental stresses in the evolution of pathogenicity. Stress is considered to play an important role in adaptive evolution in general, in particular via its effect on mutation rates [32]. For certain fungi and bacteria, including plant pathogens, stress increases the activity of transposable elements [33]–[35] and induces the SOS response and other systems involved in the modification or repair of DNA [32]. Mutations can target the ensemble of the microbial genome. However, it has been suggested that adaptation of bacteria to multiple stresses can lead, in particular, to the acquisition of virulence factors and to the emergence of pathogenic variants [36].

Adaptation to specific habitats—which involves adapting to a particular ensemble of biotic and abiotic parameters—could also influence the evolution of parasitic fitness. Available examples focus on soil-borne and rhizosphere microorganisms. The rhizosphere is a dynamic soup whose chemistry changes as plants grow, die, and degrade. Chemicals in the rhizosphere are food substrates and means of communication, antagonism, and collaboration among microorganisms, among plants, and between plants and microorganisms. To decompose dead plant material and recycle carbon, microorganisms have developed a range of cell wall–degrading enzymes, without which our planet would be quite encumbered by the accumulation of tissue from dead plants. Pectolytic, cellulolytic, and lignolytic enzymes are also well-known pathogenicity factors [37]–[39]. To hone the efficiency of these enzymes in planta, pectinolytic fungi are adept at modulating the surrounding pH. Alternaria, Penicillium, Fusarium spp., and Sclerotinia sclerotiorum also exploit these pH changes to enhance the action of these enzymes as virulence factors [40]. Streptomyces spp. are considered quintessential soil inhabitants. Their ability to degrade biopolymers, including cellulose and chitin, contributes greatly to nutrient cycling, and their vast array of antimicrobials contributes to survival and microbial communication in soil [29]. Some Streptomyces species are pathogenic to root crops and to potatoes in particular. A recently discovered virulence factor in Streptomyces, a saponinase homologue

[29], may be the result of adaptation to the rhizosphere. Saponins are plant gly-cosides that contribute to resistance against fungi and insect herbivores. Bacteria, and especially Gram-positive bacteria, can also be sensitive. Saponins are also ex-uded from the roots of some plant species where they have allelopathic as well as antimicrobial activity [41],[42].

Key vital functions, housekeeping functions, and basic life cycle processes should also be considered for their potential to give rise to pathogenicity factors. Traits fundamental to fitness and survival in general can confer or enhance patho-genic fitness. In plant pathogenic bacteria these include flagella, motility, lipo- and exo- polysaccharides, O-antigens, fimbriae, mechanisms for iron acquisition and for quorum sensing, toxin production, cell wall–degrading enzymes, and re-sistance to oxidative stress [43]. Motility, for example, is essential to dispersal and for attaining new resources. In Ralstonia solanacearum it is also essential for early stages of plant invasion and colonization during pathogenesis [44]. In the fungus Aschochyta rabiei, kinesins that are essential for polarized growth and transport of organelles are suspected to be a virulence factor [45]. An F-box protein of Giber-rella zeae has been reported to be involved in sexual reproduction and in pathoge-nicity [46]. The enzymes that allow fungi to detoxify compounds resulting from plant defense mechanisms are probably also simply means of acquiring nutrients [47]. For example, detoxification of tomatine in tomatoes by Septoria lycoper-cici and by Fusarium oxysporum f. sp. lycopersici is achieved by the deployment of glycosyl hydrolases by these fungi; Gaeumannomyces graminis detoxifies ave-nacins in oats via a beta-glucosidase [28]. Another example of adaptation of basic cellular functions into pathogenicity factors concerns elicitins. Elicitins are part of one of the most highly conserved protein families in the Phytophthora genus and are widespread throughout Phytophthora species. Elicitins of P. infestans induce hypersensitivity in plants. Recent work from Jiang and colleagues [48] suggests that a primary function of elicitins is the acquisition of sterols from the environ-ment.

Toward New Paradigms about the Evolution of Plant Pathogenicity: The Roles of Dual-Use Traits and Exaptation

How can we make sense of the processes that have led to the wide variety of pathogenicity factors in plant pathogens and that continue to drive the evolu-tion of pathogens? Bacterial plant pathogens are particularly illustrative of the differences in suites of secretion systems [43],[49],[50],[51] and of effectors [50],[51],[52],[53],[54],[55] among members of different genera, species, or strains of the same species that attack plants. Effectors are proteins secreted by plant pathogens that modulate plant defense reactions, thereby enabling the pathogen to colonize the plant tissues. It is tempting to wonder if the effectors and secretion systems have critical roles in fitness elsewhere other than in association with the host plant. The examples listed above that describe traits that play roles in both environmental fitness and virulence to plants provide a compelling incentive to expand our paradigms concerning the forces that drive evolution of plant pathogenicity. The evolutionary forces that have been described to date for plant pathogens [56] need to be extended beyond the current agro-centric paradigm.

To expand this paradigm we propose that the life cycles and life histories of plant pathogens be reconsidered. Studies of pathogen ecology, evolution, and life history should include the full range of habitats and reservoirs these organisms can inhabit. This in turn will permit testing a range of novel hypotheses about the role of ecological contexts—other than direct interaction with host plants—as forces of evolution. In Table 3 we propose some such hypotheses. For example, rates of mutation and of transposition of insertion sequences or of transposable elements including phages might be different when a microorganism inhabits nonagricultural habitats (biofilms, lake water, or inert surfaces exposed to UV, for example) than when it colonizes plants. The consequences of these mutations for pathogenicity might in turn be markedly different than for fitness in nonagricultural habitats. Likewise, the formation of spores or aggregates that can be released into the air and their survival over long distances might be highly influenced by the nature of the reservoir that the pathogen colonizes, resulting in direct effects of habitat on gene flow. Furthermore, the biotic and abiotic stresses endured in nonagricultural habitats might exert positive selection for adaptive survival traits that have dual-use as virulence factors as illustrated in the examples above. These questions are clearly pertinent for pathogens that are not obligate biotrophs. However, the complexity of the biotic and abiotic environment perceived by obligate biotrophs during colonization of plants (powdery mildews on leaf surfaces inhabited by other microorganisms, for example) or during their dissemination (survival in air or in association with vectors) are also likely to exert selection independent of that due to the host plant genotype per se. These are only some of the ways in which environmental parameters other than the host plant are expected to have a marked influence on the diversification of plant pathogens.

If nonagricultural environments can foster the evolution of traits that contribute to pathogen virulence, other scenarios are also probable where i) crop plants foster the emergence of traits antagonistic to survival outside of agricultural contexts ii) or nonagricultural environments foster the emergence of traits that are detrimental to pathogen virulence in crops. Understanding the prevalence and significance of alternative habitats to pathogen life history is crucial to determining the broad costs of virulence for pathogen fitness. The cost of virulence in terms of fitness in association with plants has been explored extensively for several obligate parasites such as rusts and powdery mildews. Work by Thrall and Burdon [57] has shown clear fitness tradeoffs between pathogen aggressiveness (capacity to induce intense disease symptoms) and dissemination (via intense spore production). For nonobligate pathogens we do not know the cost of fitness outside of agricultural habitats. The interplay between evolutionary forces and habitat has not been explored for plant pathogens and might be a key feature in the emergence of certain diseases.

Table 3. Novel hypotheses to be tested concerning the impact of substrates other than host plants on the evolutionary potential of plant pathogens.

Evolutionary Force[a]	Novel Hypothesis Arising from Expanded Paradigms about the Evolution of Plant Pathogenicity Concerning:
Mutation	*Modifications of the genome.*
	Relative to its association with cultivated plant hosts, association of the pathogen with a given nonagricultural substrate leads to:
	• a significantly greater overall mutation rate.
	• a greater rate of transposition of insertion sequences or of transposable elements.
	• more frequent mutations or transpositions that target genes involved in pathogenicity.
	• a higher probability of acquisition of alien nucleic acids.
	• genetic exchange with more phylogenetically diverse microbes.
Genetic drift	*Effective population size.*
	The effective sub-population size of a pathogen associated with a given nonagricultural (or nonplant) substrate is significantly different from that for sub-populations from cultivated host plants. This could lead to genetic and/or phenotypic differentiation of sub-populations based on substrate of origin.
Gene flow	*Dissemination.*
	The habitats occupied by the plant pathogen influence the mode(s) of dissemination, thereby influencing the distance of dissemination and the spatial and temporal scales of gene flow.
Mode of reproduction (recombination)	*Genetic recombination.*
	The frequency of recombination (via sexual cycle or other means) varies among strains of plant pathogens as a function of the habitat or substrate.
Selection	*Selective pressures and impact on fitness.*
	Strains of pathogens adapted to a broad range of habitats have the greatest parasitic fitness.

[a]The evolutionary forces listed here are those that have been considered for plant pathogens in agricultural contexts [56]. These hypotheses concern pathogens with a marked saprophytic phase or for which nonagricultural or nonplant substrates can be a notable reservoir for survival. Reservoirs can include irrigation water, natural waterways and bodies of water, biological vectors (animals, fungi, etc.), abiotic vectors (aerosols, clouds, precipitation), wild plants and weeds, soil, and physical structures in agricultural systems (greenhouse materials, tubing, plastics).

By expanding our paradigms concerning pathogen life history and the selective forces that drive plant pathogen evolution, we will enhance our understanding of how pathogens survive in the absence of hosts, how and where new pathotypes are likely to emerge, and the significance of natural habitats to agricultural epidemics. Insights will come from fundamental research to identify the mechanisms that drive the evolution of pathogenic traits and to explore the ecological significance of pathogenic traits to microbial fitness apart from the plant host. Distinguishing the role of adaptation sensu stricto in the emergence of plant pathogenicity relative to that of exaptation [58], the useful cooptation of phenotypes that have arisen under natural selection due to forces unrelated to interaction with the primary host plant, will yield critical insight into how plant pathogens evolve independently of agricultural practices. A more complete understanding of the forces that drive plant pathogen evolution will be critical to enhancing and diversifying sustainable disease control strategies, and will improve prediction of the conditions that support the emergence of novel pathogens.

Acknowledgements

We thank the three anonymous reviewers for their constructive comments and for the suggestion of additional materials to incorporate into the text. We also thank Dr. Melodie Putnam (Oregon State University, United States of America) for useful discussions about the ecology of bacterial plant pathogens.

Competing Interests

The authors have declared that no competing interests exist.

References

1. Hubálek Z (2003) Emerging human infectious diseases: anthroponoses, zoonoses and sapronoses. Emerg Infect Dis 9: 403–404.

2. Cangelosi GA, Freitag NE, Buckley MR (2004) From outside to inside: environmental microorganisms as human pathogens. A report from the American Academy of Microbiology. Washington (D.C.): American Academy of Microbiology. 18 p. Available: http://academy.asm.org/images/stories/documents/fromoutsidetoinsidecolor.pdf . Accessed 29 November 2009.

3. Casadevall A, Pirofski L (2007) Accidental virulence, cryptic pathogenesis, Martians, lost hosts, and the pathogenicity of environmental microbes. Eukaryotic Cell 6: 2169–2174.

4. Hall-Stoodley L, Stoodley P (2005) Biofilm formation and dispersal and the transmission of human pathogens. Trends Microbiol 13: 7–10.

5. Reedy JL, Bastidas RJ, Heitman J (2007) The virulence of human pathogenic fungi: notes from the south of France. Cell Host Microbe 2: 77–83.

6. Yildiz FH (2007) Processes controlling the transmission of bacterial pathogens in the environment. Res Microbiol 158: 195–202.

7. Greub G, Raoult D (2004) Microorganisms resistant to free-living amoebae. Clin Microbiol Rev 17: 413–433.

8. Casadevall A, Steenbergen JN, Nosanchuk JD (2003) 'Ready made' virulence and 'dual use' virulence factors in pathogenic environmental fungi—the Cryptococcus neoformans paradigm. Curr Opin Microbiol 6: 332–337.

9. Martinez JL, Sánchez MB, Martínez-Solano L, Hernandez A, Garmendia L, et al. (2009) Functional role of bacterial multidrug efflux pumps in microbial natural ecosystems. FEMS Microbial Rev 33: 430–449.

10. Sanchez P, Moreno E, Martinez JL (2005) The biocide triclosan selects Stenotrophomonas maltophilia mutants that overproduce the SmeDEF multidrug efflux pump. Antimicrob Agents Chemother 2: 781–782.

11. Pallen MJ, Wren MW (2007) Bacterial pathogenomics. Nature 449: 835–842.

12. Del Sorbo G, Schoonbeek H, De Waard MA (2000) Fungal transporters involved in efflux of natural toxic compounds and fungicides. Fungal Genet Biol 30: 1–15.

13. Schoonbeek HJ, Raaijmakers JM, De Waard MA (2002) Fungal ABC transporters and microbial interactions in natural environments. Mol Plant Microbe Interact 15: 1165–1172.

14. Schouten A, Maksimova O, Cuesta-Arenas Y, van den Berg G, Raaijmakers JM (2008) Involvement of the ABC transporter BcAtrB and the laccase BcLCC2 in defence of Botrytic cinerea against the broad-spectrum antibiotic 2,4-diacetylphloroglucinol. Environ Microbiol 10: 1145–1157.

15. Schoonbeek H, Del Sorbo G, De Waard MA (2001) The ABC transporter BcatrB affects the sensitivity of Botrytis cinerea to the phytoalexin resveratol and the fungicide fenpiclonil. Mol Plant Microbe Interact 14: 562–571.

16. Etzel RA (2002) Mycotoxins. JAMA 287: 425–427.

17. Idnurum A, Howlett BJ (2001) Pathogenicity genes of phytopathogenic fungi. Mol Plant Pathol 2: 241–255.

18. de Jonge R, Thomma BPHJ (2009) Fungal LysM effectors: extinguishers of host immunity? Trends Microbiol 17: 151–157.

19. Bolton MD, van Esse HP, Vossen JH, de Jonge R, Stergiopoulos I, et al. (2008) The novel Cladosporium fulvum lysin motif effector Ecp6 is a virulence factor with orthologues in other fungal species. Mol Microbiol 69: 119–136.

20. Da Silva CA, Hartl D, Liu W, Lee CG, Elias JA (2008) TLR-2 and IL-17A in chitin-induced macrophage activation and acute inflammation. J Immunol 181: 4279–4286.

21. Hwang LH, Mayfield JA, Rine J, Sil A (2008) Histoplasma requires SID1, a member of an iron-regulated siderophore gene cluster, for host colonization. PLoS Pathog 4: e1000044. doi:1000010.1001371/journal.ppat.1000044.

22. Ratledge C, Dover LG (2000) Iron metabolism in pathogenic bacteria. Annu Rev Microbiol 54: 881–941.

23. Schrettl M, Bignell E, Kragl C, Sabiha Y, Loss O, et al. (2007) Distinct roles for intra- and extracellular siderophores during Aspergillus fumigatus infection. PLoS Pathog 3: e128.

24. Liu GY, Nizet V (2009) Color me bad: microbial pigments as virulence factors. Trends Microbiol 17: 406–413.

25. Lee BN, Kroken S, Chou DYT, Robbertse B, Yoder OC, et al. (2005) Functional analysis of all nonribosomal peptide synthetases in Cochliobolus heterostrophus reveals a factor, NPS6, involved in virulence and resistance to oxidative stress. Eukaryot Cell 4: 545–555.

26. Hof C, Eisfeld K, Welzel K, Antelo L, Foster AJ, et al. (2007) Ferricrocin synthesis in Magnaporthe grisea and its role in pathogenicity in rice. Molec Plant Pathol 8: 163–172.

27. Rausch CR, Hoof I, Weber T, Wohlleben W, Huson DH (2007) Phylogenetic analysis of condensation domains in NRPS sheds light on their functional evolution. BMC Evol Biol 7: 78.

28. Berbee ML (2001) The phylogeny of plant and animal pathogens in the Ascomycota. Physiol Mol Plant Pathol 59: 165–187.

29. Loria R, Kers J, Joshi M (2006) Evolution of plant pathogenicity in Streptomyces. Annu Rev Phytopathol 44: 469–487.

30. Bender CL, Alarcon-Chaidez F, Gross DC (1999) Pseudomonas syringae phytotoxins: Mode of action, regulation and biosynthesis by peptide and polyketide synthetases. Microbiol Molec Biol Rev 63: 266–292.

31. Viaud M, Fillinger S, Liu W, Polepalli JS, Le Pêcheur P, et al. (2006) A class III histidine kinase acts as a novel virulence factor in Botrytis cinerea. Mol Plant Microbe Interact 9: 1042–1050.

32. Bjedov I, Tenaillon O, Gérard B, Souza V, Denamur E, et al. (2003) Stress-induced mutagenisis in bacteria. Science 300: 1404–1409.

33. Ikeda K, Nakayashiki H, Takagi M, Tosa Y, Mayama S (2001) Heat shock, copper sulfate and oxidative stress activate the retrotransposon MAGGY resident in the plant pathogenic fungus Magnaporthe grisea. Mol Genet Genomics 266: 318–325.

34. Ilves H, Hõrak R, Teras R, Kivisaar M (2004) IHF is the limiting host factor in transposition of Pseudomonas putida transposon Tn4652 in stationary phase. Mol Microbiol 51: 1773–1785.

35. Mes JJ, Haring MA, Cornelissen BJC (2000) Foxy: an active family of short interspersed nuclear elements from Fusarium oxysporum. Mol Gen Genet 263: 271–280.

36. Arnold DL, Jackson RW, Waterfield NR, Mansfield J (2007) Evolution of microbial virulence: the benefits of stress. Trends Genet 23: 293–300.

37. Annis SL, Goodwin PH (1997) Mini review: recent advances in the molecular genetics of plant cell wall-degrading enzymes produced by plant pathogenic fungi. Europ J Plant Pathol 103: 1–14.

38. Collmer A, Bauer DW, He SY, Lindeberg M, Kelemu S, et al. (1990) Pectic enzyme production and bacterial plant pathogenicity. In: Hennecke H, Verma DPS, editors. Advances in molecular genetics of plant-microbe interactions. Kluwer Academic Publishers. pp. 65–72.

39. Reignault P, Valette-Collet O, Boccara M (2008) The importance of fungal pectinolytic enzymes in plant invasion, host adaptability and symptom type. Eur J Plant Pathol 120: 1–11.

40. Prusky D, Yakoby N (2003) Pathogenic fungi: leading or led by ambient pH? Mol Plant Pathol 5: 509–516.

41. Bais HP, Weir TL, Perry LT, Gilroy S, Vivanco JM (2006) The role of root exudates in rhizosphere interactions with plants and other organisms. Annu Rev Plant Biol 57: 233–266.

42. Oleszek WA, Hoagland RE, Zablotwicz RM (1999) Ecological significance of plant saponins. In: Inderjit D, Foy CL, editors. Principles and practices in plant ecology: allelochemical interactions. Baton Rouge: CRC Press. pp. 451–465.

43. Toth IK, Pritchard L, Birch PRJ (2006) Comparative genomics reveals what makes an enterobacterial plant pathogen. Annu Rev Phytopathol 44: 305–336.

44. Tans-Kersten J, Huang H, Allen C (2001) Ralstonia solanacearum needs motility for invasive virulence on tomato. J Bacteriol 183: 3597–3605.

45. White D, Chen W (2007) Towards identifying pathogenic determinants of the chickpea pathogen Ascochyta rabiei. Eur J Plant Pathol 119: 3–12.

46. Han YK, Kim MD, Lee SH, Yun SH, Lee YW (2007) A novel F-box protein involved in sexual development and pathogenesis in Gibberella zeae. Molec Microbiol 63: 768–779.

47. Divon HH, Fluhr R (2007) Nutrition acquisition strategies during fungal infection of plants. FEMS Microbial Lett 266: 65–74.

48. Jiang RHY, Tyler BM, Whisson SC, Hardham AR, Govers F (2006) Ancient origin of elicitin gene clusters in Phytophthora genomes. Molec Biol Evol 23: 338–351.

49. Bell KS, Sebaihia M, Pritchard L, Holden MTG, Hyman LJ, et al. (2004) Genome sequence of the enterobacterial phytopathogen Erwinia carotovora subsp. atroseptica and characterization of virulence factors. 101: 11105–11110.

50. da Silva ACR, Ferro JA, Reinach FC, Farah CS, Furlan LR, et al. (2002) Comparison of the genomes of two Xanthomonas pathogens with differing host specificities. Nature 417: 459–463.

51. Thieme F, Koebnik R, Bekel T, Berger C, Boch J, et al. (2005) Insights into genome plasticity and pathogenicity of the plant pathogenic bacterium Xanthomonas campestris pv. vesicatoria revealed by the complete genome sequence. J Bacteriol 187: 7254–7266.

52. Lindeberg M, Myers CR, Collmer A, Schneider DJ (2008) Roadmap to new virulence determinants in Pseudomonas syringae: insights from comparative genomics and genome organization. Mol Plant Microbe Interact 21: 685–700.

53. Rohmer L, Guttman DS, Dangl JL (2004) Diverse evolutionary mechanisms shape the type III effector virulence factor repertoire in the plant pathogen Pseudomonas syringae. Genetics 167: 1341–1360.

54. Kamoun S (2007) Groovy times: filamentous pathogen effectors revealed. Curr Opinion Plant Biol 10: 358–365.

55. Setubal JC, Moreira AM, da Silva ACR (2005) Bacterial phytopathogens and genome science. Curr Opinion Microbiol 8: 595–600.

56. McDonald BA, Linde C (2002) The population genetics of plant pathogens and breeding strategies for durable resistance. Euphytica 124: 163–180.

57. Thrall PH, Burdon JJ (2003) Evolution of virulence in a plant host-pathogen metapopulation. Science 299: 1735–1737.

58. Gould SJ, Vrba E (1982) Exaptation–A missing term in the science of form. Paleobiology 8: 4–14.

59. Pruzzo C, Vezzulli L, Colwell RR (2008) Global impact of Vibrio cholerea interactions with chitin. Environ Microbiol 10: 1400–1410.

60. Reguera G, Kolter R (2005) Virulence and the environment: a novel role for Vibrio cholera toxin-coregulated pili in biofilm formation on chitin. J Bacteriol 187: 3551–3555.

61. Albert-Weissenberger C, Cazalet C, Buchrieser C (2007) Legionella pneimophila–a human pathogen that co-evolved with fresh water protozoa. Cell Mol Life Sci 84: 432–448.

62. Huber B, Feldmann F, Köthe M, Vandamme P, Wopperer J, et al. (2004) Identification of a novel virulence factor in Burkholderia cenocepacia H111 required for efficient slow killing of Caenorhabditis elegans. Infect Immun 72: 7220–7230.

63. Jarrett CO, Deak E, Isherwood KE, Oyston PC, Fischer ER, et al. (2004) Transmission of Yersinia pestis from an infectious biofilm in the flea vector. Journal of Infectious Diseases 190: 283–292.

64. Alonso A, Martínez JL (2000) Cloning and characterization of SmeDEF, a novel multidrug efflux pump from Stenotrophomonas maltophilia. Antimicrob Agents Chemother 44: 3079–3086.

65. Nikaido H (1996) Multidrug efflux pumps of Gram-negative bacteria. J Bacteriol 178: 5853–5859.

66. Peleg AY, Seifert H, Paterson DL (2008) Acinetobacter baumannii: Emergence of a successful pathogen. Clin Microbiol Rev 21: 538–582.

67. Parke JL, Gurian-Sherman D (2001) Diversity of the Burkholderia cepacia complex and implications for risk assessment of biological control strains. Annu Rev Phytopathol 39: 225–258.

68. Cother EJ, Gilbert RL (1990) Presence of Erwinia chrysanthemi in two major river systems and their alpine sources in Australia. J Appl Bacteriol 69: 729–738.

69. Franc GD (1988) Long distance transport of Erwinia carotovora in the atmosphere and surface water [PhD dissertation]. Fort Collins: Colorado State University. 131 p.

70. Harrison MD, Franc GD, Maddox DA, Michaud JE, McCarter-Zorner NJ (1987) Presence of Erwinia carotovora in surface water in North America. J Appl Bacteriol 62: 565–570.

71. McCarter-Zorner NJ, Franc GD, Harrison MD, Michaud JE, Quinn CE (1984) Soft rot Erwinia bacteria in surface and underground waters in southen Scotland and in Colorado, United-States. J Appl Bacteriol 57: 95–105.

72. Cruz AT, Cazacu AC, Allen CH (2007) Pantoea agglomerans, a plant pathogen causing human disease. J Clin Microbiol 45: 1989–1992.

73. Manulis S, Barash I (2003) Pantoea agglomerans pvs. gypsophilae and betae, recently evolved pathogens? Mol Plant Pathol 4: 307–314.

74. Morris CE, Kinkel LL, Kun X, Prior P, Sands DC (2007) A surprising niche for the plant pathogen Pseudomonas syringae. Infection, Genetics and Evolution 7: 84–92.

75. Morris CE, Sands DC, Vinatzer BA, Glaux C, Guilbaud C, et al. (2008) The life history of the plant pathogen Pseudomonas syringae is linked to the water cycle. ISME Journal 2: 321–334.

76. Larkin MJ, DeMot R, Kulakov LA, Nagy I (1998) Applied aspects of Rhodococcus genetics. Antonie van Leeuwenhoek 74: 133–153.

77. Putnam ML, Miller ML (2007) Rhodococcus fascians in herbaceous perennials. Plant Dis 91: 1064–1076.

78. Vereecke D, Cornelis K, Temmerman W, Holsters M, Goethals K (2002) Versatile persistence pathways for pathogens of animals and plants. Trends Microbiol 10: 485–488.

79. Frohlich-Nowoisky J, Pickersgill DA, Despres VR, Pöschl U (2009) High diversity of fungi in air particulate matter. Proc Nat Acad Sci 106: 12814–12819.

80. Thomma BPHJ (2003) Alternaria spp.: from general saprophyte to specific parasite. Molec Plant Pathol 4: 225–236.

81. Azmi OR, Seppelt RD (1998) The broad-scale distribution of microfungi in the Windmill Islands region, continental Antarctica. Polar Biol 19: 92–100.

82. Baakza A, Vala AK, Dave BP, Dube HC (2004) A comparative study of siderophore production by fungi from marine and terrestrial habitats. J Exp Mar Biol Ecol 311: 1–9.

83. Keller L, Bidochka MJ (1998) Habitat and temporal differences among soil microfungal assemblages in Ontario. Can J Bot 76: 1798–1805.

84. St. Leger RJ, Screen SE, Shams-Pirzadeh B (2000) Lack of host specialization in Aspergillus flavus. Appl Environ Microbiol 66: 320–324.

85. Baker R (1981) Ecology of the fungus, Fusarium: competition. In: Nelson PE, Toussoun TA, Cook RJ, editors. Fusarium, diseases, biology, and taxonomy. University Park: The Pennsylvania State University Press. pp. 245–249.

86. Burgess LW (1981) General ecology of the fusaria. In: Nelson PE, Toussoun TA, Cook RJ, editors. Fusarium, diseases, biology, and taxonomy. University Park: The Pennsylvania State University Press. pp. 225–235.

87. Mandeel QA (2006) Biodiversity of the genus Fusarium in saline soil habitats. J Basic Microbiol 46: 480–494.

88. Palmero Llamas D, de Cara Gonzalez M, Iglesias Gonzalez C, Ruíz Lopez G, Tello Marquina JC (2008) Effects of water potential on spore germination and viability of Fusarium species. J Ind Microbiol Biotechnol 35: 1411–1418.

89. Vesonder RF, Hesseltine CW (1981) Metabolites of Fusarium. In: Nelson PE, Toussoun TA, Cook RJ, editors. Fusarium, diseases, biology, and taxonomy. University Park: The Pennsylvania State University Press. pp. 350–364.

90. Rouxel T, Balesdent MH (2005) The stem canker (blackleg) fungus, Leptosphaeria maculans, enters the genomic era. Mol Plant Pathol 6: 225–241.

91. Sun BD, Yu HY, Chen AJ, Liu XZ (2008) Insect-associated fungi in soils of field crops and orchards. Crop Protect 27: 1421–1426.

92. Nechwatal J, Wielgoss A, Mendgen K (2008) Diversity, host, and habitat specificity of oomycete communities in declining reed stands (Phragmites australis) of a large freshwater lake. Mycol Res 112: 689–696.

93. Sonjak S, Frisvad JC, Gunde-Cimerman N (2006) Penicillium mycobiota in Arctic subglacial ice. Microbial Ecol 52: 207–216.

94. Castello JD, Lakshman DK, Tavantzis SM, Rogers SO, Bachand GD, et al. (1995) Detection of infectious tomato mosaic tobamovirus in fog and clouds. Phytopathology 85: 1409–1412.

95. Castello JD, Rogers SO, Starmer WT, Catrianis CM, Ma L, et al. (1999) Detection of tomato mosaic tobamovirus RNA in ancient glacial ice. Polar Biol 22: 207–212.

96. Fillhart RC, Bachand GD, Castello JD (1998) Detection of infectious tobamovirus in forest soils. Appl Environ Microbiol 64: 1430–1435.

CITATION

Originally published under the Creative Commons Attribution License. Morris CE, Bardin M, Kinkel LL, Moury B, Nicot PC, et al. (2009) Expanding the Paradigms of Plant Pathogen Life History and Evolution of Parasitic Fitness beyond Agricultural Boundaries. PLoS Pathog 5(12): e1000693. doi:10.1371/journal.ppat.1000693.

Two Plant Viral Suppressors of Silencing Require the Ethylene-Inducible Host Transcription Factor RAV2 to Block RNA Silencing

Matthew W. Endres, Brian D. Gregory, Zhihuan Gao,
Amy Wahba Foreman, Sizolwenkosi Mlotshwa, Xin Ge,
Gail J. Pruss, Joseph R. Ecker, Lewis H. Bowman
and Vicki Vance

ABSTRACT

RNA silencing is a highly conserved pathway in the network of interconnected defense responses that are activated during viral infection. As a counterdefense, many plant viruses encode proteins that block silencing, often also interfering with endogenous small RNA pathways. However, the mechanism of action of viral suppressors is not well understood and the role of host factors in the process is just beginning to emerge. Here we report that the ethylene-inducible transcription factor RAV2 is required for suppression of RNA silencing by two unrelated plant viral proteins, potyvirus HC-Pro and carmovirus P38. Using a hairpin transgene silencing system, we find that both viral suppressors require RAV2 to block the activity of primary siRNAs, whereas suppression of transitive silencing is RAV2-independent. RAV2 is also required for many HC-Pro-mediated morphological anomalies in transgenic plants, but not for the associated defects in the microRNA pathway. Whole genome tiling microarray experiments demonstrate that expression of genes known to be required for silencing is unchanged in HC-Pro plants, whereas a striking number of genes involved in other biotic and abiotic stress responses are induced, many in a RAV2-dependent manner. Among the genes that require RAV2 for induction by HC-Pro are FRY1 and CML38, genes implicated as endogenous suppressors of silencing. These findings raise the intriguing possibility that HC-Pro-suppression of silencing is not caused by decreased expression of genes that are required for silencing, but instead, by induction of stress and defense responses, some components of which interfere with antiviral silencing. Furthermore, the observation that two unrelated viral suppressors require the activity of the same factor to block silencing suggests that RAV2 represents a control point that can be readily subverted by viruses to block antiviral silencing.

AUTHOR SUMMARY

RNA silencing is an important antiviral defense in plants, and many plant viruses encode proteins that block RNA silencing. However, the mechanism of action of the viral suppressors is complex, and little is known about the role of host plant proteins in the process. Here we report the first example of a host protein that plays a required role in viral suppression of silencing—a transcription factor called RAV2 that is required for suppression of silencing by two different and unrelated viral proteins. Analysis of plant gene expression patterns shows that RAV2 is required for induction of many genes involved in other stress and defense pathways, including genes implicated as plant suppressors of silencing. Overall, the results suggest that RAV2 is an important factor in viral suppression of silencing and that the role of RAV2 is to divert host defenses toward responses that interfere with antiviral silencing.

Introduction

Plants have a complex interconnected system of defense and stress pathways [1],[2] that receives incoming stimuli, transduces the signal and initiates the appropriate response. The process is orchestrated by a variety of plant hormones and small signaling molecules, and the final shape of the response is refined by crosstalk among different pathways in the network. Evidence emerging over the last decade has made it clear that RNA silencing and endogenous small RNA pathways constitute a major response to a variety of biotic and abiotic stresses [3],[4],[5]. Surprisingly, however, although many of the components of the silencing machinery are known, little is yet known about how silencing is regulated or how it is integrated into the network of other defense and stress pathways.

RNA silencing is a sequence specific RNA degradation mechanism that serves an important antiviral role in plants [6]. Antiviral silencing is triggered by double stranded RNA (dsRNA) that arises during virus infection. The dsRNA trigger is processed by DICER-LIKE (DCL) ribonucleases into primary short interfering RNAs (siRNAs), which incorporate into an ARGONAUTE (AGO) protein-containing effector complex and guide it to complementary target RNAs. The destruction of target RNAs can be amplified via a process called transitive silencing, in which the target RNA serves as template for host RNA-dependent RNA polymerases (RDRs) to produce additional dsRNA that is subsequently processed into secondary siRNAs. In addition to these RDRs, a number of other genes, including DCL2, AGO1 and SUPPRESSOR OF GENE SILENCING 3 (SGS3), are required for transitive silencing, but not for primary silencing [7],[8],[9]. The primary and transitive silencing pathways work together to limit the accumulation of viral RNAs during both the initial and systemic phases of infection.

In addition to antiviral silencing and related pathways that target invading nucleic acids, there are endogenous small RNA pathways that regulate gene expression by directing cleavage of target RNA, inhibition of mRNA translation, or modification of chromatin structure. The best studied of the endogenous small RNAs are the microRNAs (miRNAs), which play major roles in development and in response to a variety of stresses [10],[11],[12]. Although different small RNA mediated pathways have unique genetic requirements, all make use of an overlapping set of genes for their biogenesis (four DCL genes) and function (ten AGO genes), and there is growing evidence that these pathways are interconnected and compete with one another. For example, DCL1, the Dicer that produces most miRNAs, represses antiviral silencing by down-regulation of DCL3 and DCL4 [13] and, when over-expressed, blocks silencing induced by a sense transgene [14]. In addition, many viral suppressors of RNA silencing also interfere with the biogenesis and/or function of endogenous small RNAs such as miRNAs and trans-acting small interfering RNAs (tasiRNAs) [15],[16]. However, the mechanisms that regulate and integrate the various small RNA pathways are just beginning to be elucidated.

Plant viruses have evolved a variety of effective counter-defensive strategies to suppress silencing. Numerous plant viruses encode proteins that block some aspect of RNA silencing [15],[16]. These viral proteins are highly diverse in primary sequence and protein structure, though they may share certain mechanistic features. For example, the ability to bind small RNAs is a feature of many viral suppressors of silencing, including the two used in the present work. Indeed, it has been proposed that most viral suppressors of silencing work by binding and sequestering small RNAs, thereby blocking their activity [17],[18]. However, the physiological significance of small RNA binding is not yet clear in many cases [6], and some suppressors manipulate silencing via interaction with host proteins that are either components of the silencing machinery [19],[20],[21],[22] or proposed regulators of the pathway [23]. Thus, the mechanism of action of viral suppressors is likely both diverse and complex and is not yet fully understood.

Our studies have focused on understanding the mechanism of action of HC-Pro, a potent viral suppressor of silencing that blocks both primary and transitive silencing. Our approach has been to identify host proteins that physically interact with HC-Pro and examine the effect of altering the levels of these proteins on both RNA silencing and the ability of HC-Pro to block silencing [23]. Using this approach, we find that RAV2/EDF2 (hereafter referred to as RAV2), an HC-Pro-interacting protein that is a member of the RAV/EDF family of transcription factors, is required for suppression of silencing not only by potyvirus HC-Pro, but also by carmovirus P38, the silencing suppressor from a virus family unrelated to potyviruses. Interestingly, RAV2 is required exclusively for blocking the activity of primary siRNAs, whereas suppression of transitive silencing and effects on the endogenous microRNA pathway are RAV2-independent. Whole genome tiling microarray experiments were used to characterize HC-Pro-mediated changes in host expression and identify which, if any, were RAV2-dependent. The results raise the interesting possibility that HC-Pro-suppression of silencing is not caused by decreased expression of genes that are required for silencing, but instead, by induction of stress and defense pathways that interfere with antiviral silencing.

Results

Ectopic Expression of a RAV/EDF Transcription Factor Delays the Onset of Transgene-Induced RNA Silencing in Tobacco

In previous work we used a yeast two-hybrid screen to identify Nicotiana tabacum proteins that interact with Tobacco Etch Potyvirus (TEV) HC-Pro [23]. One of the proteins identified in this way was named ntRAV because of its relatedness to the Arabidopsis thaliana RAV/EDF family of transcription factors. The RAV/EDF protein family has six members, and these are unique among transcription factors in having two unrelated DNA binding domains (AP2 and B3) [24]. Members of this family are responsive to numerous biotic and abiotic stresses [25],[26],[27],[28] and are inducible by the plant hormone ethylene [29], which controls many aspects of plant physiology, including defense against pathogens [30],[31].

In vitro pull-down experiments were used to confirm a physical interaction between TEV HC-Pro and ntRAV. 35S-methionine-labeled ntRAV produced in a coupled in vitro transcription/translation system co-purified with an HC-Pro-GST fusion protein isolated from recombinant bacteria, but not with GST alone (Fig. 1A). This result validates the HC-Pro-ntRAV interaction initially identified in the yeast two-hybrid system.

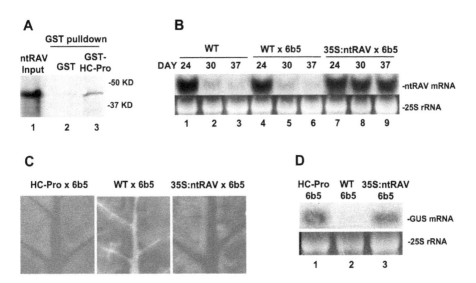

Figure 1. ntRAV Interacts with HC-Pro and Delays the Onset of Sense Transgene Silencing when Overexpressed in Tobacco. (A) Tobacco ntRAV interacts with TEV HC-Pro in in vitro pulldown experiments. 35S-labelled ntRAV co-purifies with HC-Pro-GST (lane 3), but not with GST (lane 2). Lane 1 shows the amount of input 35S-labelled ntRAV protein used in the pulldown experiments. (B) The accumulation of ntRAV mRNA at 24, 30 and 37 days after germination in whole leaves of wild type (WT) tobacco plants (lanes 1–3), plants heterozygous for the silenced 6b5 GUS transgene (WT X 6b5) (lanes 4–6), and plants heterozygous for the silenced 6b5 GUS transgene and expressing the 35S:ntRAV transgene (lanes 7–9). (C) Histochemical staining of leaves from HC-Pro X 6b5 (left panel), WT X 6b5 (center panel) and 35S:ntRAV X 6b5 leaves (right panel) at 26 days after germination. (D) GUS mRNA levels in the veins of leaves of HC-Pro X 6b5 (lane 1), WT X 6b5 (lane 2) and 35SntRAV X 6b5 plants (lane 3) at 26 days after germination.

To determine if ntRAV plays a role in RNA silencing, we evaluated the effect of ntRAV over-expression on transgene-induced silencing. In tobacco, ntRAV is normally expressed at high levels throughout fully expanded healthy leaves of young plants, but expression decreases greatly starting at about 24 days after germination (Fig. 1B, lanes 1–6). In contrast, a tobacco line that ectopically expresses ntRAV from the constitutive Cauliflower mosaic virus (CaMV) 35S promoter maintains high level expression of ntRAV (Fig. 1B, lanes 7–9). We crossed the 35S:ntRAV transgenic line, as well as wild type and HC-Pro-expressing control lines, to the well-characterized tobacco transgenic line 6b5 [32], which is post-transcriptionally silenced for a transgene encoding β-glucuronidase (GUS). Silencing of the GUS locus in line 6b5 reinitiates every generation, starting in the vascular tissue of the oldest leaves and then spreading throughout the leaf. The expression of GUS in F1 progeny of these crosses was assayed histochemically in leaves (Fig. 1C) and by northern blots of RNA from the vascular tissue (Fig. 1D) at 26 days after germination. In these young plants, ectopic expression of ntRAV blocked silencing of GUS in vascular tissue of fully expanded, healthy leaves about as well as HC-Pro (Fig. 1C and D). However, unlike HC-Pro, which completely blocks silencing over the lifetime of the plant, ectopic expression of ntRAV only delayed the onset of silencing, and GUS was eventually silenced throughout the leaf (data not shown). These results, together with those showing a physical interaction between ntRAV and TEV HC-Pro proteins, raised the possibility that ntRAV plays a role in HC-Pro-mediated suppression of silencing.

Experiments in the Model Plant, Arabidopsis Thaliana

To further investigate the role of ntRAV in HC-Pro suppression of silencing, we switched from tobacco to Arabidopsis thaliana, in order to take advantage of the numerous genetic tools available in that model system. Our experiments focused on a RAV gene family member closely related to the tobacco ntRAV, Arabidopsis RAV2 (At1g68840), which had already been cloned and characterized, and for which a validated T-DNA insertional knockout line was available [29]. The change in experimental system also necessitated a change from the HC-Pro encoded by TEV to that encoded by turnip mosaic virus (TuMV), a related potyvirus that infects Arabidopsis. Like the TEV HC-Pro transgene in tobacco, expression of the TuMV HC-Pro in transgenic Arabidopsis plants has been shown to suppress both virus- and transgene-induced RNA silencing [14],[33]. The TuMV HC-Pro transgenic line used in our experiments expresses HC-Pro at a high level and is highly phenotypic [14].

We used in vivo pull-down experiments to determine whether the TuMV HC-Pro and RAV2 proteins interact, as would be expected if RAV2 were a functional homolog of ntRAV. In these experiments, the homozygous rav2 knockout line [29] was transformed with a construct designed to express a transgene encoding FLAG-tagged RAV2. A transformant that expressed the FLAG-RAV2 transgene was crossed to our TuMV HC-Pro transgenic line [14], and expression of both transgenes in the F1 offspring was confirmed by RNA gel blot analysis (data not shown). Pull-down experiments using antiserum specific to the FLAG tag, followed by western blot analysis, showed that TuMV HC-Pro co-immunoprecipitates with the Flag-tagged RAV2 (Fig. 2), indicating that RAV2 and TuMV HC-Pro interact in planta in Arabidopsis. This result confirms that RAV2 is a functional homolog of ntRAV and also provides evidence that the interaction between potyviral HC-Pro and host RAV-like transcription factors is a conserved feature of these proteins.

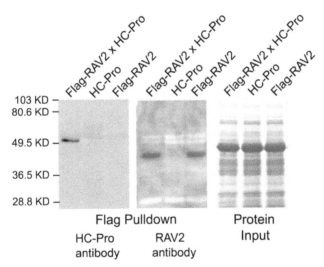

Figure 2. In vivo Interaction of RAV2 and TuMV HC-Pro in Arabidopsis. Proteins isolated from plants expressing either FLAG-tagged RAV2 (Flag-RAV2) alone, TuMV HC-Pro alone or both Flag-RAV2 and TuMV HC-Pro were incubated with anti-FLAG agarose beads. The bound protein was fractionated on acrylamide gels and subjected to western blot analysis using either HC-Pro antiserum (left panel) or RAV2 antiserum (center panel). The far right panel shows the relative input amounts of protein used in the pulldown experiments as determined by Coomassie blue staining.

RAV2 is Required for HC-Pro Suppression of Virus Induced Gene Silencing (VIGS)

Our initial experiments to examine the role of RAV2 in HC-Pro suppression of silencing focused on VIGS. These experiments used the well characterized geminivirus silencing vector, cabbage leaf curl virus (CaLCV), which carried a portion of the endogenous CHLORATA42 (CH42) gene [34]. CH42 is required for chlorophyll accumulation, and VIGS of CH42 in wild type plants results in extensive chlorosis and marked reduction in the level of CH42 mRNA. These changes are accompanied by a pronounced accumulation of 24-nt siRNAs that derive from the CH42 sequences within the viral vector [14],[34]. HC-Pro transgenic plants become infected when bombarded with the CH42 VIGS vector and, although high levels of siRNAs accumulate in the plants, the CH42 gene is not silenced as evidenced by accumulation of CH42 mRNA and the absence of chlorosis [14]. To determine if RAV2 is required for HC-Pro suppression of VIGS, plants expressing HC-Pro in either the wild type or the rav2 knockout background, along with control plants, were bombarded with the CH42 VIGS vector. Wild type control plants as well as rav2 knockout plants exhibited chlorosis of infected tissues (Fig. 3A, top two panels) accompanied by reduction in CH42 mRNA levels and the concomitant accumulation of siRNAs, as expected for VIGS (Fig. 3B, lanes 1–4). HC-Pro transgenic plants were suppressed for VIGS of CH42, remaining green (Fig. 3A, bottom left panel) and accumulating wild type levels of CH42 mRNA as previously reported (Fig. 3B, lanes 5 and 6). In contrast, HC-Pro transgenic plants in the rav2 knockout background were competent for VIGS of CH42 as evidenced by systemic chlorosis (Fig. 3A, bottom right panel) accompanied by reduction in CH42 mRNA levels (Fig. 3B, lanes 7–9). This result indicates that RAV2 is required for HC-Pro suppression of VIGS.

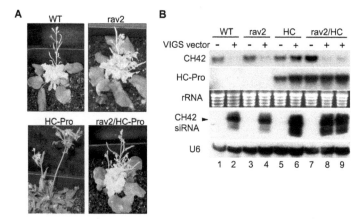

Figure 3. RAV2 is Required for HC-Pro Suppression of Virus Induced Gene Silencing (VIGS). (A) Phenotype of plants bombarded with CaLCV vector carrying a portion of the endogenous CH42 gene. VIGS of CH42 results in pronounced yellowing in wild type or rav2 knockout plants (upper left and right panels, respectively). Plants expressing HC-Pro are suppressed for VIGS and therefore remain green (lower left panel); whereas HC-Pro plants in the rav2 knockout background fail to block silencing and display yellowing typical of wild type plants (lower right panel). (B) RNA gel blot analysis of CH42 mRNA, HC-Pro and CH42 siRNA levels in wild type (lanes 1 and 2), rav2 knockout (lanes 3 and 4), HC-Pro plants (lanes 5 and 6) and HC-Pro plants in the rav2 background (lanes 7–9) either uninfected (lanes 1, 3, 5 and 7) or after bombardment with the CH42 VIGS vector (lanes 2, 4, 6, 8 and 9). Ethidium staining of rRNA is shown as the loading control for the high molecular weight blots and the hybridization signal for U6 is shown as the loading control for the small RNA blot. The migration of 24 nt siRNAs is marked by an arrow.

RAV2 is Required for HC-Pro-Suppression of the Primary, but not the Transitive, Branch of Hairpin Transgene-Induced RNA Silencing

To examine the role of RAV2 in HC-Pro-suppression of transgene silencing, we used a well-characterized system in which silencing occurs through both the primary and transitive branches of the silencing pathway [7],[35]. This system is composed of two transgenes, the 306 and 6b4 loci (Fig. 4A). The 6b4 locus encodes an expressing GUS transgene that includes the entire GUS coding sequence, while the 306 locus encodes a hairpin construct designed to silence GUS expression. The GUS sequence in the 306 locus has a 231 nucleotide deletion in the coding region (Fig. 4A, shown in green) so that RNAs originating from the 6b4 transcript can be unambiguously distinguished. The primary and transitive branches of silencing can be easily differentiated in this system. Basically, primary siRNAs derive only from the stem of the 306 hairpin transcript (Fig. 4A, shown in red, probe 1), whereas secondary siRNAs arise from either locus during an RDR6-dependent process called transitive silencing. In the case of the 306 transgene, siRNAs that arise from the loop of hairpin transcript are secondary siRNAs (Fig. 4A, shown in blue, probe 3). In contrast to the 306 hairpin transcript, the 6b4 mRNA produces only RDR6-dependent secondary siRNAs (Fig. 4A, shown in red, green and blue; [7]. Thus, in the 306/6b4 system, 6b4 mRNA can be degraded by two mechanisms. It can be targeted by a RISC complex directed by siRNAs, or it can be a substrate for RDR6, producing dsRNA that is subsequently processed by DCL to produce secondary siRNAs via transitive silencing. HC-Pro suppresses silencing in the 306/6b4 system, but has different effects on primary and secondary siRNAs: accumulation of secondary siRNAs is eliminated, as shown by the failure to detect any siRNAs when using either probe 2 or probe 3 [7]. In contrast, high levels of primary siRNAs accumulate, but are unable to mediate degradation of the 6b4 target RNA [7].

To determine if RAV2 is required for HC-Pro suppression of hairpin transgene silencing, we crossed the homozygous rav2 knockout line to a transgenic line homozygous for the 306 and 6b4 loci and hemizygous for the TuMV HC-Pro locus. F1 offspring of this cross were allowed to self-fertilize, producing an F2 population that was segregating for all four loci. F2 plants were genotyped, and individuals containing the 306/6b4/HC-Pro loci in the homozygous rav2 mutant background were identified, along with control plants containing all three loci in the wild type RAV2 background. The absence of RAV2 mRNA in rav2 knockout plants was verified by RNA gel blot analysis (Fig. 4B). Initial analysis of the 306/6b4/HC-Pro plant lines addressed the possibility of transcriptional gene silencing (TGS) of the three transgenes involved, all of which are under the control of the CaMV 35S promoter. This was especially important because it has been shown that T-DNA insertion mutants that carry 35S promoter sequences, such as the rav2 knockout line used in this work, can induce TGS of other 35S promoters in the genome [36] and because HC-Pro cannot suppress silencing at the transcriptional level [37],[38]. RNA gel blot analysis showed that the level of HC-Pro mRNA was similar in all plants carrying the HC-Pro transgene (Fig. 4B), arguing against transcriptional silencing of 35S promoter sequences in the plants. In addition, the presence of siRNAs that derive from the GUS transcripts (Fig. 4C) indicates that the observed silencing of the GUS transgenes is at the post-transcriptional rather than the transcriptional level.

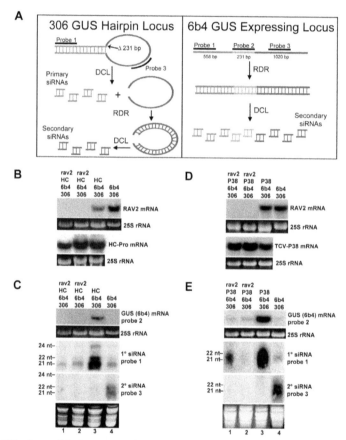

Figure 4. RAV2 is Required for Suppression of Hairpin Transgene Silencing by Two Unrelated Viral Suppressors. (A) Diagrams showing the structures of the 6b4 and 306 transgene loci. The 6b4 locus is an expressing locus which encodes a functional GUS protein. The 306 locus produces a GUS hairpin RNA that acts in trans to silence the 6b4 locus. The locations of the hybridization probes used in parts B, C and D are indicated. (B and D) The accumulation of RAV2, TCV-P38 and/or TuMV HC-Pro mRNA in plants of the genotypes indicated at the top of the lanes. (C and E) The top panel of each shows the accumulation of 6b4 GUS mRNA in plants of the genotypes indicated at the top of the lanes, and the bottom two panels show the accumulation of primary and secondary siRNAs in the same samples. The size of 21-, 22- and 24-nt marker RNAs are indicated to the left of the small RNA panels and the probes used are indicated to the right of each panel.

The role of RAV2 in HC-Pro suppression of hairpin transgene silencing was assayed using northern blot analysis to measure the accumulation of 6b4 GUS target mRNA as well as that of GUS primary and secondary siRNAs (Fig. 4C). As previously reported [7], HC-Pro blocked target RNA degradation when 306/6b4/HC-Pro transgenic plants were wild type for RAV2, showing the characteristic absence of secondary siRNAs accompanied by high levels of nonfunctional primary siRNAs (Fig. 4C, compare lanes 3 and 4). In contrast, HC-Pro failed to prevent degradation of the 6b4 GUS mRNA target in the rav2 knockout background (Fig. 4C, lanes 1 and 2). In addition, accumulation of GUS primary siRNAs was reduced in the rav2 compared to the RAV2 background and was similar to that in 306/6b4 plants without HC-Pro (Fig. 4C, lanes 1–4). Accumulation of secondary siRNAs, which are diagnostic of transitive silencing, was suppressed in HC-Pro transgenic plants even in the rav2 knockout background (Fig. 4C, lanes 1–3), suggesting that HC-Pro-suppression of transitive silencing is RAV2-independent. In this experiment, however, we cannot rule out the possibility that the rav2 knockout itself eliminates accumulation of secondary siRNAs. Therefore, our results suggest that RAV2 is required for the HC-Pro-mediated block in primary siRNA activity, but not for HC-Pro suppression of transitive silencing.

RAV2 is Required for Suppression of Hairpin Transgene-Induced Silencing by the Carmovirus Suppressor of Silencing, P38

To determine if RAV2 plays a general role in viral suppression of silencing, we used the 306/6b4 hairpin transgene silencing system to investigate whether Turnip Crinkle Virus (TCV) P38, a viral suppressor of silencing from a different virus family than TuMV HC-Pro [39], requires RAV2 to block silencing. The rav2 knockout line was crossed to a 306/6b4 line that expresses P38, and the resultant F1 plants were allowed to self-fertilize. F2 plants were genotyped, and individuals containing the 306/6b4/P38 loci in the homozygous rav2 mutant background were identified along with control plants containing all three loci in the RAV2 background.

We used northern blot analysis to confirm the expected pattern of expression of RAV2 and P38 in these two sets of plants (Fig. 4D) and to examine suppression of silencing by P38 in the presence and absence of RAV2. Previous experiments showed that P38 behaves much like HC-Pro in the 306/6b4 transgene silencing system, blocking silencing and allowing 6b4 GUS mRNA to accumulate, even though high levels of GUS primary siRNAs also accumulate [7]. Similar to HC-Pro, P38 also blocks transitive silencing in this system as indicated by the absence of GUS secondary siRNAs [7]. In the current work, P38 transgenic 306/6b4 plants with at least one copy of the wild type RAV2 locus replicated those earlier results, showing P38 suppression of silencing, with a concomitant increase in accumulation of GUS primary siRNAs and elimination of GUS secondary siRNAs (Fig. 4E, compare lanes 3 and 4). In contrast, P38 suppression of silencing was strongly diminished in the rav2 knockout background (Fig. 4E, lanes 1 and 2). Similar to our results with HC-Pro, accumulation of primary siRNAs in plants expressing P38 was much reduced in the rav2 compared to the RAV2 background, whereas secondary siRNA accumulation was unaffected by the loss of RAV2 and remained undetectable (Fig. 4E, compare lanes 1 and 2 with lane 3). The variability in accumulation of primary siRNAs observed in the rav2 background (Fig. 4E, lanes 1 and 2) probably reflects the facts that individual plants were tested and accumulation of primary siRNAs is greatly reduced, but not eliminated in the absence of RAV2. Altogether our results indicate that RAV2 plays similar roles in suppression of silencing by P38 and HC-Pro. Interestingly, in both cases, RAV2 function is required for suppression of primary siRNA-directed target degradation, but dispensable for the block to transitive silencing.

RAV2 is Required for Some of the Phenotypic Defects Induced by HC-Pro, but not for HC-Pro-Mediated Defects in the miRNA Pathway

Arabidopsis plants expressing TuMV HC-Pro display a number of developmental anomalies: the plants are dwarfed with serrated leaves and have abnormal flower morphology associated with severely reduced fertility (Fig. 5A; [14],[33]). The phenotype of homozygous rav2 knockout plants, however, is indistinguishable from that of wild type plants (data not shown). To determine if RAV2 is required for any of the HC-Pro associated developmental anomalies, we compared the phenotype of HC-Pro plants in the wild type RAV2 background to that of plants expressing approximately equal levels of HC-Pro mRNA, but in the rav2 knockout background. The HC-Pro-mediated defects in flower morphology and fertility are completely alleviated in the absence of RAV2 (Fig. 5 and data not shown). In addition, both the dwarfing and serrated leaf phenotypes are mitigated–but not eliminated–in the rav2 knockout background, resulting in an intermediate phenotype that is most visible when the plants are young (Fig. 5A), but becomes less distinguishable from that of wild type after the plants have flowered (Fig. 3A, 5A, and data not shown). These observations indicate that RAV2 is required for HC-Pro-mediated flower and fertility defects and contributes to the defects in plant size and leaf shape.

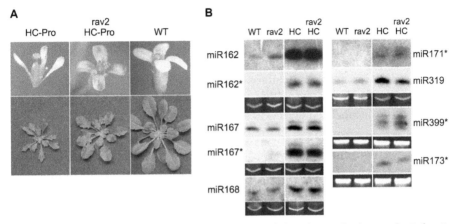

Figure 5. RAV2 is Required for Many HC-Pro-associated Morphological Anomalies but not for Defects in MicroRNA Biogenesis. (A) Flower morphological defects in HC-Pro transgenic plants (top left panel) are rescued in the rav2 knockout background (top middle panel) resulting in flower phenotype indistinguishable from wild type (top right panel). Rosette dwarfing and leaf serration in transgenic plants (bottom left panel) are partially rescued in the rav2 knockout background (bottom middle panel) resulting in a phenotype intermediate between wild type (bottom right panel) and Hc-Pro plants. (B) The accumulation of the indicated miRNAs and miRNA*s was determined from RNA gel blot analysis of low molecular weight RNA from wild type (WT), rav2 knockout plants (rav2), HC-Pro plants (HC) and HC-Pro plants in the rav2 knockout background (rav2, HC). Ethidium bromide (EtBr) staining of the predominant RNA species in the low molecular weight fraction is shown as a loading control.

In addition to its role in suppression of silencing, HC-Pro also causes defects in the biogenesis and function of certain endogenous small RNAs, including miRNAs, a class of small regulatory RNAs that plays critical roles in development. MiRNAs arise by processing of stem-loop primary transcripts by a Dicer-like enzyme, usually DCL1. The initial product is a 21-nt duplex, composed of the mature miRNA and the imperfectly complementary opposite strand, which is called miRNA*. The two strands separate and the mature miRNA binds to an AGO protein, forming the core of the miRNA effector complex. In HC-Pro transgenic plants, the level of many miRNAs is increased, often dramatically [33],[40]. Despite the increased level of the miRNA in the HC-Pro plants, the miRNA-targeted messenger RNAs also show an increased accumulation, suggesting that the miRNAs have reduced function [33],[41]. In addition, the miRNA* strand, which is unstable and fails to accumulate in wild type plants, characteristically accumulates to high levels in HC-Pro transgenic plants [33]. Together these results have led to the idea that HC-Pro impedes the proper separation of the strands of the miRNA:miRNA* duplex, leading to reduced association of the mature miRNA with AGO and thereby reducing miRNA function.

Because RAV2 is required for HC-Pro effects on the biogenesis and function of primary siRNAs, as well as for many of the HC-Pro-associated developmental anomalies, we hypothesized that RAV2 might also be required for HC-Pro-me-diated defects in the miRNA pathway. To address the role of RAV2 in HC-Pro-associated defects in miRNA biogenesis, we compared the levels of a variety of miRNAs and their corresponding miRNA* strands in HC-Pro plants in the pres-ence and absence of RAV2. In all cases, the levels of miRNA and miRNA* were independent of RAV2 (Fig. 5B). These results indicate that RAV2 is not required for the HC-Pro-associated defects in miRNA biogenesis.

To determine if RAV2 is involved in HC-Pro-associated defects in miRNA function, we compared the levels of a set of known miRNA-targeted messen-ger RNAs in RAV2/HC-Pro plants to those in rav2/HC-Pro plants using whole genome tiling microarray data (see following section for details of the tiling ar-ray experiments). Because HC-Pro interferes with the activity of some miRNAs [33],[41], we expected the tiling array data to show increased expression of at least some miRNA-targeted genes in HC-Pro plants. The tiling array data supported this expectation. Specifically, out of 146 verified miRNA targets [42],[43],[44],[45], we found that 39 showed altered expression in the HC-Pro transgenic line com-pared to the wild type control. Of these, 35 had increased expression, and only one of these was up-regulated in HC-Pro/RAV2 versus HC-Pro/rav2 plants, sug-gesting that RAV2 does not play a general role in HC-Pro inhibition of miRNA activity. Altogether, the results suggest that, although RAV2 is required for many of the morphological anomalies in HC-Pro transgenic plants, it is not required for the HC-Pro-mediated defects in either the biogenesis or function of miRNAs.

Whole genome tiling analysis links HC-Pro suppression of silencing to the network of host defense pathways

Because RAV2 is a transcription factor, we expected that it might be required for some HC-Pro-mediated changes in gene expression and that identifying these genes could provide insight into the role of RAV2 in HC-Pro suppression of silencing. To address this idea, we employed whole genome tiling microarray ex-periments to determine if the global pattern of gene expression is altered in HC-Pro transgenic plants and, if so, whether any of the changes are dependent on RAV2 function. Arabidopsis plants with four different genotypes were used in this experiment: 1) a rav2 mutant line, 2) an HC-Pro expressing line, 3) the rav2 mutant line expressing HC-Pro, and 4) the wild type (Columbia ecotype) control. We grew all four genotypes under identical conditions, extracted total RNA from plants just before bolting and used poly-A RNA to generate probes for hybrid-ization to the Arabidopsis tiling arrays as previously described [46],[47]. Tile-Map [48] was used to identify genes that are significantly up- or down-regulated in each line as compared to wild type plants, as well as to compare the pattern of gene expression in RAV2/HC-Pro plants versus rav2/HC-Pro plants (Tables S2–S9). To check the tiling results, the expression of ten genes in these plant lines was additionally examined using real-time quantitative PCR (RT qPCR). This analysis confirmed the relative levels of expression of these genes determined by the tiling array in 33 of 40 two-way comparisons between the four genotypes (Fig. 6A and B).

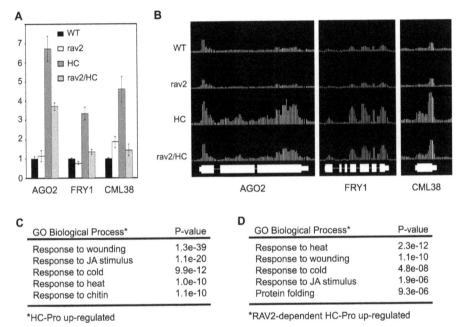

C

GO Biological Process*	P-value
Response to wounding	1.3e-39
Response to JA stimulus	1.1e-20
Response to cold	9.9e-12
Response to heat	1.0e-10
Response to chitin	1.1e-10

*HC-Pro up-regulated

D

GO Biological Process*	P-value
Response to heat	2.3e-12
Response to wounding	1.1e-10
Response to cold	4.8e-08
Response to JA stimulus	1.9e-06
Protein folding	9.3e-06

*RAV2-dependent HC-Pro up-regulated

Figure 6. Tiling Microarray Analysis and RT qPCR Show RAV2-dependent Up-regulation of Silencing-associated Genes by HC-Pro. (A) The mRNA levels for AGO2 (At1g31280), FRY1 (At5g63980) and CML38 (At1g76650) in rav2 knockout plants (rav2), HC-Pro transgenic plants (HC), HC plants in the rav2 knockout background (rav2/HC) and wild type control plants (WT) were determined by oligo(dT)-primed RT qPCR analysis. Error bars, ±SD. (B) The mRNA levels for the same genes shown in (A) were determined by Arabidopsis whole-genome tiling microarray expression analysis. The top four tracks show the level of these mRNAs in the genotypes indicated to the left of the track. The bottom track indicates the annotated gene models for the three loci. (C) Gene ontology (GO) analysis results for genes that are up-regulated in HC-Pro transgenic plants as compared to wild type plants. The top five over-represented biological processes categories and the associated hypergeometric distribution P-values are shown. (D) GO analysis results for genes that are up-regulated by HC-Pro in a RAV2-dependent manner. The top five over-represented biological processes categories and the associated hypergeometric distribution P-values are shown.

One of the first questions we addressed was whether genes involved in antiviral silencing and other small RNA pathways were affected by HC-Pro and RAV2. Unexpectedly, none of the genes encoding components of the silencing machinery or otherwise known to be required for silencing were down-regulated in the HC-Pro plants. Expression of RAV2 itself was also not altered in HC-Pro plants. However, a number of silencing-associated genes were up-regulated in HC-Pro plants. The up-regulated genes included three of the ten Arabidopsis

AGO family members, AGO2, AGO3, and AGO4. AGO4 is required for some kinds of transcriptional silencing. The roles of AGO2 and AGO3 are unknown, but neither has been associated with antiviral silencing [49],[50]. Interestingly, two genes implicated as endogenous suppressors of silencing were also up-regulated in HC-Pro: Arabidopsis FIERY1 (FRY1), which negatively regulates transitive silencing [51], and CML38 (At1g76650), which is a likely Arabidopsis homolog of rgsCaM, an endogenous suppressor of antiviral silencing in tobacco [23]. Like RAV2, rgsCaM was originally identified as an HC-Pro interacting protein [23]; however, it is not yet known whether rgsCaM is required for HC-Pro to suppress silencing. RT qPCR confirmed the relative expression levels of AGO2, FRY1, and CML38 in the HC-Pro expressing line compared to wild type plants (compare Fig. 6A and 6B). The RT qPCR data also showed that increases in both FRY1 and CML38 expression required RAV2, whereas the increase in AGO2 expression was only partially dependent on RAV2 (Fig. 6A). These results argue that the mechanism for HC-Pro suppression of silencing does not involve down-regulation of genes required for silencing, but rather a RAV2-dependent up-regulation of genes that potentially antagonize antiviral silencing.

The tiling array analysis was used to identify global HC-Pro-mediated changes in gene expression and determine which, if any, depended on RAV2. A significant number of genes were differentially regulated in the HC-Pro plants; 2580 were up-regulated and 2060 were down-regulated. Many fewer genes were differentially affected in RAV2/HC-Pro compared to rav2/HC-Pro plants (Tables S4 and S5). Of 265 genes that showed dependence on RAV2 for up-regulation by HC-Pro, only a small number showed changed expression in rav2 mutant plants in the absence of HC-Pro as compared to wild type (20 of 265 were up-regulated; 17 of 265 were down-regulated). Similarly, of 433 genes that showed dependence on RAV2 for down-regulation by HC-Pro, a relatively small number showed changed expression in the rav2 knockout plants in the absence of HC-Pro as compared to wild type (15 of 433 were up-regulated; 98 of 433 were down-regulated). Together, these results suggest that HC-Pro causes major changes in global gene expression patterns, some of which are mediated by RAV2. Interestingly, based on comparison of the set of genes with altered expression in rav2 mutant plants with the set altered by HC-Pro in a RAV2-dependent manner, it appears that HC-Pro changes the scope and spectrum of genes that are controlled by RAV2.

Gene Ontology (GO) term analysis was used to give a functional characterization of the tiling array results [52]. A key finding of this analysis was that multiple stress and defense responses were induced in HC-Pro expressing plants. The top four biological process categories that were over-represented among genes up-regulated in HC-Pro compared to wild type plants were: response to wounding (67 of 119 genes), response to jasmonic acid (JA) stimulus (48 of 119 genes), cold stress (49 of 197 genes) and heat stress (33 of 109 genes) (Fig. 6C). Strikingly, genes in these same four categories were also over-represented among the genes that are up-regulated by HC-Pro in a RAV2-dependent manner (Fig. 6D). Tables showing the specific genes that are up-regulated by HC-Pro in each of these GO categories, as well as the subsets that require RAV2 for HC-Pro up-regulation results indicate that RAV2 plays a role in altered expression of stress and defense pathways in HC-Pro plants. Interestingly, FRY1 and CML38, both of which have been implicated as suppressors of silencing [23],[51] and are induced by HC-Pro in a RAV2-dependent manner (Fig. 6B), have GO annotations of response to cold and wounding, respectively, suggesting a link between silencing and other stress and defense pathways.

Discussion

It has been over a decade since the first plant viral suppressors of RNA silencing were reported [53],[54],[55], providing an early clue that silencing serves as an anti-viral defense in plants and leading to the identification of many other such silencing suppressors [56]. However, the mechanisms by which these viral proteins manipulate silencing have remained largely elusive. Here we report the identification of a host protein, the transcription factor RAV2, that is required for suppression of silencing mediated by two unrelated viral proteins, potyviral HC-Pro and carmoviral P38. RAV2 is part of a gene family that comprises six members, two of which (RAV1; At1g13260 and RAV2-like; At1g25560) are very closely related to RAV2. Surprisingly, however, neither of these related genes is able to compensate for the loss of RAV2 with respect to suppression of silencing mediated by either HC-Pro or P38. This result indicates that RAV2 provides a unique function in suppression of silencing. The identification of RAV2 as an important element in viral suppression of silencing provides a handle for identifying additional host partners and thereby unraveling the pathway of host involvement in that process.

The discovery that plant viruses from many unrelated families encode suppressors of silencing has underscored the importance of silencing in antiviral defense. Similarly, we expect our finding that viral suppressors from two unrelated viruses have evolved independently to require RAV2 underscores the importance of host proteins in viral counter-defense. In addition, it suggests that RAV2 represents an effective and readily subverted control point—either for suppression of silencing in general or for a subset of suppressors with some mechanistic features in common. It will be interesting to see how general the requirement for RAV2 is in viral suppression of silencing.

How could a transcription factor such as RAV2 be used to suppress silencing? Two reports have identified RAV2 as a repressor of at least some target genes [57],[58]. Therefore, it seemed reasonable to hypothesize that the role of RAV2 in HC-Pro suppression of silencing is to repress transcription of genes that encode components of the silencing machinery for the anti-viral branch of the silencing pathway. However, our global analysis of genome expression indicates that the expression of genes known to be required for RNA silencing is unchanged in HC-Pro transgenic plants as compared to wild type controls. Instead, our data shows that RAV2 is required for HC-Pro-mediated up-regulation of some stress and defense response genes. Earlier work showing that induction of both biotic and abiotic stresses interferes with RNA silencing induced by a viral amplicon in tobacco is consistent with a mechanism in which induction of other defense responses can divert the host from antiviral silencing [59]. The observation that RAV2 is induced by the ethylene defense pathway and is also required for viral suppression of silencing emphasizes the importance of crosstalk among defense pathways and supports the idea that RAV2 constitutes an important control point for the integration of defense responses during virus infection.

One puzzle raised by the observation that HC-Pro, which is a cytoplasmic protein [60],[61], interacts with a host transcription factor is: How and where do the two proteins have the opportunity to meet? Although HC-Pro has been shown to accumulate in nuclear inclusions in certain potyviral infections, it is thought that such inclusions represent storage of excess protein [61]. Thus, it seems more likely that HC-Pro and RAV2 interact in the cytoplasm. Sequestering transcription factors in the cytoplasm is a common mechanism used in eukaryotic organisms for controlling the activity of such proteins [62],[63]. The interaction of HC-Pro with RAV2 in the cytoplasm could either reflect a direct involvement of RAV2 itself in suppression of silencing or interference by HC-Pro in the cellular control of RAV2—either to block activation or promote inappropriate activation—thereby changing host gene expression in such a way that promotes suppression of silencing. Elucidating these issues, as well as examining whether P38 also physically interacts with RAV2, is likely to be a fruitful area of research.

Another particularly interesting aspect of our results is the differential requirement for RAV2 in suppression of different small RNA-mediated processes. Both HC-Pro and P38 suppress transitive silencing in the absence of RAV2; yet, both suppressors require RAV2 for suppression of target degradation via the activity of primary siRNAs. Furthermore, although HC-Pro requires RAV2 to block the activity of primary siRNAs, RAV2 is not required for HC-Pro-mediated defects in miRNA activity. Our present work does not distinguish whether these differential requirements for RAV2 indicate a fundamental difference in the mechanisms responsible for suppression of these processes or simply a difference in the cofactor requirements of a common mechanism.

One current model for viral suppression of small RNA pathways posits a general mechanism in which small RNA duplexes are bound by the suppressor, thereby blocking the incorporation of one strand of the duplex into an active effector complex [17],[64]. Our data showing a role for RAV2 in suppression of silencing does not directly support this proposed mechanism, but is also not inconsistent with it. Indeed, it has been shown that small RNA binding by HC-Pro in vitro is enhanced by unknown cellular factors [17],[64]. Thus, RAV2 might be one such factor, acting either directly or indirectly to enhance small RNA binding.

Expression of HC-Pro in transgenic plants causes a set of morphological anomalies that have been attributed to defects in the biogenesis and function of endogenous miRNAs [33]. However, there is emerging evidence that suggests that the phenotypic changes are largely independent of the miRNA pathway [14],[15],[20],[65]. In support of this notion, the data we have presented here indicate that many of the HC-Pro-mediated morphological anomalies are RAV2-dependent whereas the defects in the miRNA pathway are RAV2-independent, arguing against a causative role for miRNAs in most HC-Pro-associated morphological anomalies

Although the mechanism by which HC-Pro uses RAV2 to suppress silencing is not yet clear, the results of our tiling array analysis suggest two interesting, though speculative, possibilities. The first of these relates to the induction of AGO2 and a subset of other AGO genes in HC-Pro transgenic plants, an effect that is only partially dependent on RAV2. The AGO genes that are up-regulated by HC-Pro are not required for post-transcriptional gene silencing (PTGS). These results suggest that an alteration of the mix of AGO proteins in the cell might tip the balance away from PTGS towards other small RNA pathways that are not directly involved in anti-viral defense. The recent demonstration that changing the 5' nucleotide of a miRNA so as to favor binding to AGO2 instead of AGO1 inactivates that miRNA [66] supports the idea that an overabundance of the wrong AGO proteins could contribute to suppression of silencing. The second interesting possibility suggested by our tiling data concerns the result that HC-Pro requires RAV2 to induce expression of FRY1 and CML38, both of which have been implicated as endogenous suppressors of silencing and both of which are associated with stress or defense responses. Induction of endogenous suppressors of silencing may be more widespread than we know because most have probably not yet been identified [51]. It is tempting to speculate that the induction of stress and defense pathways by HC-Pro might have the counter-productive result—from the plant's perspective—of inducing a set of endogenous suppressors of antiviral silencing.

Materials and Methods

Plant Material and Transgenic Lines

The tobacco 6b5 [32] and Arabidopsis TuMV HC-Pro [CT25 [14]], TCV-P38 [39], 306 and 6b4 [35] lines have been previously described. The Arabidopsis rav2/edf2 (At1g68840) T-DNA insertion line (SALK_070847) was used and did not express detectable levels of RAV2 mRNA as assayed by northern analysis.

GUS Histochemical Staining

Histochemical staining for GUS activity was carried out as described [53].

VIGS Silencing Assays

The silencing of endogenous CH42 expression using the geminivirus CaLCV vector was performed exactly as described previously [14].

RNA Isolation and Northern Analysis

RNA isolation and RNA gel blot analysis of high and low molecular weight RNA were performed exactly as previously described [14],[40],[67]. Probes for detection of TuMV HC-Pro, TCV-P38 and 6b4 mRNAs, miRNA as well as those for primary and secondary siRNAs from the 6b4/306 transgene silencing system were previously described [7]. The RAV2 probe was generated using the primer set (5′ primer-TTGGAAAGTTCGGTCTGGTC and 3′ primer-TAATACGACTCAC-TATAGGGACCGCAAACATATCATCAACATCTC), which generate a 152 bp fragment from the 3′ end of the gene. The 3′ RAV2 primer contains T7 promoter sequences and a 4 nucleotide spacer at its 5′ end to facilitate synthesis of the probe using T7 polymerase.

GST Pulldown Assays

To determine if HC-Pro-GST and ntRAV interact, approximately equimolar amounts of GST or HC-Pro-GST fusion protein were added to 20 μl of glutathione sepharose 4B beads (GE Healthcare) in GLB buffer (50 mM Tris-HCl, pH 8.0, 150 mM NaCl, 1 mM EDTA, and 1 mM PMSF) supplemented to contain 100 μg/ml BSA and 0.1% NP-40 (Roche) and shaken gently for 1 hour at 4°C. After rinsing with supplemented GLB, an equal amount of 35S-methionine labeled ntRAV was added to each sample, shaken gently at 4°C for 2 hours and rinsed again with supplemented GLB. Bound protein was eluted from the beads with Laemmli sample buffer, resolved by SDS-PAGE, and transferred to PVDF membrane. 35S-methionine labeled ntRAV was visualized by autoradiography.

Co-Immunoprecipitation of RAV2 and HC-Pro

Protein was extracted from 0.5 g of Arabidopsis rosette leaf tissue by the following procedure. Tissue was frozen in liquid nitrogen, ground into powder with a mortar and pestle, homogenized in 4 ml of protein extraction buffer (40 mM Tris-Cl, pH 8.0, 200 mM NaCl, 2.5 mM EDTA, 1% Triton X-100, 0.1% NP-40) containing protease inhibitor cocktail (Roche), and centrifuged (12,000 g at 4°C). The supernatant was incubated with 100 µl pre-washed anti-FLAG M2 agarose beads (Sigma F2426) at 4°C for two hours. Agarose beads containing protein complexes were washed three times with extraction buffer, boiled in SDS sample buffer, resolved on a 10% SDS polyacrylamide gel, and subjected to western blotting. The presence of RAV2 protein was detected using a rabbit anti-RAV2 peptide antibody generated from the peptide GGKRSRDVDDMFALRC, and a rabbit anti-HC-Pro peptide antibody generated from the peptide KEFT-KVVRDKLVGE was used to detect HC-Pro. Both RAV2 and HC-Pro peptide antibodies were produced by Sigma-Genosys.

Tiling Microarray Analysis

Total RNA was isolated as described above from the above ground portions of six week old plants that had not yet bolted. Generation of probes to poly-A RNA and hybridization to the tiling arrays were performed as described previously [46],[47]. The data was analyzed using the program TileMap with a posterior probability of 0.8 [48]. The TileMap program identifies sequences that have significant changes in expression compared to controls, but does not provide fold-differences in expression levels. GO analysis was performed using ProfCom [68].

Competing Interests

The authors have declared that no competing interests exist.

Acknowledgements

We thank Jim Carrington for seeds of the TCV-P38 transgenic line and Herve Vaucheret for seeds of the 6b4 and 306 lines.

Authors' Contributions

Conceived and designed the experiments: JRE LHB VV. Performed the experiments: MWE BDG ZG AWF SM XG. Analyzed the data: MWE BDG GJP LHB VV. Contributed reagents/materials/analysis tools: JRE. Wrote the paper: MWE GJP VV.

References

1. Fujita M, Fujita Y, Noutoshi Y, Takahashi F, Narusaka Y, et al. (2006) Crosstalk between abiotic and biotic stress responses: a current view from the points of convergence in the stress signaling networks. Curr Opin Plant Biol 9: 436–442.

2. Kunkel BN, Brooks DM (2002) Cross talk between signaling pathways in pathogen defense. Current Opinion in Plant Biology 5: 325–331.

3. Sunkar R, Zhu JK (2004) Novel and stress-regulated microRNAs and other small RNAs from Arabidopsis. Plant Cell 16: 2001–2019. Epub 2004 Jul 2016.

4. Hamilton AJ, Baulcombe DC (1999) A species of small antisense RNA in posttranscriptional gene silencing in plants. Science 286: 950–952.

5. Ratcliff F, Harrison BD, Baulcombe DC (1997) A Similarity Between Viral Defense and Gene Silencing in Plants. Science 276: 1558–1560.

6. Ding SW, Voinnet O (2007) Antiviral immunity directed by small RNAs. Cell 130: 413–426.

7. Mlotshwa S, Pruss GJ, Peragine A, Endres MW, Li J, et al. (2008) DICER-LIKE2 plays a primary role in transitive silencing of transgenes in Arabidopsis. PLoS ONE 3: e1755.

8. Fagard M, Boutet S, Morel JB, Bellini C, Vaucheret H (2000) AGO1, QDE-2, and RDE-1 are related proteins required for post-transcriptional gene silencing in plants, quelling in fungi, and RNA interference in animals. Proc Natl Acad Sci USA 97: 11650–11654.

9. Mourrain P, Beclin C, Elmayan T, Feuerbach F, Godon C, et al. (2000) Arabidopsis SGS2 and SGS3 genes are required for posttranscriptional gene silencing and natural virus resistance. Cell 101: 533–542.

10. Chuck G, Candela H, Hake S (2009) Big impacts by small RNAs in plant development. Curr Opin Plant Biol 12: 81–86.

11. Mallory AC, Vaucheret H (2006) Functions of microRNAs and related small RNAs in plants. Nat Genet 38: S31–36.

12. Sunkar R, Chinnusamy V, Zhu J, Zhu JK (2007) Small RNAs as big players in plant abiotic stress responses and nutrient deprivation. Trends Plant Sci 12: 301–309.

13. Qu F, Ye X, Morris TJ (2008) Arabidopsis DRB4, AGO1, AGO7, and RDR6 participate in a DCL4-initiated antiviral RNA silencing pathway negatively regulated by DCL1. Proc Natl Acad Sci USA 105: 14732–14737.

14. Mlotshwa S, Schauer SE, Smith TH, Mallory AC, Herr JM Jr, et al. (2005) Ectopic DICER-LIKE1 expression in P1/HC-Pro Arabidopsis rescues phenotypic anomalies but not defects in microRNA and silencing pathways. Plant Cell 17: 2873–2885.

15. Diaz-Pendon JA, Ding SW (2008) Direct and indirect roles of viral suppressors of RNA silencing in pathogenesis. Annu Rev Phytopathol 46: 303–326.

16. Burgyan J (2008) Role of silencing suppressor proteins. Methods Mol Biol 451: 69–79.

17. Lakatos L, Csorba T, Pantaleo V, Chapman EJ, Carrington JC, et al. (2006) Small RNA binding is a common strategy to suppress RNA silencing by several viral suppressors. Embo J 25: 2768–2780.

18. Merai Z, Kerenyi Z, Kertesz S, Magna M, Lakatos L, et al. (2006) Double-stranded RNA binding may be a general plant RNA viral strategy to suppress RNA silencing. J Virol 80: 5747–5756.

19. Baumberger N, Tsai CH, Lie M, Havecker E, Baulcombe DC (2007) The Polerovirus silencing suppressor P0 targets ARGONAUTE proteins for degradation. Curr Biol 17: 1609–1614.

20. Deleris A, Gallego-Bartolome J, Bao J, Kasschau KD, Carrington JC, et al. (2006) Hierarchical action and inhibition of plant Dicer-like proteins in antiviral defense. Science 313: 68–71.

21. Glick E, Zrachya A, Levy Y, Mett A, Gidoni D, et al. (2008) Interaction with host SGS3 is required for suppression of RNA silencing by tomato yellow leaf curl virus V2 protein. Proc Natl Acad Sci USA 105: 157–161.

22. Zhang X, Yuan YR, Pei Y, Lin SS, Tuschl T, et al. (2006) Cucumber mosaic virus-encoded 2b suppressor inhibits Arabidopsis Argonaute1 cleavage activity to counter plant defense. Genes Dev 20: 3255–3268.

23. Anandalakshmi R, Marathe R, Ge X, Herr JM Jr, Mau C, et al. (2000) A calmodulin-related protein that suppresses posttranscriptional gene silencing in plants. Science 290: 142–144.

24. Kagaya Y, Ohmiya K, Hattori T (1999) RAV1, a novel DNA-binding protein, binds to bipartite recognition sequence through two distinct DNA-binding domains uniquely found in higher plants. Nucleic Acids Res 27: 470–478.

25. Sohn KH, Lee SC, Jung HW, Hong JK, Hwang BK (2006) Expression and functional roles of the pepper pathogen-induced transcription factor RAV1 in bacterial disease resistance, and drought and salt stress tolerance. Plant Mol Biol 61: 897–915.

26. Kim SY, Kim YC, Lee JH, Oh SK, Chung E, et al. (2005) Identification of a CaRAV1 possessing an AP2/ERF and B3 DNA-binding domain from pepper leaves infected with Xanthomonas axonopodis pv. glycines 8ra by differential display. Biochim Biophys Acta 1729: 141–146.

27. Kagaya Y, Hattori T (2009) Arabidopsis transcription factors, RAV1 and RAV2, are regulated by touch-related stimuli in a dose-dependent and biphasic manner. Genes Genet Syst 84: 95–99.

28. Fowler S, Thomashow MF (2002) Arabidopsis transcriptome profiling indicates that multiple regulatory pathways are activated during cold acclimation in addition to the CBF cold response pathway. Plant Cell 14: 1675–1690.

29. Alonso JM, Stepanova AN, Leisse TJ, Kim CJ, Chen H, et al. (2003) Genome-wide insertional mutagenesis of Arabidopsis thaliana. Science 301: 653–657.

30. Broekaert WF, Delaure SL, De Bolle MF, Cammue BP (2006) The role of ethylene in host-pathogen interactions. Annu Rev Phytopathol 44: 393–416.

31. Guo H, Ecker JR (2004) The ethylene signaling pathway: new insights. Curr Opin Plant Biol 7: 40–49.

32. Elmayan T, Vaucheret H (1996) Expression of single copies of a strongly expressed 35S transgene can be silenced post-transcriptionally. Plant Journal 9: 787–797.

33. Kasschau KD, Xie Z, Allen E, Llave C, Chapman EJ, et al. (2003) P1/HC-Pro, a viral suppressor of RNA silencing, interferes with Arabidopsis development and miRNA function. Dev Cell 4: 205–217.

34. Turnage MA, Muangsan N, Peele CG, Robertson D (2002) Geminivirus-based vectors for gene silencing in Arabidopsis. Plant J 30: 107–114.

35. Beclin C, Boutet S, Waterhouse P, Vaucheret H (2002) A branched pathway for transgene-induced RNA silencing in plants. Curr Biol 12: 684–688.

36. Daxinger L, Hunter B, Sheikh M, Jauvion V, Gasciolli V, et al. (2008) Unexpected silencing effects from T-DNA tags in Arabidopsis. Trends Plant Sci 13: 4–6.

37. Marathe R, Smith TH, Anandalakshmi R, Bowman LH, Fagard M, et al. (2000) Plant viral suppressors of post-transcriptional silencing do not suppress transcriptional silencing. Plant J 22: 51–59.

38. Mette MF, Matzke AJ, Matzke MA (2001) Resistance of RNA-mediated TGS to HC-Pro, a viral suppressor of PTGS, suggests alternative pathways for dsRNA processing. Curr Biol 11: 1119–1123.

39. Chapman EJ, Prokhnevsky AI, Gopinath K, Dolja VV, Carrington JC (2004) Viral RNA silencing suppressors inhibit the microRNA pathway at an intermediate step. Genes Dev 18: 1179–1186.

40. Mallory AC, Reinhart BJ, Bartel D, Vance VB, Bowman LH (2002) From the Cover: A viral suppressor of RNA silencing differentially regulates the accumulation of short interfering RNAs and micro-RNAs in tobacco. Proc Natl Acad Sci USA 99: 15228–15233.

41. Dunoyer P, Lecellier CH, Parizotto EA, Himber C, Voinnet O (2004) Probing the microRNA and small interfering RNA pathways with virus-encoded suppressors of RNA silencing. Plant Cell 16: 1235–1250.

42. Allen E, Xie Z, Gustafson AM, Carrington JC (2005) microRNA-directed phasing during trans-acting siRNA biogenesis in plants. Cell 121: 207–221.

43. German MA, Pillay M, Jeong DH, Hetawal A, Luo S, et al. (2008) Global identification of microRNA-target RNA pairs by parallel analysis of RNA ends. Nat Biotechnol 26: 941–946.

44. Fahlgren N, Howell MD, Kasschau KD, Chapman EJ, Sullivan CM, et al. (2007) High-throughput sequencing of Arabidopsis microRNAs: evidence for frequent birth and death of MIRNA genes. PLoS ONE 2: e219.

45. Howell MD, Fahlgren N, Chapman EJ, Cumbie JS, Sullivan CM, et al. (2007) Genome-wide analysis of the RNA-DEPENDENT RNA POLYMERASE6/DICER-LIKE4 pathway in Arabidopsis reveals dependency on miRNA- and tasiRNA-directed targeting. Plant Cell 19: 926–942.

46. Gregory BD, O'Malley RC, Lister R, Urich MA, Tonti-Filippini J, et al. (2008) A link between RNA metabolism and silencing affecting Arabidopsis development. Dev Cell 14: 854–866.

47. Chekanova JA, Gregory BD, Reverdatto SV, Chen H, Kumar R, et al. (2007) Genome-wide high-resolution mapping of exosome substrates reveals hidden features in the Arabidopsis transcriptome. Cell 131: 1340–1353.

48. Ji H, Wong WH (2005) TileMap: create chromosomal map of tiling array hybridizations. Bioinformatics 21: 3629–3636.

49. Mallory AC, Elmayan T, Vaucheret H (2008) MicroRNA maturation and action–the expanding roles of ARGONAUTEs. Curr Opin Plant Biol 11: 560–566.

50. Vaucheret H (2008) Plant ARGONAUTES. Trends Plant Sci 13: 350–358.

51. Gy I, Gasciolli V, Lauressergues D, Morel JB, Gombert J, et al. (2007) Arabidopsis FIERY1, XRN2, and XRN3 are endogenous RNA silencing suppressors. Plant Cell 19: 3451–3461.

52. Thomas PD, Mi H, Lewis S (2007) Ontology annotation: mapping genomic regions to biological function. Curr Opin Chem Biol 11: 4–11.

53. Anandalakshmi R, Pruss GJ, Ge X, Marathe R, Mallory AC, et al. (1998) A viral suppressor of gene silencing in plants. Proc Natl Acad Sci USA 95: 13079–13084.

54. Brigneti G, Voinnet O, Li WX, Ji LH, Ding SW, et al. (1998) Viral pathogenicity determinants are suppressors of transgene silencing in Nicotiana benthamiana. Embo J 17: 6739–6746.

55. Kasschau KD, Carrington JC (1998) A counterdefensive strategy of plant viruses: suppression of posttranscriptional gene silencing. Cell 95: 461–470.

56. Roth BM, Pruss GJ, Vance VB (2004) Plant viral suppressors of RNA silencing. Virus Res 102: 97–108.

57. Castillejo C, Pelaz S (2008) The balance between CONSTANS and TEMPRANILLO activities determines FT expression to trigger flowering. Curr Biol 18: 1338–1343.

58. Ikeda M, Ohme-Takagi M (2009) A novel group of transcriptional repressors in Arabidopsis. Plant Cell Physiol 50: 970–975.

59. Taliansky M, Kim SH, Mayo MA, Kalinina NO, Fraser G, et al. (2004) Escape of a plant virus from amplicon-mediated RNA silencing is associated with biotic or abiotic stress. Plant J 39: 194–205.

60. Mlotshwa S, Verver J, Sithole-Niang I, Gopinath K, Carette J, et al. (2002) Subcellular location of the helper component-proteinase of Cowpea aphidborne mosaic virus. Virus Genes 25: 207–216.

61. Riedel D, Lesemann DE, Maiss E (1998) Ultrastructural localization of nonstructural and coat proteins of 19 potyviruses using antisera to bacterially expressed proteins of plum pox potyvirus. Arch Virol 143: 2133–2158.

62. Garcia AV, Parker JE (2009) Heaven's Gate: nuclear accessibility and activities of plant immune regulators. Trends Plant Sci 14: 479–487.

63. Lee Y, Lee HS, Lee JS, Kim SK, Kim SH (2008) Hormone- and light-regulated nucleocytoplasmic transport in plants: current status. J Exp Bot 59: 3229–3245.

64. Silhavy D, Burgyan J (2004) Effects and side-effects of viral RNA silencing suppressors on short RNAs. Trends Plant Sci 9: 76–83.

65. Diaz-Pendon JA, Li F, Li WX, Ding SW (2007) Suppression of antiviral silencing by cucumber mosaic virus 2b protein in Arabidopsis is associated with drastically reduced accumulation of three classes of viral small interfering RNAs. Plant Cell 19: 2053–2063.

66. Mi S, Cai T, Hu Y, Chen Y, Hodges E, et al. (2008) Sorting of small RNAs into Arabidopsis argonaute complexes is directed by the 5′ terminal nucleotide. Cell 133: 116–127.

67. Mlotshwa S, Yang Z, Kim Y, Chen X (2006) Floral patterning defects induced by Arabidopsis APETALA2 and microRNA172 expression in Nicotiana benthamiana. Plant Mol Biol 61: 781–793.

68. Antonov AV, Schmidt T, Wang Y, Mewes HW (2008) ProfCom: a web tool for profiling the complex functionality of gene groups identified from high-throughput data. Nucleic Acids Res 36: W347–351.

CITATION

Originally published under the Creative Commons Attribution License. Endres MW, Gregory BD, Gao Z, Foreman AW, Mlotshwa S, et al. (2010) Two Plant Viral Suppressors of Silencing Require the Ethylene-Inducible Host Transcription Factor RAV2 to Block RNA Silencing. PLoS Pathog 6(1): e1000729. doi:10.1371/journal.ppat.1000729.

Enhanced Disease Susceptibility 1 and Salicylic Acid Act Redundantly to Regulate Resistance Gene-Mediated Signaling

Srivathsa C. Venugopal, Rae-Dong Jeong, Mihir K. Mandal,
Shifeng Zhu, A. C. Chandra-Shekara, Ye Xia, Matthew Hersh,
Arnold J. Stromberg, DuRoy Navarre, Aardra Kachroo
and Pradeep Kachroo

ABSTRACT

Resistance (R) protein–associated pathways are well known to participate in defense against a variety of microbial pathogens. Salicylic acid (SA) and its associated proteinaceous signaling components, including enhanced disease susceptibility 1 (EDS1), non–race-specific disease resistance 1 (NDR1), phytoalexin deficient 4 (PAD4), senescence associated gene 101 (SAG101), and EDS5, have been identified as components of resistance derived from many R proteins. Here, we show that EDS1 and SA fulfill redundant functions in defense signaling mediated by R proteins, which were thought to function independent of EDS1 and/or SA. Simultaneous mutations in EDS1 and the SA–synthesizing enzyme SID2 compromised hypersensitive response and/or resistance mediated by R proteins that contain coiled coil domains at their N-terminal ends. Furthermore, the expression of R genes and the associated defense signaling induced in response to a reduction in the level of oleic acid were also suppressed by compromising SA biosynthesis in the eds1 mutant background. The functional redundancy with SA was specific to EDS1. Results presented here redefine our understanding of the roles of EDS1 and SA in plant defense.

AUTHOR SUMMARY

Salicylic acid and enhanced disease susceptibility 1 are important components of resistance gene-mediated defense signaling against diverse pathogens in a variety of plants. Present understanding of plant defense signaling pathways places salicylic acid and enhanced disease susceptibility 1 downstream of resistant protein activation. In addition, enhanced disease susceptibility 1 is primarily thought to function in the signaling initiated via Toll-interleukin 1-receptor type of resistance proteins. Here, we show that salicylic acid and enhanced disease susceptibility 1 serve redundant functions in defense signaling mediated by coiled-coil-domain containing resistance proteins that were thought to function independent of enhanced disease susceptibility 1. Furthermore, resistance signaling induced under low oleic acid conditions also requires enhanced disease susceptibility 1 and salicylic acid in a redundant manner, but these components are required upstream of resistance gene expression. Together, these results show that the functional redundancy between salicylic acid and enhanced disease susceptibility 1 has precluded their detection as required components of many resistance protein–signaling pathways.

Introduction

Plants have evolved highly specific mechanisms to resist pathogens. One of the common ways to counter pathogen growth involves the deployment of resistant (R) proteins, which confer protection against specific races of pathogens carrying corresponding avirulence (Avr) genes [1]. Following recognition of the pathogen, one or more signal transduction pathways are induced in the host plant and these lead to the prevention of colonization by the pathogen. Induction of defense responses is often accompanied by localized cell death at the site of pathogen entry. This phenomenon, termed the hypersensitive response (HR), is one of the earliest visible manifestations of induced defense reactions and resembles programmed cell death in animals [1]–[6]. Concurrent with HR development, defense reactions are triggered in both local and distant parts of the plant and accompanied by a local and systemic increase in endogenous salicylic acid (SA) levels and the upregulation of a large set of defense genes, including those encoding pathogenesis-related (PR) proteins [7]–[9].

The SA signal transduction pathway plays a key role in plant defense signaling (see reviews in [10]–[12]). Arabidopsis mutants that are impaired in SA responsiveness, such as npr1 (Nonexpressor of PR; [13]–[15]), or are defective in pathogen-induced SA accumulation, such as eds1 (Enhanced Disease Susceptibility 1; [16]), eds5 (Enhanced Disease Susceptibility 5; [17]), sid2 (isochorishmate synthase; [18]) and pad4 (Phytoalexin Deficient 4; [19]), exhibit enhanced susceptibility to pathogen infection and show impaired PR gene expression. The EDS1, EDS5, PAD4, NPR1 and SID2 proteins participate in both basal disease resistance to virulent pathogens as well as R protein-mediated resistance to avirulent pathogens [20]. Defense signaling mediated via a majority of R proteins, which contain Toll-interleukin1-like (TIR) domains at their N-terminal ends, is dependent on EDS1 [21]. Conversely, the NDR1 (Non-race-specific Disease Resistance) protein is required for many R proteins that contain coiled-coil (CC) domains at their N-terminal ends. However, several CC-nucleotide binding site (NBS)-leucine rich repeat (LRR) type of R proteins, including RPP8, RPP13-Nd, HRT, and RPP7, signal resistance via a pathway(s) that is independent of NDR1 [21], [22]–[24]. Strikingly, the CC-NBS-LRR gene HRT, which confers resistance to Turnip Crinkle Virus (TCV), is dependent on EDS1 [23]. Besides HRT, the only other CC domain-containing R protein that utilizes an EDS1-dependent pathway is RPW8, which confers broad-spectrum resistance to powdery mildew [25]. However, RPW8 is not a typical NBS-LRR type of R protein; it contains an N-terminal transmembrane domain in addition to the CC domain. Although several components contributing to resistance against pathogens have been identified, the molecular signaling underlying R gene-mediated resistance still remains obscure. Furthermore, potential relationship(s) among different downstream components and how they relay information leading to resistance remains unknown.

The EDS1 and PAD4 proteins are structurally related to lipase/esterase-like proteins although their lipase-like biochemical functions have not been demonstrated [16],[19]. EDS1 interacts with PAD4 and SAG (senescence associated gene) 101 and the combined activities of these proteins are required for HR formation and to restrict the growth of virulent bacterial strains [26]. PAD4 and SAG101 also restrict the post-invasive growth of non-pathogenic fungi in Arabidopsis [27].

In addition to the major phytohormone-mediated defense pathways, fatty acid (FA)-derived signals have emerged as important mediators of defense signaling [28]–[35]. The Arabidopsis SSI2/FAB2-encoded stearoyl-acyl carrier protein-desaturase (SACPD) converts stearic acid (18:0) to oleic acid (18:1). A mutation in SSI2 results in the accumulation of 18:0 and a reduction in 18:1 levels. The mutant plants show stunting, spontaneous lesion formation, constitutive PR gene expression, and enhanced resistance to bacterial and oomycete pathogens [29],[36]. Characterization of ssi2 suppressor mutants has shown that the altered defense-related phenotypes are the result of the reduction in the levels of the unsaturated FA, 18:1 [30], [31], [35], [37]–[40]. The altered defense-related phenotypes in ssi2 plants can be rescued by restoring the 18:1 levels via second site mutations in genes encoding a glycerol-3-phosphate (G3P) acyltransferase [ACT1, 30], a G3P dehydrogenase [GLY1, 31], and an acyl carrier protein [ACP4, 35]. A mutation in act1 disrupts the acylation of G3P with 18:1 resulting in the increased accumulation of 18:1, thereby restoring wild-type (wt) phenotypes in ssi2 plants. ACT1 preferentially utilizes 18:1 conjugated to the ACP4 isoform in Arabidopsis [35]. Thus, a mutation in acp4 produces similar phenotypes as the act1 mutant and suppresses ssi2-mediated signaling by increasing 18:1 levels [35]. A mutation in GLY1 also restores 18:1 levels in ssi2 gly1 plants because it disrupts the formation of G3P from dihydroxyacetone phosphate [31]. Reduced availability of G3P in turn impairs the ACT1-catalyzed reaction resulting in accumulation of 18:1 in ssi2 gly1 plants. Concurrently, increasing the endogenous G3P levels via exogenous application of glycerol reduces 18:1 levels and induces ssi2-like phenotypes in wt plants [31],[40]. This effect of glycerol is highly specific because ssi2-associated phenotypes are not induced upon glycerol treatment of act1 (defective in the acylation of G3P with 18:1) or gli1 (defective in the phosphorylation of glycerol to G3P) mutants [40].

Recently, we showed that a reduction in 18:1 levels upregulates the expression of several R genes in an SA-independent manner [37]. Furthermore, we showed that pathogen resistance induced via this mode bypasses the requirement for components that are normally required for signaling downstream of R protein activation. For example, resistance to TCV mediated by the R gene HRT (HR to TCV), requires the recessive locus rrt (regulates resistance to TCV), SA, EDS1 and PAD4 [23]. Exogenous application of SA induces the expression of HRT and overcomes the requirement for rrt. However, exogenous SA is unable to induce HRT or confer resistance in pad4 background [23]. Interestingly, even though a reduction in 18:1 levels also upregulates HRT expression to confer resistance to TCV, this mode of resistance is independent of PAD4, SA, EDS1 and EDS5, which are required for HRT-mediated resistance to TCV [37]. Remarkably, induction of R genes in response to reduced 18:1 is conserved in plants as diverse as Arabidopsis and soybean [41]. Furthermore, this low 18:1-mediated induction of defense responses was also demonstrated in rice recently [42]. Together, these studies strengthen the conserved role of 18:1 in plant defense signaling.

Here, we show that R gene expression induced in response to a reduction in 18:1 levels and the associated defense signaling can be suppressed by simultaneous mutations in EDS1 and the genes governing synthesis of SA. We also show that EDS1 and SA function redundantly in R gene-mediated resistance against bacterial, viral and oomycete pathogens and that EDS1 also regulates signaling mediated by CC domain containing R proteins.

Results

EDS1 and SA are Essential but Redundant Components Required for R Gene Expression Induced in Response to a Reduction in 18:1 Levels

Signaling mediated by many R genes is known to require EDS1 and/or NDR1. Previously, we have shown that ssi2 eds1 plants continue to express R genes at high levels, including those that are dependent on EDS1 for their signaling [37]. To determine if NDR1 played a role in ssi2-triggered phenotypes, we generated ssi2 ndr1 plants. The double-recessive plants segregated in a Mendelian fashion and all ssi2 ndr1 plants showed ssi2-like morphology in the F2, F3 and F4 generations (Figure 1A). Although the ssi2 ndr1 plants accumulated significantly less SA/SAG (Figure 1C), compared to ssi2 plants, they showed ssi2-like PR-1 and R gene expression (Figure 1D and 1E). Exogenous glycerol application, which reduces 18:1 levels, also induced R gene expression in eds1 and ndr1 plants (data not shown). Together, these results suggest that R gene expression induced by low 18:1 levels does not require EDS1 or NDR1.

The SA/SAG levels in ssi2 eds1 and ssi2 ndr1 plants were significantly higher compared to those in wt plants (Figure 1C). To determine whether high SA in these genotypes was responsible for increased R gene expression, we generated ssi2 eds1 sid2 and ssi2 ndr1 sid2 plants. Interestingly, only the ssi2 eds1 sid2 plants showed wt-like morphology and did not develop visible or microscopic cell death (Figure 1A and 1B). In contrast, ssi2 sid2, ssi2 ndr1, ssi2 ndr1 sid2 or ssi2 eds1 plants exhibited ssi2-like phenotypes. PR-1 gene expression was restored to wt-like levels in the ssi2 eds1 sid2 and ssi2 ndr1 sid2 plants, due to the sid2-derived

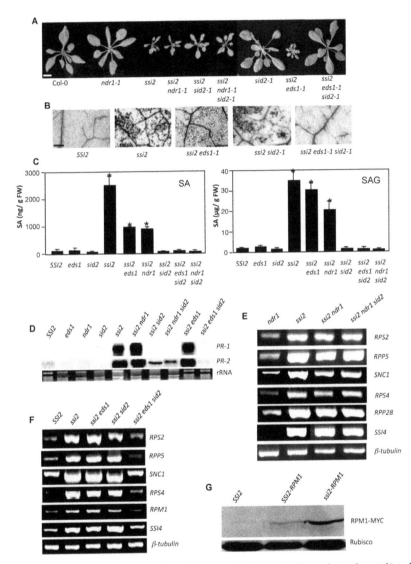

Figure 1. Morphological, molecular, and defense phenotypes of ssi2 ndr1-1 sid2-1 and ssi2 eds1-1 sid2-1 plants. (A) Comparison of the morphological phenotypes displayed by 3-week-old soil-grown plants (scale, 0.5 cm). (B) Microscopy of trypan blue-stained leaves from wt (SSI2, Col-0 ecotype), ssi2, ssi2 eds1-1, ssi2 sid2-1 and ssi2 eds1-1 sid2-1 plants (scale bars, 270 microns). (C) SA and SAG levels in indicated genotypes. The error bars indicate SD. Asterisks indicate data statistically significant from wt Nö ecotype (SSI2) (P<0.05, n = 4). (D) Expression of PR-1 and PR-2 genes in indicated genotypes. Total RNA was extracted from 4-week-old plants and used for RNA gel-blot analysis. Ethidium bromide staining of rRNA was used as the loading control. The PR-1 transcript levels in EDS1 SID2 F2 plants were similar to those of wt plants (data not shown). (E) RT-PCR analysis of various R genes in indicated genotypes. The level of β-tubulin was used as an internal control to normalize the amount of cDNA template. (F) RT-PCR analysis of various R genes in indicated genotypes. The level of β-tubulin was used as an internal control to normalize the amount of cDNA template. The expression of R genes in EDS1 SID2 F2 plants was similar to that of wt plants (data not shown). (G) Levels of Myc-tagged RPM1 protein in indicated genotypes. Levels of Rubisco were used as the loading control.

reduction in SA levels (Figure 1D). In contrast, expression of the SA-independent PR-2 gene was restored to basal levels only in ssi2 eds1 sid2 [43], but not in ssi2 sid2 or ssi2 ndr1 sid2 plants (Figure 1D). Most importantly, ssi2 eds1 sid2 showed basal expression of R genes, unlike ssi2 ndr1 sid2 plants (Figure 1E and 1F). R gene induction was further confirmed by comparing the transcript profiles of 162 NBS-LRR genes in ssi2 sid2 with that of wt plants using Affymetrix ATH1 GeneChips arrays. Twenty-one NB-LRR genes were specifically expressed at 2-fold or higher levels in ssi2 sid2 plants as compared to wt (Col-0) or eds1 plants (P<0.05). All 21 NB-LRR genes were expressed at low levels in ssi2 eds1 sid2 plants, further confirming the results from the RT-PCR analysis. Transcriptional profiling performed using Affymetrix arrays showed that the induction of several R genes (RPM1, RPS2, RPP5, RPS4) was lower than 2-fold in ssi2 or ssi2 sid2 compared to wt plants. To determine if this low-level induction translated to a significant increase in R protein levels, we analyzed the levels of RPM1 in ssi2 plants. Indeed, ssi2 plants accumulated significantly higher levels of the RPM1-Myc protein (Figure 1G).

To rule out the effects of the varied ecotypes of the ssi2 sid2 eds1 (Nössen, Col-0, Ler) plants we introduced eds1-1 (Ws-0 ecotype) and eds1-2 (Ler ecotype) alleles in ssi2 sid2 and ssi2 nahG (Nössen ecotype) backgrounds. All combinations of ssi2 with eds1-1/eds1-2 and sid2/nahG produced similar phenotypes (data not shown). FA profiling showed that the ssi2 eds1 sid2 plants contained low 18:1 levels, similar to ssi2 plants. We thus concluded that EDS1 and SA function downstream of 18:1 levels, but upstream of R gene expression. Furthermore, ssi2 eds1 sid2 plants were wt-like, even though neither ssi2 eds1 nor ssi2 sid2 were restored for defense signaling. Therefore, EDS1 and SA likely fulfill redundant functions in defense signaling induced in response to a reduction in 18:1 levels.

To further test the redundancy for EDS1 and SA, ssi2 eds1 sid2 plants were treated with SA or its active analog benzo(1,2,3)thiadiazole-7-carbothioic acid (BTH). Application of SA or BTH induced lesion formation on ssi2 eds1 sid2 plants but not on wt, eds1, sid2, eds1 sid2 or EDS1 SID2 F2 plants (Figure 2A and 2B, data not shown for eds1 sid2 and EDS1 SID2). Also, application of SA or BTH induced R gene expression in ssi2 eds1 sid2 plants (Figure 2C). Thus, application of SA restored ssi2-like phenotypes in ssi2 eds1 sid2 plants. Since glycerol application mimics the effects of the ssi2 mutation, we generated eds1 sid2 plants and evaluated them for their ability to induce R genes in response to glycerol. Exogenous application of glycerol lowered 18:1 levels in all genotypes, but induced the expression of R genes only in wt, eds1, sid2 and EDS1 SID2 F2 plants (Figure 2D). Only a marginal or no increase in R gene expression was observed in the eds1 sid2 plants (Figure 2D). These results confirmed that EDS1 and SA function redundantly downstream of signaling induced by low 18:1 levels, but upstream of R gene expression.

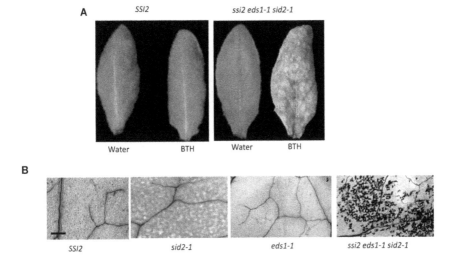

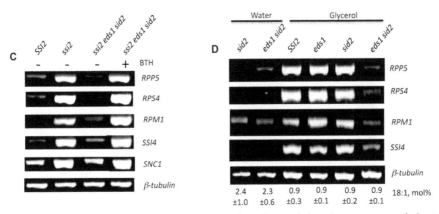

Figure 2. Restoration of ssi2 phenotypes in ssi2 eds1-1 sid2-1 plants and glycerol responsiveness of eds1-1 sid2-1 plants. (A) Visual phenotypes of water- or BTH–treated wt (SSI2; Col-0 ecotype) and ssi2 eds1-1 sid2-1 plants. The plants were photographed at 2 days post treatment (dpt). (B) Microscopy of trypan blue-stained leaves from BTH–treated wt (SSI2; Col-0 ecotype), sid2, eds1-1 and ssi2 eds1-1 sid2-1 plants. The plants were treated with BTH and stained at 2 dpt (scale bars, 270 microns). (C) RT–PCR analysis of R genes in water- or BTH-treated ssi2 eds1-1 sid2-1 plants. Untreated wt (SSI2; Col-0 ecotype) and ssi2 plants were used as controls. The expression of R genes in EDS1 SID2 F2 plants was similar to that of wt plants (data not shown). The level of β-tubulin was used as an internal control to normalize the amount of cDNA template. (D) RT–PCR analysis of various R genes in water- or glycerol-treated sid2-1 and eds1-1 sid2-1 plants. The glycerol-treated wt (SSI2; Col-0 ecotype) and eds1-1 were included as additional controls. The expression of R genes in water- or glycerol-treated EDS1 SID2 F2 plants was similar to that of water- or glycerol-treated wt plants, respectively (data not shown). The expression of R genes in wt and eds1-1 plants was similar to that seen in sid2-1 or eds1-1 sid2-1 plants. The plants were treated with water or glycerol for three days and analyzed for 18:1 levels and R gene expression. The level of β-tubulin was used as an internal control to normalize the amount of cDNA template. The 18:1 content of each genotype is shown as mol%±SD.

EDS1 and SA Function Redundantly in Pathogen Resistance Induced in Response to Reduction in 18:1 Levels

We next evaluated the effect of simultaneous mutations in EDS1- and SA-signaling pathways on resistance to TCV in the ssi2 background. We reported previously that resistance to TCV is dependent on the R gene, HRT, and a recessive locus rrt [23]. However, the ssi2 mutation overcomes the requirement for rrt in HRT-containing plants [23],[37]. Furthermore, the ssi2 mutation only confers resistance to TCV when HRT is present (Figure 3A). The ssi2 mutation also overrides a requirement for EDS1 and SA and consequently ssi2 HRT eds1 as well as ssi2 HRT sid2 plants exhibit resistance to TCV [37] (Figure 3A). Unlike HRT ssi2, HRT ssi2 eds1 or HRT ssi2 sid2 plants, the HRT ssi2 eds1 sid2 plants showed susceptibility to TCV; ~85% HRT ssi2 eds1 sid2 plants were susceptible to TCV as against ~2–4% of HRT ssi2 sid2 or HRT ssi2 eds1 plants (Figure 3A). TCV-induced expression of PR-1 is also independent of EDS1 and SA. However, TCV inoculation failed to induce PR-1 expression in HRT ssi2 eds1 sid2 plants, unlike in HRT ssi2 sid2 plants (Figure 3B). These results showed that both EDS1 and SA have redundant functions in ssi2-mediated resistance to TCV in HRT plants.

EDS1 and SA Function Redundantly in Signaling Mediated by HRT, RPS2, and RPP8 Genes that Encode CC-NBS-LRR Proteins

To determine the redundancy of EDS1 and SA in signaling mediated by CC-NBS-LRR R proteins, we tested the effects of mutations in EDS1- and/or SID2 on HR to TCV. Earlier, we showed that HRT-mediated HR to TCV and PR-1 gene expression is not affected by mutations in the EDS1 or SID2 genes [23]. Consistent with previous results, Di-17 (HRT-containing resistant ecotype), HRT sid2 and HRT eds1 plants revealed discrete and similar-sized HR lesions on TCV-inoculated leaves (Figure 3C and 3D). In comparison, HR in HRT eds1 sid2 plants was diffused and formed larger lesions (Figure 3C and 3D). Increased lesion size in HRT eds1 sid2 plants correlated with increased accumulation of the TCV coat protein (CP) and TCV CP transcript (Figure 3E and 3F). Analysis of PR-1 and PR-2 gene expression indicated that TCV-inoculated HRT eds1 sid2 plants accumulated lower levels of PR-1 and PR-2 transcripts, unlike Di-17, HRT eds1 or HRT sid2 plants (Figure 3G and 3H). In contrast to PR, HRT expression remained unaltered in HRT eds1 sid2 plants (Figure 3H). Together, these results suggested that EDS1 and SA function redundantly in HRT-mediated signaling leading to HR formation and expression of PR-1. The functional redundancy with SA was specific to EDS1 and did not extend to PAD4; HRT pad4 sid2 plants showed normal replication of the virus and wt-like HR and PR-1 gene expression (Figure 3C–3G).

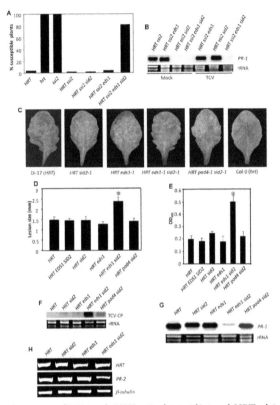

Figure 3. Interaction phenotypes of TCV with HRT ssi2 eds1-1 sid2-1 and HRT eds1-1 sid2-1 plants. (A) Percentage TCV susceptible plants. HRT and hrt indicate resistant and susceptible ecotypes Di-17 and Col-0, respectively. Approximately 70–100 plants were scored for each genotype three-weeks post inoculation and all susceptible plants showed crinkling phenotype and drooping of the bolt [23]. (B) Expression of PR-1 gene in indicated genotypes after mock- or TCV-inoculation. Total RNA was extracted from inoculated leaves at 3 dpi. Ethidium bromide staining of rRNA was used as the loading control. (C) HR formation in indicated genotypes at 3 dpi. The HR response in TCV-inoculated HRT SID2 F2 plants was similar to that seen in TCV-inoculated Di-17, HRT sid2-1 or HRT eds1-1 plants. Plants lacking HRT (Col-0, Nö ecotypes or EDS1 SID2 F2's) did not show any HR. (D) Lesion size in indicated genotypes at 3 dpi. Lesion size was determined from ~23 individual leaves from each genotype. Statistical significance was determined using Students t-test. Asterisks indicate data statistically significant from those of HRT, HRT EDS1 SID2, HRT sid2-1 or HRT eds1-1 plants (P<0.05, n = 23). The error bars indicate SD. (E) ELISA showing levels of TCV CP in the inoculated leaves of indicated genotypes at 3 dpi. Asterisks indicate data statistically significant from results for HRT (Di-17 ecotype) plants (P<0.05, n = 4). The error bars indicate SD. (F) Transcript levels of TCV CP in the inoculated leaves of indicated genotypes at 3 dpi. Ethidium bromide staining of rRNA was used as the loading control. (G) Expression of PR-1 gene in indicated genotypes. Total RNA was extracted from inoculated leaves at 3 dpi. Ethidium bromide staining of rRNA was used as the loading control. The PR-1 gene expression in TCV-inoculated HRT EDS1 SID2 F2 plants was similar to that observed in TCV-inoculated HRT, HRT eds1-1 or HRT sid2-1 plants (data not shown). (H) RT–PCR analysis showing HRT and PR-2 transcript levels in indicated genotypes. The plants were inoculated with TCV and leaf samples were harvested 24 h post inoculation. The level of β-tubulin was used as an internal control to normalize the amount of cDNA template.

A majority of CC-domain containing R proteins, including RPS2, have been reported as not requiring EDS1 for resistance signaling [21]. To determine the effect of simultaneous mutations in EDS1 and SID2 on RPS2-mediated resistance, we compared defense phenotypes produced in single or double mutant plants with that of plants lacking a functional RPS2 gene. Since different alleles of RPS2 confer varying levels of resistance to Pseudomonas syringae (containing AvrRPT2) [44], we screened and isolated an EDS1 knockout (KO) mutant (designated eds1-22) in the Col-0 background and crossed it into the sid2 background (Col-0 ecotype). Inoculation with P. syringae expressing AvrRPT2 induced severe chlorosis on eds1-22 sid2 leaves (Figure 4A). Similar results were obtained when

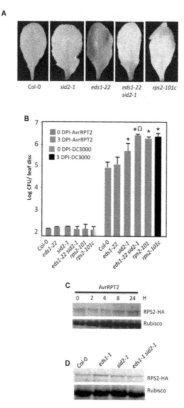

Figure 4. Interaction phenotypes of virulent or AvrRPT2-expressing P. syringae with eds1 sid2 plants. (A) Photograph showing phenotypes produced upon infiltration of 105 CFU/ml bacteria (AvrRPT2). All genotypes were in the Col-0 background. The leaves were photographed at 3 days post inoculation (dpi). The pathogen-inoculated EDS1 SID2 F2 plants showed absence of any visible symptoms in response to bacterial inoculations, similar to Col-0 plants (data not shown). (B) Growth of virulent or avirulent (expressing AvrRPT2) P. syringae on indicated genotypes. The error bars indicate SD. Asterisks and omega symbols indicate data statistically significant from wt (Col-0) or sid2 (P<0.05, n = 4), respectively. All genotypes are in the Col-0 background. (C) Levels of HA-tagged RPS2 protein at 0, 2, 4, 8, and 24 h post inoculation with P. syringae expressing AvrRPT2. Levels of Rubisco were used as the loading control. (D) Levels of HA-tagged RPS2 protein in indicated genotypes. Levels of Rubisco were used as the loading control.

P. syringae expressing AvrRPT2 was inoculated into eds1-1 sid2 double mutant plants. Interestingly, these phenotypes were very similar to those produced on plants lacking a functional RPS2 (rsp2-101c), while eds1 and sid2 showed no or very mild symptoms, respectively (Figure 4A). The appearance of symptoms correlated with bacterial growth; eds1-22 sid2 plants and the rps2 mutant supported maximum growth of the pathogen, followed by sid2 plants (Figure 4B). Similarly, the eds1-1 sid2 double mutant plants supported more pathogen growth compared to eds1-1 or sid2 plants (data not shown). Together, these data suggest that the simultaneous loss of EDS1- and SA-dependent signals is required to mimic a phenotype produced by the loss of the cognate R gene, RPS2.

To determine if the loss of both EDS1- and SA-dependent signaling impaired resistance by affecting the RPS2 protein, we analyzed R protein levels in eds1-1 and sid2 single and eds1-1 sid2 double mutant plants. Analysis of RPS2 tagged with HA epitope at various times did not detect any significant changes in RPS2 levels in response to inoculation with P. syringae expressing AvrRPT2 (Figure 4C). Therefore, RPS2 levels in mutant plants were analyzed at only 12 and 24 h post-pathogen inoculation. The RPS2-HA levels in eds1-1, sid2 or eds1-1 sid2 plants were similar to that in wt plants (Figure 4D). These results suggested that abrogation of resistance in eds1 sid2 double mutants was not due to a defect in the accumulation of the R protein.

We next evaluated the effects of mutations in EDS1 and SID2 on RPP8-mediated resistance to Hyalopernospora arabidopsidis biotype Emco5 encoding Atr8. RPP8 (encodes a CC-NBS-LRR type R protein)-mediated resistance signaling was previously reported to be independent of both EDS1 and SA [21],[24]. As expected, RPP8 plants (ecotype Ler) inoculated with the Emco5 isolate showed localized HR and did not support growth of the pathogen (Figure 5A). Consistent with earlier reports [21],[24], RPP8 eds1-2 plants also did not support the growth of Emco5, although they did develop trailing necrosis (Figure 5A and 5B). The presence of the nahG transgene did not alter HR formation or pathogen response in the RPP8 nahG plants (Ler ecotype). In contrast, eds1-2 nahG plants were affected in both HR as well as resistance; eds1-2 nahG plants not only showed extensive trailing necrosis but also supported growth and sporulation of the pathogen (Figure 5A–5C). Although RPP8 EDS1 nahG and RPP8 eds1-2 nahG plants showed contrasting phenotypes (Figure 5A–5C), we still wanted to rule out the possibility that susceptibility of eds1 nahG plants was not due to the accumulation of catechol, which is formed upon degradation of SA by NAHG. Estimation of SA levels in Emco5 inoculated RPP8 (Ler) plants showed marginal increase in SA and no significant increase in SAG levels compared to mock-inoculated plants (data not shown). This suggests that Emco5 inoculated nahG plants are unlikely to show a significant increase in catechol levels. In addition to this, we tested two independent lines of RPP8 eds1-2 sid2 (in the ssi2 background) plants and both showed increased susceptibility to Emco5 (Figure 5D). In comparison, RPP8 eds1-2 or RPP8 sid2 genotypes did not support any growth or sporulation of the pathogen (Figure 5D). Taken together, these results show that EDS1 and SA have redundant functions in RPP8-mediated resistance to H. arabidopsidis Emco5.

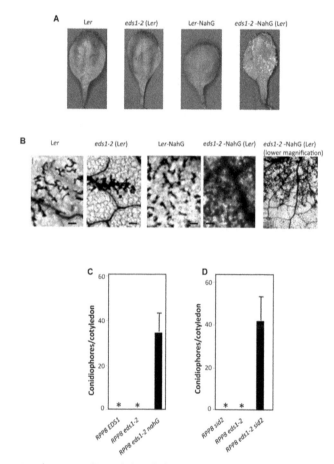

Figure 5. Interaction phenotypes of H. arabidopsidis biotype Emco5 expressing Atr8 with RPP8 eds1-2 nahG or RPP8 eds1-2 sid2-1 plants. (A) Whole leaf pictures showing growth of Emco5 on the cotyledons from indicated genotypes. All genotypes were in the Ler background. Cotyledons were photographed 10 days after inoculation. (B) Trypan blue stained leaf showing microscopic HR on Ler and Ler nahG leaves, and trailing necrosis on eds1-2 and eds1-2 nahG leaves (scale bars, 270 microns). Both high (100×) and low magnification (100×) images of eds1-2 nahG leaf are shown. Pathogen inoculations were carried out in F2, F3, and F4 generations with consistent results. The F2 plants showing wt genotype at the mutant locus were resistant to pathogen infection (data not shown). (C) Quantification of pathogen growth on RPP8 EDS1, RPP8 eds1-2 and RPP8 eds1-2 nahG plants. Approximately, 40–60 cotyledons were assayed for each genotype. Asterisks indicate absence of spores. All genotypes were in the Ler background. (D) Quantification of pathogen growth on RPP8 sid2, RPP8 eds1-2, and RPP8 eds1-2 sid2-1 plants. All genotypes were in the ssi2 background. Approximately, 40–60 cotyledons were assayed for each genotype. Asterisks indicate absence of spores.

Exogenous SA and Overexpression of EDS1 Have Additive Effects on Pathogen Resistance in Wild-Type Plants

To determine the relation between EDS1- and SA-derived signaling, we compared PR-1 gene expression and resistance in plants that were either overexpressing EDS1 or were pretreated with SA. EDS1 overexpression was achieved by expressing EDS1 (At3g48090 from the Col-0 ecotype) under control of the CaMV 35S promoter in Col-0 plants (Figure 6A). The 35S-EDS1 plants analyzed in the T2 and T3 generations showed wt-like morphology (data not shown), wt-like expression of the PR-1 gene (Figure 6A) and accumulated wt-like levels of SA/SAG (data not shown). In comparison, exogenous application of SA induced PR-1 and EDS1 gene expression [data not shown; 16].

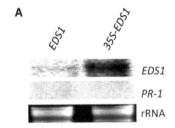

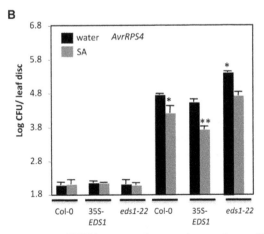

Figure 6. Effect of SA pretreatment and EDS1 overexpression on pathogen resistance. (A) Expression of EDS1 and PR-1 in EDS1 (Col-0) and 35S-EDS1 (Col-0) plants. Total RNA was extracted from 4-week-old plants and ethidium bromide staining of rRNA was used as the loading control. (B) Growth of P. syrinage AvrRPS4 on indicated genotypes (all in Col-0 background). Single asterisks indicate data statistically significant from results for water-treated wt (Col-0) (P<0.05, n = 4). Two asterisks indicate data statistically significant from results for SA–treated wt (Col-0) (P<0.05, n = 4). The error bars indicate SD.

Analysis of RPS4 (encodes a TIR-NBS-LRR type R protein)-mediated resistance showed that exogenous application of SA enhanced resistance to P. syringae (expressing AvrRPS4) in wt as well as eds1-22 plants, although wt plants were more resistant to AvrRPS4 bacteria than the eds1-22 plants (Figure 6B). Overexpression of EDS1, on the other hand, did not alter the response to AvrRPS4 bacteria. Strikingly, exogenous application of SA on 35S-EDS1 plants enhanced resistance even more than in the SA-treated wt or eds1-22 plants. Together, these results suggest that EDS1- and SA-derived signaling contribute additively towards pathogen resistance.

Simultaneous Defects in EDS1 and SA Biosynthesis Do Not Additively Lower Basal Defense

We next evaluated the effect of the eds1 sid2 mutations on basal resistance to virulent P. syringae, since both EDS1 and SID2 are known to contribute to basal defense as well. The eds1-1, eds1-22, sid2 and eds1 sid2 plants all showed enhanced susceptibility to virulent bacteria as compared to the respective wt ecotypes (Figure 7A). Interestingly, unlike in the case of the avirulent bacteria, growth of virulent bacteria was similar in eds1 sid2 double mutant plants as compared to that in eds1 or sid2 single mutant plants. These results suggested that loss-of-function mutations in EDS1 and SID2 do not additively reduce basal resistance to virulent P. syringae. Similar to the results obtained with the bacterial pathogen, the loss of both EDS1- and SA-dependent signals did not additively lower basal resistance to TCV either (Figure 7B). This further suggested that the redundant functions of EDS1 and SA might be relevant only for R gene-mediated signaling.

Mutations in FAD7 FAD8 and EDS5 Restore Altered Defense Signaling in ssi2 eds1 Plants

Besides SID2, mutations in FAD7 FAD8, which catalyze desaturation of 18:2 to 18:3 on membrane glycerolipids, also lower the SA levels in ssi2 plants [40]. To test if fad7 or fad7 fad8 mutations produced a similar effect as sid2, these mutations were mobilized into the ssi2 eds1 background. The ssi2 eds1 fad7 and ssi2 eds1 fad7 fad8 plants were bigger in size compared to ssi2 fad7 or ssi2 fad7 fad8 plants. The ssi2 eds1 fad7 fad8 were wt-like in morphology and showed no or greatly reduced cell death lesions PR-1 expression was greatly reduced or abolished in ssi2 eds1 fad7 and ssi2 eds1 fad7 fad8 plants, respectively and correlated with their endogenous SA/SAG levels; the ssi2 eds1 fad7 and ssi2 eds1 fad7 fad8 plants showed greatly reduced or basal levels of SA and SAG, respectively. Expression of some R genes (SSI4, RPS2, RPP5) was nominally or moderately reduced in ssi2 eds1 fad7 plants. By comparison, all R genes tested were expressed at basal levels in ssi2 eds1 fad7 fad8 plants. These results showed that presence of fad7 fad8 mutations restored the altered defense phenotypes of ssi2 eds1 plants. FA profiling did not detect any significant increase in 18:1 levels in ssi2 eds1 fad7 and ssi2 eds1 fad7 fad8 plants, compared to ssi2 fad7 and ssi2 fad7 fad8, respectively. This suggested that restoration of defense phenotypes in ssi2 eds1 fad7 fad8 was not the result of restored 18:1 levels, but rather the reduction of SA levels in the eds1 background.

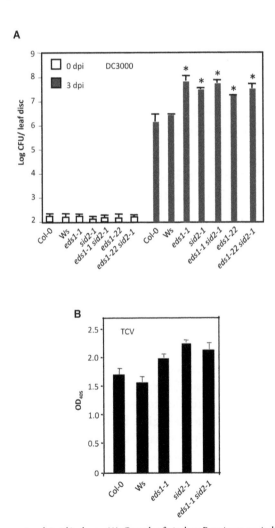

Figure 7. Basal resistance in eds1 sid2 plants. (A) Growth of virulent P. syringae on indicated genotypes. The error bars indicate SD. Asterisks indicate data statistically significant from wt (Col-0 or Ws) (P<0.05, n = 4). The eds1-1 and eds1-22 are in Ws and Col-0 ecotypic backgrounds, respectively. (B) ELISA showing levels of TCV CP in the inoculated leaves of indicated genotypes at 3 dpi. The error bars indicate SD (n = 4).

Mutations in EDS5 and PAD4 also lower SA/SAG levels in ssi2 plants [40]. To determine if mutations in these can substitute for sid2 triple mutants containing ssi2 eds1 pad4 and ssi2 eds1 eds5 were generated. The ssi2 eds1 pad4 plants were morphologically similar to ssi2 eds1 or ssi2 pad4 plants and showed spontaneous cell death and increased expression of PR-1 gene (Figure 8A–8C). In comparison, ssi2 eds1 eds5 showed wt-like morphology, greatly reduced cell death and basal expression of PR-1 gene (Figure 8A–8C). Quantification of endogenous SA levels showed that both ssi2 eds1 eds5 and ssi2 eds1 pad4 accumulated lower SA/SAG levels compared to ssi2 eds5 and ssi2 pad4, respectively (Figure 8D and 8E). However, while ssi2 eds1 eds5 plants accumulated basal levels of SA/SAG, the ssi2 eds1 pad4 accumulated significantly higher levels of SA/SAG compared to wt, ssi2 sid2 and ssi2 eds1 eds5 plants (Figure 8D and 8E). Analysis of R gene expression showed greatly reduced levels in ssi2 eds1 eds5 plants but the ssi2 eds1 pad4 expressed ssi2-like levels of R genes (Figure 8F). Taken together, these results suggest that the suppression of SA levels was required for the normalization of defense phenotypes in the ssi2 eds1 background.

PAD4, SAG101, and EDS5 are not Functionally Redundant with SA in Low 18:1-Mediated Signaling

Besides EDS1, the SA signaling pathway is also regulated by PAD4 and EDS5 and via the physical association of EDS1 with SAG101 and PAD4 [17],[19],[45]. To determine if PAD4, SAG101 or EDS5 also function redundantly with SA, we introduced the pad4, sag101 and eds5 mutations in the ssi2 and ssi2 sid2 backgrounds.

The ssi2 sag101, ssi2 pad4 and ssi2 eds5 plants showed ssi2-like morphology, visible and microscopic cell death and constitutive PR-1 gene expression. Consistent with these phenotypes, the ssi2 sag101, ssi2 pad4, ssi2 eds5 plants showed increased expression of R genes and accumulated elevated levels of SA and SAG. Notably, the SA levels in ssi2 sag101 plants were ~6-fold lower than in ssi2 plants, suggesting that SAG101 contributed to the accumulation of SA in ssi2 plants. To determine if the reduced SA in the sag101 background could restore wt-like phenotypes in ssi2 eds1 plants, triple mutant ssi2 eds1 sag101 plants were generated. Although the ssi2 eds1 sag101 plants accumulated significantly lower levels of SA/SAG, these plants were only slightly bigger than ssi2 eds1 or ssi2 sid2 plants, showed spontaneous cell death and expressed PR-1 and R genes constitutively. We next analyzed the triple mutant ssi2 sag101 sid2, ssi2 pad4 sid2 and ssi2 eds5 sid2 plants. All the triple mutants contained wt-like levels of SA and SAG. The ssi2 sag101 sid2 plants were morphologically similar to ssi2 plants, showed spontaneous cell death and expressed R genes constitutively. In comparison, the ssi2 pad4 sid2 and ssi2 eds5 sid2 plants were bigger in morphology. However, plants of both genotypes showed cell death and expressed R genes constitutively. Together, these data suggest that the functional redundancy with SA was specific only to EDS1 and did not extend to PAD4, SAG101 or EDS5.

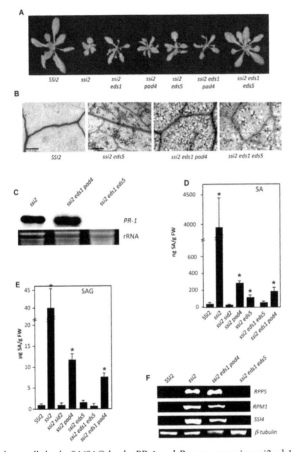

Figure 8. Morphology, cell death, SA/SAG levels. PR-1 and R gene expression ssi2 eds1-2 pad4-1 and ssi2 eds1-2 eds5-1 plants. (A) Comparison of the morphological phenotypes displayed by 4-week-old soil-grown wt (SSI2), ssi2, ssi2 eds1, ssi2 pad4, ssi2 eds5, ssi2 eds1 pad4, and ssi2 eds1 eds5 plants. (B) Microscopy of trypan blue-stained leaves from indicated genotypes. (C) Expression of PR-1 gene in indicated genotypes. Total RNA was extracted from 4-week-old plants and used for RNA gel-blot analysis. Ethidium bromide staining of rRNA was used as the loading control. (D) Endogenous SA levels in the leaves of 4-week-old soil-grown plants. Values are presented as mean of three replicates and the error bars represent SD. Statistical significance was determined using Students t-test. Asterisks indicate data statistically significant compared to SSI2 (Col-0) plants (P<0.05, n = 5). (E) Endogenous SAG levels in the leaves of 4-week-old soil-grown plants. Values are presented as mean of three replicates and the error bars represent SD. Asterisks indicate data statistically significant compared to SSI2 (Col-0) plants (P<0.05, n = 5). (F) RT–PCR analysis of R genes in indicated genotypes. The level of β-tubulin was used as an internal control to normalize the amount of cDNA template. The SSI2 EDS1, SSI2 PAD4, SSI2 EDS1 PAD4, and SSI2 EDS1 EDS5 F2 plants showed wt–like morphology, accumulated basal levels of SA and showed basal level expression of PR-1 and R genes (data not shown).

Discussion

SA is long known as an essential modulator of R gene-derived signaling in pathogen defense. Several molecular components, including EDS1, have been identified as essential effectors of SA-derived signaling [23],[26],[45]. Since SA upregulates expression of EDS1, both SA and EDS1 are thought to function in a positive feedback loop and EDS1 is widely considered an upstream effector of SA [16],[19],[23],[45]. Recent data has shown that EDS1 signals resistance via both SA-dependent as well as SA-independent pathways [46]. Strikingly, EDS1-dependent but SA-independent branch of EDS1 pathway still requires SA pathway for full expression of resistance [46]. In this study, we have characterized the relationship between EDS1 and SA. We show that the two components act in a redundant, and not necessarily sequential manner to regulate R gene expression induced in response to a reduction in the levels of the FA 18:1. Furthermore, EDS1 and SA also function redundantly in R gene-mediated defense against viral, bacterial and oomycete pathogens. It appears that the redundant functions of EDS1 and SA may have prevented their identification as required components for signaling mediated by CC-NBS-LRR R proteins. Indeed, RPS2-mediated signaling is fully compromised only in eds1 sid2 and not in the single mutant plants. Similarly, HRT-mediated signaling leading to HR formation and PR-1 gene expression is only affected in eds1 sid2 plants, while eds1 or sid2 plants behave similar to wt plants. Furthermore, RPP8-mediated resistance, which was previously reported not to require EDS1 or SA [21],[24], is compromised in plants lacking both EDS1 and SA. In contrast to their effect on R gene-mediated resistance, loss of both EDS1- and SA-dependent signals did not additively lower basal resistance to P. syringae or TCV. Together, these data suggests that the redundant functions of EDS1 and SA might be relevant only for R gene-mediated signaling.

In contrast to SA application, overexpression of EDS1 was unable to confer increased resistance to the avirulent pathogen P. syringae. Furthermore, unlike SA, overexpression of EDS1 was not associated with the induction of PR-1 gene expression. These findings, together with the observation that SA was able to induce EDS1 expression and that SA application on wt plants resulted in higher resistance than that in eds1, suggests that SA feedback regulates EDS1-derived signaling in a unidirectional manner (Figure 9B). Thus, SA application induces both SA- and EDS1-derived signaling, the additive effects of which enhance resistance in wt plants much more than in eds1-22 plants. Furthermore, the combined effects of SA pretreatment and EDS1 overexpression induced much better resistance than the individual effects of each. This is consistent with a previous report that 35S-EDS1 plants induce rapid and stronger expression of PR-1 in response to pathogen inoculation [47]. The additive effects of EDS1 and SA was also supported by the observation that eds1 sid2 plants showed pronounced chlorosis upon inoculation with AvrRPS4 expressing pathogen, which is recognized by a TIR-NBS-LRR protein RPS4. Since mutations in SA-independent branch of EDS1 pathway and sid2 have additive effects on R gene-mediated resistance [46], it is possible that overexpression of EDS1 triggers signaling via both SA-dependent and/or -independent branches of EDS1 pathway.

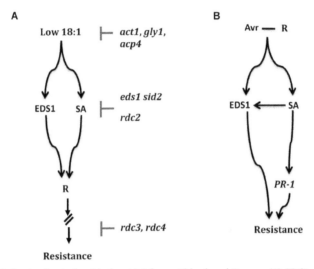

Figure 9. Models for signaling induced by low 18:1 fatty acid levels and R genes. (A) EDS1 and SA function upstream of R genes and regulate expression of R genes induced by low 18:1 fatty acid levels. Mutations in EDS1 and SA-synthesizing enzyme, encoded by SID2, abolish constitutive upregulation of R genes and associated enhanced resistance in genetic backgrounds containing low 18:1 levels. Similar to EDS1/SA, restored in defective crosstalk (RDC) 2 acts downstream of signaling induced by low levels of 18:1 but upstream of R gene expression. Signaling induced by low 18:1 fatty acid levels can also be suppressed by mutations in ACT1-encoded G3P acyltransferase [30], GLY1-encoded G3P dehydrogenase [31], or ACP4-encoded acyl carrier protein 4 [35], which normalize 18:1 levels, or by blocking steps downstream of R gene expression (rdc3 and rdc4). Upregulation of R genes induced by low 18:1 fatty acid levels does not require PAD4, SAG101, or EDS5, which are components of the resistance signaling pathway(s) initiated upon R-Avr interaction. (B) Direct or indirect interaction between host-encoded R and pathogen-encoded Avr products initiate resistance signaling, which requires EDS1 and SA. Exogenous application of SA induces expression of PR-1 and EDS1 genes but overexpression of EDS1 does not induce PR-1 expression or increase SA levels. The EDS1– and SA–dependent pathways have additive effects.

Although the Col-0 ecotype is thought to contain two functional alleles of EDS1 [26], a KO mutation in At3g48090 was sufficient to compromise both basal and R gene (RPS4)-mediated resistance. However, the Col-0 eds1-22 mutant consistently supported less growth of virulent or avirulent pathogens compared to eds1-1 or eds1-2 plants. This suggests that the second EDS1 allele in the Col-0 ecotype might also contribute towards the resistance response. This is consistent with another study where constitutive defense phenotypes due to the overexpression of the SNC1 gene, encoding a TIR-NBS-LRR R protein, are not completely suppressed by a mutation in eds1 in the Col-0 background but restored by the eds1 mutation in the Ws background [48].

The inability to accumulate SA together with a mutation in EDS1 was also required to suppress constitutive defense signaling resulting from the overexpression of R genes induced in response to reduced 18:1 levels. Although eds1 or sid2 plants were entirely competent in inducing R gene expression in response to a reduction in 18:1, eds1 sid2 plants were not. Thus, ssi2 eds1 sid2 as well as glycerol-treated eds1 sid2 plants showed wt-like expression of R genes while ssi2 eds1, ssi2 sid2 and glycerol-treated eds1 or sid2 plants showed increased expression of R genes. Moreover, treatment of ssi2 eds1 sid2 plants with exogenous SA restored R transcript induction and cell death in these plants. The fact that glycerol treatment is unable to induce R gene expression in eds1 sid2 plants supports the possibility that EDS1 and SA function upstream of, and not merely serve as a feedback loop in, R gene induction. Signaling induced by low 18:1 levels continues to function in the absence of SA, suggesting a novel SA-independent role for EDS1 in defense signaling.

Since ssi2 eds1 sid2 plants contain a mixed ecotypic background (Nö, Ws/ Ler, Col-0, ecotypes), it is possible that ecotypic variations in various genetic backgrounds resulted in the restoration of ssi2-triggered defense phenotypes. Indeed, phenotypic variations amongst different Arabidopsis ecotypes have been associated with many physiological processes [48]–[51]. Moreover, certain alleles can express themselves only in specific ecotypic backgrounds [48],[51]. However, since ssi2 EDS1 SID2, ssi2 EDS1 sid2 or ssi2 eds1 SID2 plants (F2 population) always exhibited ssi2-like phenotypes, it is highly unlikely that ecotypic variations resulted in the restoration of phenotypes in ssi2 eds1 sid2 plants. The effect of ecotypic variations on the observed phenotypes can be further ruled out for the following reasons. First, the effects of different mutations were assessed in multiple backgrounds. For example, we used both eds1-1 (Ws-0 ecotype) and eds1-2 (Ler ecotype) alleles in ssi2 sid2 (Nö, Col-0 ecotypes) and ssi2 nahG (Nö ecotype) backgrounds and all combinations of ssi2 with eds1-1/eds1-2 and sid2/ nahG produced similar phenotypes. Second, all defense phenotypes were assessed over three generations using multiple progeny. Third, similar results were obtained when different ecotypic backgrounds were evaluated for their response to different pathogens. For example, eds1 nahG or eds1 sid2 backgrounds conferred increased susceptibility to H. arabidopsidis, P. syringae and TCV, even though only the genotypes used for TCV were of mixed ecotypic backgrounds. Fourth, F2 plants containing wild-type alleles behaved like wild-type parents. Finally, the effects of various mutant backgrounds on ssi2 phenotypes were also confirmed by glycerol application on individual mutants.

Although glycerol treatment failed to induce R gene expression in eds1 sid2 plants, it did induce cell death. This is in contrast to the absence of a cell death phenotype in ssi2 eds1 sid2 leaves. One possibility is that the glycerol-triggered cell death is not due to a reduction in 18:1 levels. However, significant overlap between ssi2- and exogenous glycerol-triggered signaling pathways lessens such a possibility [40]. An alternate possibility is that, while EDS1 affects a majority of the responses induced by low 18:1 levels, the cell death phenotype is also governed by some additional molecular factor(s). This is supported by the fact that ssi2 pad4 sid2 plants exhibit improved morphology and reduced cell death even though they are not restored for other defense-related phenotypes.

Since the overexpression of R genes can initiate defense signaling in the absence of a pathogen [48],[52], it is possible that the induced defense responses in ssi2 plants are the result of increased R gene expression. This idea is supported by the fact that ssi2-related phenotypes can be normalized by restoring R gene expression to wt-like levels, irrespective of their 18:1 levels. Thus, wt-like defense phenotypes are restored in suppressors containing high 18:1 levels, such as ssi2 act1, ssi2 gly1 or ssi2 acp4 [30],[31],[35], as well as in suppressor containing low 18:1 levels, such as ssi2 eds1 sid2 (this work) and restored in defective crosstalk (rdc) 2 (unpublished data) (Figure 9A). We have also characterized additional ssi2 suppressors that show wt-like phenotypes even though they contain low 18:1 levels and express R genes constitutively (rdc3, rdc4). Together, these results suggest that the ssi2-associated phenotypes can be restored by normalizing R gene expression to wt-like levels either by increasing 18:1 levels, impairing factors downstream of signaling induced by low 18:1 levels, or impairing events downstream of R gene expression induced by low 18:1 levels.

In addition to 18:1 levels or R gene expression, ssi2-related defense signaling could also be normalized by altering some factor(s) that function downstream of R gene induction. Indeed, our preliminary characterizations have identified additional ssi2 suppressors that yield wt-like phenotypes with regards to defense signaling but continue to express R genes at high levels. Reduced 18:1 levels may induce defense signaling by directly regulating the transcription of activators or suppressors of defense gene expression. This is supported by the fact that 18:1-mediated activation of a transcription factor induces the expression of genes required for neuronal differentiation [53]. Similarly, in Sacharromyces cerevisiae as well as mammalian cells, binding of 18:1 to specific transcription factors induces the transcription of genes carrying 18:1 responsive elements in their promoters [54],[55]. On the other hand, expression of the oncogene HER2 is inhibited via the 18:1-upregulated expression of its transcriptional repressor [56]. Reduced 18:1 might also directly activate/inhibit/alter protein activities. For example, 18:1 is known to activate the Arabidopsis phospholipase D [57] and inhibit glucose-6-phosphate transporter activity in Brassica embryos [58]. Indeed, we have also identified several Arabidopsis proteins for which enzymatic activities are inhibited upon binding to 18:1 (unpublished data).

In conclusion, results presented here redefine the currently accepted pathway for SA-mediated signaling by showing that EDS1 and SA play a redundant role in plant defense mediated by R proteins and in signaling induced by low 18:1 fatty acid levels. Further biochemical characterization should help determine if 18:1 binds to EDS1 and if cellular levels of 18:1 modulate the as yet undetected lipase activity of EDS1.

Materials and Methods

Plant Growth Conditions and Genetic Analysis

Plants were grown in MTPS 144 Conviron (Winnipeg, MB, Canada) walk-in-chambers at 22°C, 65% relative humidity and 14 hour photoperiod. The photon flux density of the day period was 106.9 μmoles m^{-2} s^{-1} and was measured using a digital light meter (Phytotronic Inc, Earth city, MO). All crosses were performed by emasculating the flowers of the recipient genotype and pollinatng with the pollen from the donor. In most cases, single, double, or triple mutant plants were obtained from more than one combination of crosses and showed similar mor-phological, molecular and biochemical phenotypes. F2 plants showing the wt genotype at the mutant locus were used as controls in all experiments. The wt and mutant alleles were identified by PCR, CAPS, or dCAPS analysis and/or based on the FA profile [30],[31],[38],[40]. The EDS1 KO mutant in At3g48090 was, isolated by screening SALK_071051 insertion line, obtained from ABRC. The EDS1 KO was designated eds1-22, based on the previous designation assigned to SALK_071051 T-DNA KO line [48]. The At3g48090 gene showed 98.8% identity at amino acid level to EDS1 allele from Ler ecotype. The homozygous insertion lines were verified by sequencing PCR products obtained with primers specific for the T-DNA left border in combination with an EDS1-specific primer. The eds1-22 lines did not show any detectable expression of EDS1.

RNA Extraction and Northern Analyses

Small-scale extraction of RNA from one or two leaves was performed with the TRIzol reagent (Invitrogen, CA), following the manufacturer's instructions. Northern blot analysis and synthesis of random-primed probes for PR-1 and PR-2 were carried out as described previously [29].

Reverse Transcription–PCR

RNA quality and concentration were determined by gel electrophoresis and determination of A260. Reverse transcription (RT) and first strand cDNA synthesis were carried out using Superscript II (Invitrogen, CA). Two-to-three independent RNA preparations were used for RT-PCR and each of these were analyzed at least twice by RT–PCR. The RT–PCR was carried out for 35 cycles in order to determine absolute levels of transcripts. The number of amplification cycles was reduced to 21–25 in order to evaluate and quantify differences among transcript levels before they reached saturation. The amplified products were quantified using ImageQuant TL image analysis software (GE, USA).

Trypan-Blue Staining

The leaves were vacuum-infiltrated with trypan-blue stain prepared in 10 mL acidic phenol, 10 mL glycerol, and 20 mL sterile water with 10 mg of trypan blue. The samples were placed in a heated water bath (90°C) for 2 min and incubated at room temperature for 2–12 h. The samples were destained using chloral hydrate (25 g/10 mL sterile water; Sigma), mounted on slides and observed for cell death with a compound microscope. The samples were photographed using an AxioCam camera (Zeiss, Germany) and images were analyzed using Openlab 3.5.2 (Improvision) software.

Pathogen Infections

The asexual conidiospores of H. arabidopsidis Emco5 expressing Atr8 were maintained on the susceptible host Nössen (Nö) or Nö NahG. The spores were removed by agitating the infected leaves in water and suspended to a final concentration of 10^5 spores/mL. Two-week-old seedlings were sprayed with spore suspension and transferred to a MTR30 reach-in chamber (Conviron, Canada) maintained at 17°C, 98% relative humidity and 8 h photoperiod. Plants were scored at ~10–14 dpi and the conidiophores were counted under a dissecting microscope.

The bacterial strain DC3000 derivatives containing pVSP61 (empty vector), AvrRpt2 or AvrRps4 were grown overnight in King's B medium containing rifampicin (Sigma, MO). The bacterial cells were harvested, washed and suspended in 10 mM $MgCl_2$. The cells were diluted to a final density of 10^5 to 10^7 CFU/mL (A600) and used for infiltration. The bacterial suspension was injected into the abaxial surface of the leaf using a needle-less syringae. Three leaf discs from the inoculated leaves were collected at 0 and 3 dpi. The leaf discs were homogenized in 10 mM $MgCl_2$, diluted 10^3 or 10^4 fold and plated on King's B medium.

Transcripts synthesized in vitro from a cloned cDNA of TCV using T7 RNA polymerase were used for viral infections [59],[60]. For inoculations, the viral transcript was suspended at a concentration of 0.05 μg/μL in inoculation buffer, and the inoculation was performed as described earlier [56]. After viral inoculations, the plants were transferred to a Conviron MTR30 reach-in chamber maintained at 22°C, 65% relative humidity and 14 hour photoperiod. HR was determined visually three-to-four days post-inoculation (dpi). Resistance and susceptibility was scored at 14 to 21 dpi and confirmed by northern gel blot analysis. Susceptible plants showed stunted growth, crinkling of leaves and drooping of the bolt.

Transcriptional Profiling

Total RNA isolated from four-week-old plants using TRIZOL as outlined above. The experiment was carried out in triplicate and a separate group of plants was used for each set. RNA was processed and hybridized to the Affimetric Arabidopsis ATH1 genome array GeneChip following the manufacturers instructions (http://www.affymetrix.com/Auth/support/downloads/manuals/expression_analysis_technical_manual.pdf). All probe sets on the Genechips were assigned hybridization signal above background using Affymetrix Expression Console Software v1.0 (http://www.affymetrix.com/Auth/support/downloads/manuals/expression_console_userguide.pdf). Data was analyzed by one-way Anova followed by post hoc two sample t-tests. The P values were calculated individually and in pair-wise combination for each probe set. The identities of 162 NBS-LRR genes were obtained from the Arabidopsis information resource (TAIR; www.arabidopsis.org) and disease resistance gene homolog databases (http://niblrrs.ucdavis.edu/).

Fatty Acid Profiling

FA analysis was carried out as described previously [61]. For FA profiling, one or few leaves of four-week-old plants were placed in 2 ml of 3% H_2SO_4 in methanol containing 0.001% butylated hydroxytoluene (BHT). After 30 minutes incubation at 80°C, 1 mL of hexane with 0.001% BHT was added. The hexane phase was then transferred to vials for gas chromatography (GC). One-microliter samples were analyzed by GC on a Varian FAME 0.25 mm×50 m column and quantified with flame ionization detection. The identities of the peaks were determined by comparing the retention times with known FA standards. Mole values were calculated by dividing peak area by molecular weight of the FA.

SA and SAG Quantification

SA and SAG quantifications were carried out from ~300 mg of leaf tissue as described before [23].

Chemical Treatment of Plants

SA treatments were carried out by spraying or subirrigating 3-week-old plants with 500 μM SA or 100 μM BTH. For glycerol treatment, plants were sprayed with 50 mM solution prepared in sterile water.

Enzyme Linked Immuno-Sorbent Assay and Western Analysis

Total protein was extracted in buffer containing 50 mM Tris pH 8.0, 1 mM EDTA, 12 mM β-mercaptoethanol and 10 μg ml^{-1} phenylmethylsulfonyl fluoride. Proteins were fractionated on a 10–12% SDS-PAGE to confirm the quality. An antigen-coated enzyme-linked immunosorbent assay was used to determine levels of TCV CP in the infected plants as described before [62].

For protein gel blot analysis, leaf tissue from 4-week-old plants was extracted with a buffer containing 50 mM Tris-HCl, pH 7.5, 10% glycerol, 150 mM NaCl, 10 mM MgCl2, 5 mM EDTA, 5 mM DTT, and 1× proteinase inhibitor (Sigma). Protein concentrations were determined by the Bradford assay (Bio-Rad, CA). For immunodetection, 10–50-μg protein samples were electrophoresed on 10–15% polyacrylamide gels and run in the presence of 0.38 M Tris and 0.1% SDS. Proteins were transferred from the gels to polyvinylidene difluoride membranes by electroblotting, incubated with primary anti-HA antibody (Sigma) and alkaline phosphatase-conjugated secondary antibody (Sigma). Immunoblots were developed using color detection.

Competing Interests

The authors have declared that no competing interests exist.

Acknowledgements

We thank David Smith for critical comments; John Johnson for help with gas chromatography; Ludmila Lapchyk, Thomas Muse, and Lev Orlov for help with fatty acid extractions and genotyping; and Amy Crume for maintaining the growth facility. We thank Jeff Dangl for rps2-101c, RPM1-MYC, and RPS2-HA seeds; Keiko Yoshioka for Ler NahG; and Jane Parker for sag101 seeds. We thank Walter Grassmann for the Pseudomonas syringae strain containing AvrRPS4 and Barbara Kunkle for the AvrRPT2 strain. We thank Jack Morris for providing anti-TCV CP antisera.

Authors' Contributions

Conceived and designed the experiments: SCV RDJ MKM SZ ACCS AK PK. Performed the experiments: SCV RDJ MKM SZ ACCS YX DN AK PK. Analyzed the data: SCV RDJ MKM SZ ACCS YX MH AJS DN AK PK. Contributed reagents/materials/analysis tools: SCV RDJ MKM SZ ACCS YX MH AJS DN AK PK. Wrote the paper: AK PK.

References

1. Flor H (1971) Current status of gene-for-gene concept. Annu Rev Phytopathol 9: 275–296.

2. Greenberg JT, Guo A, Klessig DF, Ausubel FM (1994) Programmed cell death in plants: a pathogen-triggered response activated coordinately with multiple defense functions. Cell 77: 551–563.

3. Dangl JL, Dietrich RA, Richberg MH (1996) Death don't have no mercy: cell death programs in plant-microbe interactions. Plant Cell 8: 1793–1807.

4. Hammond-Kosack KE, Jones JDJ (1996) Resistance gene-dependent plant defense responses. Plant Cell 8: 1773–1791.

5. Jabs T, Dietrich RA, Dangl JL (1996) Initiation of runaway cell death in an Arabidopsis mutant by extracellular superoxide. Science 273: 1853–1856.

6. Gray WM (2002) Plant defence: a new weapon in the arsenal. Curr Biol 12: R352–R354.

7. Ward ER, Uknes SJ, Williams SC, Dincher SS, Wiederhold DL, et al. (1991) Coordinate gene activity in response to agents that induce systemic acquired resistance. Plant Cell 3: 1085–1094.

8. Gaffney T, Friedrich L, Vernooij B, Negrotto D, Nye G, et al. (1993) Requirement of salicylic acid for the induction of systemic acquired resistance. Science 261: 754–756.

9. Uknes S, Winter AM, Delaney T, Vernooij B, Morse A, et al. (1993) Biological induction of systemic acquired resistance in Arabidopsis. Mol Plant-Microbe Interact 6: 692–698.

10. Durrant WE, Dong X (2004) Systemic acquired resistance. Annu Rev Phytopathol 42: 185–209.

11. Kachroo A, Kachroo P (2006) Salicylic Acid-, Jasmonic Acid- and Ethylene-Mediated Regulation of Plant Defense Signaling. In: Setlow Jane, editor. Genetic Regulation of Plant Defense Mechanisms. 28Springer pubs. pp. 55–83.

12. Kachroo P, Chandra-Shekara AC, Klessig D (2006) Plant signal transduction. and defense against viral pathogens. In: Maramososch K, Shatkin AJ, editors. Advances in Viral Research. 66: 161–191.

13. Cao H, Glazebrook J, Clarke JD, Volko S, Dong X (1997) The Arabidopsis NPR1 gene that controls systemic acquired resistance encodes a novel protein containing ankyrin repeats. Cell 88: 57–63.

14. Ryals JA, Weymann K, Lawton K, Friedrich L, Ellis D, et al. (1997) The Arabidopsis NIM1 protein shows homology to the mammalian transcription factor inhibitor IκB. Plant Cell 9: 425–439.

15. Shah J, Tsui F, Klessig DF (1997) Characterization of a salicylic acid-insensitive mutant (sai1) of Arabidopsis thaliana, identified in a selective screen utilizing the SA-inducible expression of the tms2 gene. Mol Plant-Microbe Interact 1: 69–78.

16. Falk A, Feys BJ, Frost LN, Jones JD, Daniels MJ, et al. (1999) EDS1, an essential component of R gene-mediated disease resistance in Arabidopsis has homology to eukaryotic lipases. Proc Natl Acad Sci USA 96: 3292–3297.

17. Nawrath C, Heck S, Parinthawong N, Metraux JP (2002) EDS5, an essential component of salicylic acid-dependent signaling for disease resistance in Arabidopsis, is a member of the MATE transport family. Plant Cell 1: 275–286.

18. Wildermuth MC, Dewdney J, Wu G, Ausubel FM (2001) Isochorismate synthase is required to synthesize salicylic acid for plant defense. Nature 414: 562–565.

19. Jirage D, Tootle TL, Reuber TL, Frost LN, Feys BJ, et al. (1999) Arabidopsis thaliana PAD4 encodes a lipase-like gene that is important for salicylic acid signaling. Proc Natl Acad Sci USA 96: 13583–13588.

20. Glazebrook J (2001) Genes controlling expression of defense responses in Arabidopsis –2001 status. Curr Opin Plant Biol 4: 301–308.

21. Aarts N, Metz M, Holub E, Staskawicz BJ, Daniels MJ, et al. (1998) Different requirements for EDS1 and NDR1 by disease resistance genes define at least two R gene-mediated signaling pathways in Arabidopsis. Proc Natl Acad Sci USA 95: 10306–10311.

22. Bittner-Eddy PD, Beynon JL (2001) The Arabidopsis downy mildew resistance gene, RPP13-Nd, functions independently of NDR1 and EDS1 and does not require the accumulation of salicylic acid. Mol Plant-Microbe Interact 14: 416–421.

23. Chandra-Shekara AC, Navarre D, Kachroo A, Kang H-G, Klessig DF, et al. (2004) Signaling requirements and role of salicylic acid in HRT- and rrt-mediated resistance to turnip crinkle virus in Arabidopsis. Plant J 40: 647–659.

24. McDowell JM, Cuzick A, Can C, Beynon J, Dangl JL, et al. (2000) Downy mildew (Peronospora parasitica) resistance genes in Arabidopsis vary in functional requirements for NDR1, EDS1, NPR1 and salicylic acid accumulation. Plant J 6: 523–529.

25. Xiao S, Calis O, Patrick E, Zhang G, Charoenwattana P, et al. (2005) The atypical resistance gene, RPW8, recruits components of basal defence for powdery mildew resistance in Arabidopsis. Plant J 42: 95–110.

26. Feys BJ, Wiermer M, Bhat RA, Moisan LJ, Medina-Escobar N, et al. (2005) Arabidopsis SENESCENCE-ASSOCIATED GENE101 stabilizes and signals within an ENHANCED DISEASE SUSCEPTIBILITY1 complex in plant innate immunity. Plant Cell 9: 2601–2613.

27. Lipka V, Dittgen J, Bednarek P, Bhat R, Wiermer M, et al. (2005) Pre- and postinvasion defenses both contribute to nonhost resistance in Arabidopsis. Science 310: 1180–1183.

28. Vijayan P, Shockey J, Levesque CA, Cook RJ, Browse J (1998) A role for jasmonate in pathogen defence of Arabidopsis. Proc Natl Acad Sci USA 95: 7209–7214.

29. Kachroo P, Shanklin J, Shah J, Whittle EJ, Klessig DF (2001) A Fatty acid desaturase modulates the activation of defense signaling pathways in Plants. Proc Natl Acad Sci USA 98: 9448–9453.

30. Kachroo A, Lapchyk L, Fukushigae H, Hildebrand D, Klessig D, et al. (2003) Plastidial fatty acid signaling modulates salicylic acid- and jasmonic acid-mediated defense pathways in the Arabidopsis ssi2 mutant. Plant Cell 15: 2952–2965.

31. Kachroo A, Venugopal SC, Lapchyk L, Falcone D, Hildebrand D, et al. (2004) Oleic acid levels regulated by glycerolipid metabolism modulate defense gene expression in Arabidopsis. Proc Natl Acad Sci USA 101: 5152–5157.

32. Weber H (2002) Fatty acid derived signals in plants. Curr Opin Plant Biology 7: 217–224.

33. Li C, Liu G, Xu C, Lee GI, Bauer P, et al. (2003) The Tomato Suppressor of prosystemin-mediated responses2 Gene Encodes a Fatty Acid Desaturase Required for the Biosynthesis of Jasmonic Acid and the Production of a Systemic Wound Signal for Defense Gene Expression. Plant Cell 15: 1646–1661.

34. Yaeno T, Matsuda O, Iba K (2004) Role of chloroplast trienoic fatty acids in plant disease defense responses. Plant J 40: 931–941.

35. Xia Y, Gao Q-M, Yu K, Navarre D, Hildebrand D, et al. (2009) An intact cuticle in distal tissues is essential for the induction of systemic acquired resistance in plants. Cell Host & Microbe 5: 151–165.

36. Shah J, Kachroo P, Nandi A, Klessig DF (2001) A loss-of-function mutation in the Arabidopsis SSI2 gene confers SA- and NPR1-independent expression of PR genes and resistance against bacterial and oomycete pathogens. Plant J 25: 563–574.

37. Chandra-Shekara AC, Venugopal SC, Barman SR, Kachroo A, Kachroo P (2007) Plastidial fatty acid levels regulate resistance gene-dependent defense signaling in Arabidopsis. Proc Natl Acad Sci USA 104: 7277–7282.

38. Kachroo P, Kachroo A, Lapchyk L, Hildebrand D, Klessig D (2003a) Restoration of defective cross talk in ssi2 mutants; Role of salicylic acid, jasmonic acid and fatty acids in SSI2-mediated signaling. Mol Plant-Microbe Interact 11: 1022–1029.

39. Kachroo A, Shanklin J, Lapchyk L, Whittle E, Hildebrand D, et al. (2007) The Arabidopsis stearoyl-acyl carrier protein-desaturase family and the contribution of leaf isoforms to oleic acid synthesis. Plant Mol Biol 63: 257–271.

40. Kachroo P, Venugopal SC, Navarre DA, Lapchyk L, Kachroo A (2005) Role of salicylic acid and fatty acid desaturation pathways in ssi2-mediated signaling. Plant Physiol 139: 1717–1735.

41. Kachroo A, Daqi F, Havens W, Navarre R, Kachroo P, et al. (2008) Virus-induced gene silencing of stearoyl-acyl carrier protein-desaturase in soybean results in constitutive defense and enhanced resistance to pathogens. Mol Plant-Microbe Interact 21: 564–575.

42. Jiang CJ, Shimono M, Maeda S, Inoue H, Mori M, et al. (2009) Suppression of the rice fatty-acid desaturase gene OsSSI2 enhances resistance to blast and leaf blight diseases in rice. Mol Plant-Microbe Interact. In press.

43. Nawrath C, Métraux JP (1999) Salicylic acid induction-deficient mutants of Arabidopsis express PR-2 and PR-5 and accumulate high levels of camalexin after pathogen inoculation. Plant Cell 11: 1393–1404.

44. Chen Z, Kloek AP, Boch J, Katagiri F, Kunkel BN (2000) The Pseudomonas syringae avrRpt2 gene product promotes pathogen virulence from inside plant cells. Mol Plant-Microbe Interact 13: 1312–1321.

45. Wiermer M, Feys BJ, Parker JE (2005) Plant immunity: the EDS1 regulatory node. Curr Opin Plant Biol 8: 383–389.

46. Bartsch M, Gobbato E, Bednarek P, Debey S, Schultze JL, et al. (2006) Salicylic acid-independent ENHANCED DISEASE SUSCEPTIBILITY1 signaling in Arabidopsis is regulated by the monooxygenase FMO1 and the nudix hydrolase NUDT7. Plant Cell 18: 1038–1051.

47. Xing D, Chen Z (2006) Effects of mutations and constitutive overexpression of EDS1 and PAD4 on plant resistance to different types of microbial pathogens. Plant Sci 171: 251–262.

48. Yang S, Hua J (2004) A haplotype-specific resistance gene regulated by BONZAI1 mediates temperature-dependent growth control in Arabidopsis. Plant Cell 16: 1060–1071.

49. Leeuwen HV, Kliebenstein DJ, West MAL, Kim K, Poecke RV, et al. (2007) Natural variation among Arabidopsis thaliana accessions for transcriptome response to salicylic acid. Plant Cell 19: 2099–2110.

50. Bouchabke O, Chang F, Simon M, Voisin R, Pelletier G, et al. (2008) Natural variation in Arabidopsis thaliana as a tool for highlighting differential drought responses. PLoS ONE 3(2): e1705. doi:10.1371/journal.pone.0001705.

51. Sandra SL, Amasino RM (1996) Ecotype-specific expression of a flowering mutant phenotype in Arabidopsis thaliana. Plant Physiol 111: 641–644.

52. Stokes TL, Kunkel BN, Richards EJ (2002) Epigenetic variation in Arabidopsis disease resistance. Genes Dev 16: 171–182.

53. Rodriguez-Rodriguez RA, Tabernero A, Velasco A, Lavado EM, Medina JM (2004) The neurotrphic effect of oleic acid includes dendritic differentiation and the expression of the neuronal basic helix-loop-helix transcription factor. J Neurochem 88: 1041–1051.

54. Stremmel W, Strohmeyer G, Borchard F, Kochwa S, Berk PD (1985) Isolation and partial characterization of a fatty acid binding protein in rat liver plasma membrane. Proc Natl Acad Sci USA 82: 4–8.

55. Svanborg C, Agerstam H, Aronson A, Bjerkvig R, Düringer C, et al. (2003) HAMLET kills tumor cells by an apoptosis-like mechanism-cellular, molecular, and therapeutic aspects. Adv Cancer Res 88: 1–29.

56. Menendez JA, Lupu R (2006) Oncogenic properties of the endogenous fatty acid metabolism: molecular pathology of fatty acid synthase in cancer cells. Curr Opin Clin Nutr Metab Care 9: 346–357.

57. Wang C, Wang X (2001) A novel phospholipase D of Arabidopsis that is activated by oleic acid and associated with the plasma membrane. Plant Physiol 127: 1102–1112.

58. Fox SR, Hill LM, Rawsthorne S, Hills MJ (2000) Inhibition of the glucose-6-phosphate transporter in oilseed rape (Brassica napus L.) plastids by acyl-CoA thioesters reduces fatty acid synthesis. Biochem J 352: 525–532.

59. Dempsey DA, Wobbe KK, Klessig DF (1993) Resistance and susceptible responses of Arabidopsis thaliana to turnip crinkle virus. Phytopathology 83: 1021–1029.

60. Oh J-W, Kong W, Song C, Carpenter CD, Simon AE (1995) Open reading frames of turnip crinkle virus involved in satellite symptom expression and incompatibility with Arabidopsis thaliana ecotype Dijon. Mol Plant-Microbe Interact 8: 979–987.

61. Dahmer ML, Fleming PD, Collins GB, Hildebrand DF (1989) A Rapid screening for determining the lipid composition of soybean seeds. J Am Oil Chem 66: 534–538.

62. Ghabrial SA, Schultz FG (1983) Serological detection of bean pod mottle virus in bean leaf beetles. Phytopathology 73: 480–483.

CITATION

Originally published under the Creative Commons Attribution License. Venugopal SC, Jeong R-D, Mandal MK, Zhu S, Chandra-Shekara AC, et al. (2009). Enhanced Disease Susceptibility 1 and Salicylic Acid Act Redundantly to Regulate Resistance Gene-Mediated Signaling. PLoS Genet 5(7): e1000545. doi:10.1371/journal.pgen.1000545.

Strategies of *Nitrosomonas europaea* 19718 to Counter Low Dissolved Oxygen and High Nitrite Concentrations

Ran Yu and Kartik Chandran

ABSTRACT

Background

Nitrosomonas europaea is a widely studied chemolithoautotrophic ammonia oxidizing bacterium. While significant work exists on the ammonia oxidation pathway of N. europaea, its responses to factors such as dissolved oxygen limitation or sufficiency or exposure to high nitrite concentrations, particularly at the functional gene transcription level are relatively sparse. The principal goal of this study was to investigate responses at the whole-cell activity and gene transcript levels in N. europaea 19718 batch cultures, which were cultivated at different dissolved oxygen and nitrite concentrations. Transcription of genes coding for principal metabolic pathways including ammonia oxidation (amoA), hydroxylamine oxidation (hao), nitrite reduction (nirK) and nitric oxide reduction (norB) were quantitatively measured during batch growth, at a range of DO concentrations (0.5, 1.5 and 3.0 mg O2/L). Measurements were also conducted during growth at 1.5 mg O2/L in the presence of 280 mg-N/L of externally added nitrite.

Results

Several wide ranging responses to DO limitation and nitrite toxicity were observed in N. europaea batch cultures. In contrast to our initial hypothesis, exponential phase mRNA concentrations of both amoA and hao increased with decreasing DO concentrations, suggesting a mechanism to metabolize ammonia and hydroxylamine more effectively under DO limitation. Batch growth in the presence of 280 mg nitrite-N/L resulted in elevated exponential phase nirK and norB mRNA concentrations, potentially to promote utilization of nitrite as an electron acceptor and to detoxify nitrite. This response was in keeping with our initial hypothesis and congruent with similar responses in heterotrophic denitrifying bacteria. Stationary phase responses were distinct from exponential phase responses in most cases, suggesting a strong impact of ammonia availability and metabolism on responses to DO limitation and nitrite toxicity. In general, whole-cell responses to DO limitation or nitrite toxicity, such as sOUR or nitrite reduction to nitric oxide (NO) did not parallel the corresponding mRNA (nirK) profiles, suggesting differences between the gene transcription and enzyme translation or activity levels.

Conclusions

The results of this study show that N. europaea possesses specific mechanisms to cope with growth under low DO concentrations and high nitrite concentrations. These mechanisms are additionally influenced by the physiological growth state of N. europaea cultures and are possibly geared to enable more efficient substrate utilization or nitrite detoxification.

Background

Nitrosomonas europaea is a widely studied chemolithoautotrophic ammonia oxidizing bacterium (AOB) that catalyzes the aerobic oxidation of ammonia (NH_3) to nitrite (NO_2^-) using carbon dioxide (CO_2) as the preferred assimilative carbon source [1]. Bacteria closely related to N. europaea have been found in various natural and engineered environments indicating that they can proliferate under different growth conditions, by effectively utilizing growth substrates such as NH_3 and oxygen [2-4].

The oxidative catabolic pathway of N. europaea involves NH_3 oxidation to hydroxylamine (NH_2OH) by membrane bound ammonia monooxygenase (AMO) and NH2OH oxidation to NO_2^- by periplasmic hydroxylamine oxidoreductase (HAO) (Figure 1) [5]. In addition, autotrophic denitrification by N. europaea has also been shown [6-8]. It is believed that denitrification by N. europaea is especially favored during growth under low dissolved oxygen (DO) concentrations or high nitrite concentrations [9] and results in the production of nitric oxide (NO) or nitrous oxide (N_2O) [10,11]. However, little information exists on the mechanisms driving the responses of N. europaea to DO limitation and possible NO_2^- toxicity [12]. For instance, it is as yet unknown whether responses to DO limitation and NO_2^- toxicity at the whole-cell level are ultimate manifestations of changes in gene transcription and expression.

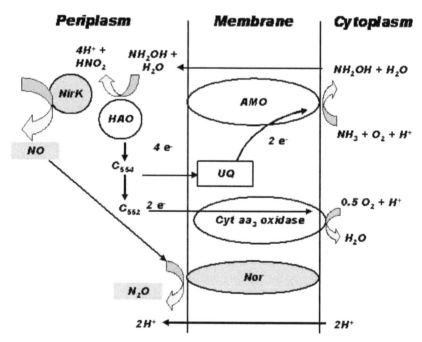

Figure 1. Schematic of oxidative (unshaded enzymes) and reductive (gray shaded enzymes) nitrogen transformations in N. europaea (modified after [5]).

In this study, the ability of N. europaea to transcribe four key genes involved in its catabolic pathway as a function of batch growth conditions (NH_3 sufficiency and starvation, DO limitation and NO_2^- toxicity) was evaluated. It was hypothesized that DO limitation and NO_2^- toxicity would result in lower transcription of genes coding for NH_3 and NH_2OH oxidation (amoA and hao, respectively), given that these are the main steps leading to energy generation in N. europaea [5]. Furthermore, given that low DO and high NO_2^- concentrations are two main triggers for expression of denitrification genes in heterotrophic bacteria [13], it was hypothesized that decreasing DO concentrations and high NO_2^- concentrations would similarly induce progressively higher transcription of NO_2^- and NO reductase genes in N. europaea (nirK and norB, respectively).

The specific objectives of this study were to (i) quantitatively measure the transcription of amoA, hao, nirK and norB, four genes involved in redox N transformations, in N. europaea during batch growth at different DO and initial NO_2^- concentrations; and (ii) compare gene transcription level responses to DO limitation and NO_2^- inhibition with 'whole-cell' responses related to activity, biokinetics and production of gaseous NO- the first product of sequential NO_2^- reduction by N. europaea.

Results

Impact of Reactor DO on N Speciation, Biokinetics and Functional Gene Transcription

Batch cultivation of N. europaea cultures at different DO concentrations (0.5, 1.5 and 3.0 mg O_2/L) led to several differences at the nitrogen speciation, biokinetics and gene transcription levels. Based on a studentized t-test, the degree of NH_3-N conversion to NO_2^- -N at DO = 0.5 mg O_2/L (76 ± 16%) was significantly lower (p < 0.05) than at DO = 1.5 mg O2/L, (90 ± 10%) or DO = 3.0 mg O2/L (89 ± 15%), respectively, (Figure 2, A1-C1). The final cell concentrations were relatively uniform for all three DO concentrations (Figure 2, A2-C2). However, the lag phase at DO = 0.5 mg O2/L was one day longer than at DO = 1.5 or 3.0 mg O_2/L pointing to the impact of electron acceptor limitation on the cell synthesizing machinery of N. europaea (Figure 2, A2-C2). Estimates of the maximum specific growth rate (obtained via non-linear estimation [14]) at DO = 0.5 mg O2/L (0.043 ± 0.005 h^{-1}), 1.5 mg O_2/L (0.057 ± 0.012 h^{-1}) and 3.0 mg O_2/L (0.060 ± 0.011 h^{-1}) were not statistically different at α = 0.05. At all three DO concentrations tested, low levels of NH_2OH transiently accumulated in the growth medium during the exponential phase, in keeping with its role as an obligate intermediate of NH_3 oxidation [5] (Figure 2, A1-C1). The initial increase in NH2OH concentrations at DO = 0.5 mg O_2/L, was the slowest, due to the longer lag-phase (Figure 2, A1). The peak NH2OH concentration at DO = 0.5 mg O2/L was also lower than at DO = 1.5 or 3.0 mg O_2/L (Figure 2, A1-C1).

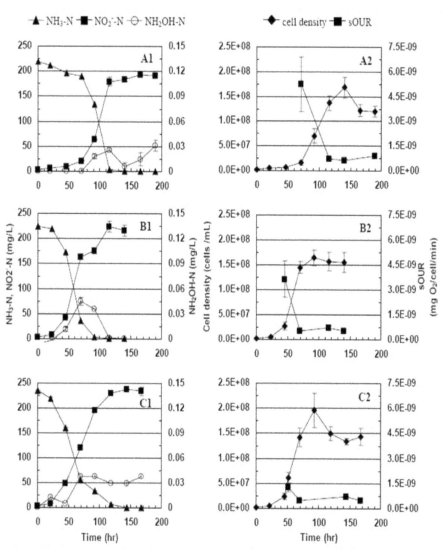

Figure 2. NH3-N, NO2--N, and NH2OH-N, (A1-C1), cell density and sOUR (A2-C2) profiles during N. europaea batch growth at DO = 0.5 mg/L (A), 1.5 mg/L (B) and 3 mg/L (C).

The peak 'potential' biokinetics of NH_3 oxidation (expressed as sOUR, and measured under non-limiting DO and ammonia concentrations) varied inversely with reactor DO concentrations (Figure 2, A2-C2). sOUR values consistently peaked during early exponential growth phase followed by a significant decrease during stationary phase (Figure 2, A2-C2), in good correspondence with recent results [15]. Additional sOUR assays could not be conducted during the lag phase, owing to low cell concentrations, which would have consequently necessitated removal of excessively high sampling volumes.

Headspace NO concentrations peaked during the exponential phase and significantly diminished upon NH_3 exhaustion in the stationary phase (Figure 3, A3-C3). An increasing trend in peak headspace NO concentrations was observed with increasing DO concentrations. NO formation was strictly biological and was not observed in cell-free controls (data not shown). At all DO concentrations tested, intracellular NO was detected in the majority of sampled cells (Figure 3, A3-C3). NO 'positive' cell concentrations were highest especially during late exponential and stationary phases when NO_2^-, the likely substrate for NO production, concentrations were the highest (Figure 3, A3-C3). The more gradual increase in the proportion of NO positive cells at DO = 0.5 mgO_2/L paralleled the trend in peak headspace NO concentrations (Figures 2, 3).

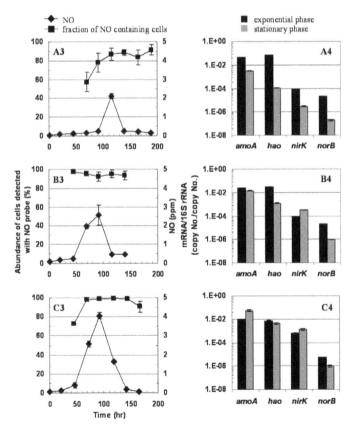

Figure 3. NO profiles and fraction of NO containing cells (A3-C3), and gene expression (A4-C4) during exponential phase and stationary phase at DO = 0.5 mg/L (A), 1.5 mg/L (B) and 3 mg/L (C) for cultures shown in Figure 2.

The impact of operating DO concentrations on gene transcript profiles, determined using primer sets described in Table 1, was dependent upon the physiological growth phase. In exponential phase cell samples, amoA and hao relative mRNA concentrations statistically decreased with increasing reactor DO concentrations (Figure 3, A4-C4, Table 2). A systematic impact of growth phase on nirK and norB relative mRNA concentrations was not observed during exponential phase. The relative mRNA concentrations for both genes during exponential phase were statistically similar for DO = 0.5 and 1.5 mg O_2/L and statistically higher (for nirK) or lower (for norB) at DO = 3.0 mg O2/L (Figure 3, A4-C4, Table 2). In direct contrast, during stationary phase, the relative mRNA concentrations of amoA, hao and nirK all statistically increased with increasing DO concentrations. Additionally, the relative mRNA concentrations of norB at DO = 1.5 mg O_2/L were statistically higher than at DO = 0.5 mg O_2/L, but statistically similar to those at DO = 3.0 mg O2/L (Table 2).

Table 1. Endpoint and real-time PCR primers employed in this study

Primer	Sequence (5'-3')	Position	Target gene	Reference
		Endpoint PCR		
A189	GGHGACTGGGAYTTCTGG	151-168	amoA	[36,37]
amoA2R'	CCTCKGSAAAGCCTTCTTC	802-820		
HAO1F	TCAACATAGGCACGGTTCATCGGA	203-226	hao	[38]
HAO1R	ATTTGCCGAACGTGAATCGGAACG	1082-1105		
NirK1F	TGCTTCCGGATCAGCGTCATTAGT	31-54	nirK	[38]
NirK1R	AGTTGAAACCGATGTGGCCTACGA	809 832		
NorB1F	CGGCACTGATGTTCCTGTTTGCTT	479-502	norB	[38]
NorB1R	AGCAACCGCATCCAGTAGAACAGA	1215-1238		
KNO50F	TNANACATGCAAGTCGAICG	49-68	Eubacterial 16S rRNA gene	[39]
KNO51R	GGYTACCTTGTTACGACTT	1492-1510		
		Quantitative PCR		
amoAFq	GGACTTCACGCTGTATCTG	408-426	amoA	[15]
amoARq	GTGCCTTCTACAACGATTGG	524-543		
HAO1Fq	TGAGCCAGTCCAACGTGCAT	266-285	hao	[38]
HAO1Rq	AAGGCAACAACCCTGCCTCA	331-350		
NirK1Fq	TGCAGGGCATACTGGACGTT	182-201	nirK	[38]
NirK1Rq	AGGTGAACGGGTGCGCATTT	291-310		
NorB1Fq	ACACAAATCACTGCCGCCCA	958-977	norB	[38]
NorB1Rq	TGCAGTACACCGGCAAAGGT	1138-1157		
EUBF	TCCTACGGGAGGCAGCAGT	339-357	Eubacterial 16S rRNA gene	[34]
EUBR	GGACTACCAGGGTATCTAATCCTGTT	780-805		

Table 2. Statistical comparison of the impact of DO concentrations on relative mRNA concentrations in exponential (E) and stationary (S) phase cultures (p values < 5.0 × 10-2 indicate statistically significant differences).

DO (mg O_2/L) comparison	p =							
	amoA		hao		nirK		norB	
	E	S	E	S	E	S	E	S
0.5 – 1.5	1.32×10^{-4}	1.64×10^{-5}	4.86×10^{-5}	3.3×10^{-5}	9.48×10^{-1}	2.9×10^{-5}	6.29×10^{-1}	4.63×10^{-6}
1.5 – 3.0	1.2×10^{-5}	1.98×10^{-3}	2.26×10^{-11}	2.16×10^{-3}	1.22×10^{-5}	1.78×10^{-3}	1.83×10^{-7}	7.52×10^{-1}

Underlined text indicates statistically similar results, bold text indicates statistical increase and regular text indicates decrease.

At DO = 0.5 mg O_2/L, the transition from exponential phase to stationary phase resulted in a systematic decrease in relative mRNA concentrations of all four genes (Figure 3A4-C4 and Table 3). At DO = 1.5 mg O_2/L, this trend was valid for amoA, hao and norB. However, the stationary phase nirK relative mRNA concentrations were statistically higher than during exponential phase. At DO = 3.0 mg O_2/L, only hao and norB displayed a decrease in relative mRNA concentrations upon transition from exponential to stationary phase (Figure 3A4-C4, Table 3). In contrast, relative mRNA concentrations of amoA and nirK increased during stationary phase (Figure 3A4-C4, Table 3). Additionally, except at DO = 1.5 mg O_2/L for nirK, the relative retention of amoA mRNA concentrations during stationary phase relative to exponential phase was consistently the highest (Figure 3 B4-C4). The retention factors averaged across all three DO concentrations were 1.98:1, 0.21:1, 1.86:1 and 0.08:1 for amoA, hao, nirK and norB, respectively (where a retention factor > 1) suggests relative increase during stationary phase).

Table 3. Statistical comparison of relative mRNA concentrations and sOUR in exponential and stationary phase cultures grown at different DO concentrations (p values < 5.0×10^{-2} indicate statistically significant differences).

DO (mg O_2/L)	p =				
	amoA	hao	nirK	norB	sOUR
0.5	5.0×10^{-5}	1.1×10^{-5}	3.2×10^{-6}	8.0×10^{-6}	5.0×10^{-1}
1.5	5.5×10^{-6}	6.4×10^{-8}	7.7×10^{-5}	3.9×10^{-6}	1.5×10^{-3}
3.0	1.5×10^{-3}	6.3×10^{-4}	5.1×10^{-3}	1.0×10^{-6}	1.2×10^{-1}

Underlined text indicates statistically similar results, bold text indicates statistical increase and regular text indicates decrease.

Impact of Growth in the Presence of Added Nitrite on N Speciation, Biokinetics and Gene Transcription

Cell growth was not detected at an initial NO_2^- concentration of 560 mg-N/L and DO = 1.5 mg O_2/L, even after 2 weeks of incubation (data not shown). An initial NO_2^- concentration of 280 mg NO_2^--N/L and DO = 1.5 mg O2/L, resulted in a lag phase one day longer than that in the initial absence of nitrite (Figure 4 D1-D2 and Figure 2, B1-B2, respectively). However, the overall cell yield was not impacted. The extent of NH_3 oxidized to NO_2^- in the presence of 280 mg NO_2^--N/L (88 ± 5%, n = 2) was not significantly different (α = 0.05) than in the absence of nitrite (90 ± 10%, n = 2). NH2OH accumulation was observed during the extended lag phase suggesting initial inhibition of NH_2OH oxidation by NO_2^- (Figure 4, D1). Lower NO production was observed in the presence of added NO_2^- (Figure 4, D3). In parallel, a substantial reduction in the fraction of cells with detectable intracellular NO was also observed, from $98.3 \pm 2.1\%$ during exponential phase to $66.6 \pm 10.4\%$ during stationary phase (Figure 4, D3). sOUR values were not significantly different ($\alpha = 0.05$) in the presence or absence of added NO_2^--N/L (Figure 4, D2, Figure 2, B2, respectively). Exponential phase relative mRNA concentrations of amoA and hao were statistically lower during growth in the presence of 280 mg NO_2^--N/L than in the absence of added nitrite (Figure 4, D4, Table 4). However, exponential phase transcription of nirK and norB was significantly higher in the presence of 280 mg NO_2^--N/L than in the absence of added nitrite (Figure 4, D4 and Figure 3, B4, Table 4). During stationary phase, amoA, hao, nirK and norB relative mRNA concentrations were all statistically lower in the presence of 280 mg NO_2^--N/L than in the absence of added nitrite (Figure 3, B4 and Figure 4, D4, Table 4).

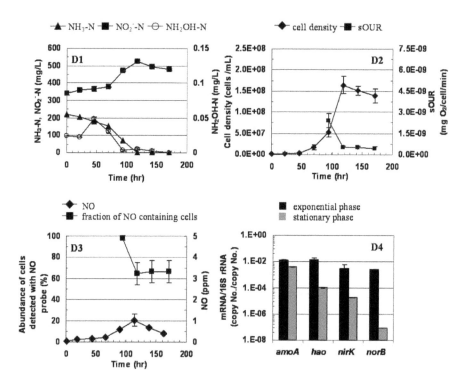

Figure 4. Profiles of NH_3-N, NO_2^--N, and NH_2OH-N (D1), cell density and sOUR (D2), NO and fraction of NO containing cells (D3) and gene expression (D4) during exponential phase and stationary phase at DO = 1.5 mg/L in the presence of added 280 mg NO_2^--N/L.

Table 4. Statistical comparison of relative mRNA concentrations and sOUR in exponential (E) and stationary (S) phase cultures grown in the presence and absence of nitrite (p values < 5.0 × 10-2 indicate statistically significant differences).

Growth phase	p =				
	amoA	hao	nirK	norB	sOUR
E	7.9×10^{-4}	1.2×10^{-3}	1.3×10^{-3}	2.8×10^{-3}	7.0×10^{-3}
S	5.1×10^{-5}	3.2×10^{-5}	3.2×10^{-5}	4.6×10^{-5}	2.0×10^{-1}

Underlined text indicates statistically similar results, bold text indicates statistical increase and regular text indicates decrease.

Discussion

Functional Gene Transcription and N Profiles During Batch Growth of N. europaea

In addition to its well-studied NH_3 oxidation pathway, the genome of N. europaea contains genes coding for several denitrification steps, including NO_2^- and NO reduction [16]. While significant work exists on expression analysis of amoA and to an extent, hao, [17-22], quantitative transcription patterns for nirK and norB are relatively less characterized. The significance of this study therefore lies in elucidating the co-transcription patterns of amoA, hao, nirK and norB under varying degree of DO and NO_2^- exposure during batch growth of N. europaea.

The general overall reduction in amoA transcription during the stationary phase, at DO = 0.5 and 1,5 mg O_2/L (Figure 3, A4-B4), can be linked to dwindling energy resources for N. europaea [15,23] or toxicity of accumulating NO_2^- concentrations [21]. The higher amoA relative mRNA concentrations during the stationary phase at DO = 3.0 mg O_2/L were not expected and likely due to the opposing trends in exponential phase and stationary phase responses to increasing DO concentrations (Figure 3, B4-D4), as discussed below.

The retention of relatively higher amoA mRNA concentrations during stationary phase compared to those for hao, nirK and norB points to the capability of N. europaea to sustain and rapidly increase NH_3 oxidation during a transition from a starvation state (as in stationary phase) to when NH_3 becomes available. Since NH_3 oxidation is the very first step in energy generation for N. europaea, it is indeed advantageous to retain the capability (by retaining amoA mRNA) for this step to a certain extent compared to downstream steps. These results are consistent with the higher retention of amoA mRNA concentrations relative to those for other genes coding for carbon dioxide fixation for growth, ion transport, electron transfer and DNA replication [23]. In fact, an actual increase in NH_3

transport genes during NH_3 starvation in stationary phase has also been observed [23].

The increasing trend in relative mRNA concentrations of amoA and hao and sOUR with decreasing DO concentrations during exponential growth reflect a possible strategy of N. europaea to (partially) make up for low DO concentrations by enhancing the ammonia and hydroxylamine oxidizing machinery. One possible means to enhance substrate utilization rates at reduced DO concentrations could be to increase the capacity for oxygen transfer into the cell itself.

An alternate means could be by enhancing the ammonia or hydroxylamine oxidizing machinery (mRNA, proteins and or protein activity). The volumetric ammonia oxidation rate depends upon the mathematical product of AMO (or HAO) protein concentrations, their activity and DO concentrations (as given by the multiplicative Monod model [24]). Therefore, potentially similar ammonia oxidation rates could be maintained at lower DO concentrations by increasing the catalytic protein concentrations (or those of their precursors, such as mRNA) or activities (as measured by sOUR assays). Such an enhancement might be manifested in higher 'potential' oxygen uptake rates, measured under non-limiting DO concentrations. Notwithstanding increased 'potential' NH3 or NH2OH oxidation activity from cells exposed to sustained lower DO concentrations, actual 'extant' activity is indeed expected to be lower under stoichiometric DO limitation, resulting in lower rates of batch cell growth or nitrite accumulation (Figure 2, A2-C2). Based on a recent study, N. europaea cultures demonstrated similar increases in amoA transcription and sOUR when subject to NH3 limitation in chemostats, relative to substrate sufficient batch cultures [15].

While it is documented that NirK is involved in NH3 oxidation by facilitating intermediate electron transport [25], the specific role of the Nor cluster in NH_3 metabolism and exclusivity in N_2O production is unclear [7]. Both NirK and Nor act upon products of upstream AMO and HAO. Thus, the lack of systematic trends in relative mRNA concentrations of nirK or norB with changing DO concentrations possibly point to less stringent regulation of these two genes during exponential growth in the overall catabolic pathways of N. europaea.

In contrast to exponential phase, the statistical increase in relative mRNA concentrations with increasing DO concentrations for all four genes during stationary phase is clearly intriguing. These trends highlight the impact of starvation on responses to different DO concentrations. Although the unique responses of N. europaea to starvation [23] and oxygen concentrations (via Fnr [26]) have been documented, the mechanisms of combined NH3 and DO based gene regulation in N. europaea are not well understood. It is well documented that ammonia oxidizing bacteria, such as N. europaea, are commonly subject to cycling between anoxic and oxic conditions and a wide range of NH3 concentrations in engineered and natural environments such as wastewater treatment plants or soils [24,27,28]. The specific responses observed herein might be part of a coordinated strategy of N. europaea to maintain active or latent substrate metabolic machinery to counter such varying environments and clearly merit further study.

The differences in observed transient accumulation of NH_2OH could also be explained at the transcription and protein activity levels. The decrease in exponential phase hao relative mRNA concentrations with increasing DO was more rapid than for amoA (Figure 3, A4-C4). This decrease coupled with a decrease in sOUR (a composite measure of AMO and HAO activity) with increasing DO, could have resulted in the observed trends in NH_2OH concentrations. Although it has been shown that N. europaea can retain high levels of HAO protein and activity under ammonia starvation [29], the impact of DO concentrations on HAO activity has not been specifically identified. While the gene transcript data provide good insights into possible responses of N. europaea to different DO concentrations, protein activity data is crucial to explain profiles of intermediates such as NH_2OH.

The parallel profiles of exponential phase nirK relative mRNA concentrations and headspace NO concentrations at different DO concentrations (Figure 3) suggest a possible link between nirK transcription and NO generation. However, the loss of this parallel in the presence of added NO2- (higher nirK gene transcription but lower NO concentrations, Figure 4) suggests the possible presence of NO generation pathways that are distinct from NO2- reduction, as pointed out previously [26] or even post-transcriptional effects. Indeed, there is still no consensus about the source of NO produced by AOB, such as N. europaea, and the potential roles of nirK, hao and a multicopper oxidase of the nirK operon have all been implicated [26].

Impact of Exposure to High Nitrite Concentrations on Gene Transcription

High NO_2^- concentrations have been implicated as the principal trigger for high NirK protein activity in N. europaea [9], which has a fundamental grounding in the similar trends observed in this study at the nirK gene mRNA level during exponential growth (Figure 4 D4). Increased nirK transcription is the result of the regulatory activity of the NsrR repressor protein, which is present in the genome of N. europaea [16]. NsrR is responsible for sensing NO and NO_2^- concentrations and is supposedly involved in the transcriptional regulation of several operons including the nirK gene cluster of N. europaea [9]. Although N. europaea contains norB, alternate pathways are possibly involved in the production of N_2O [7], the increased transcription of norB, shown in this study cannot be unequivocally reconciled with functional N2O production. Nevertheless, the increased transcription of both nirK and norB in response to high nitrite concentrations is in keeping with one of our initial hypotheses.

The uniformly lower transcript concentrations upon growth with added 280 mg NO_2^- -N/L could be a result of energy resources channeled towards mitigation of nitrite toxicity rather than its utilization as an electron acceptor during stationary phase. In general, it could be argued that in response to nitrite toxicity during ammonia starvation, there is little incentive to increase transcription of putative nitrite and nitric oxide reduction pathways. However, it should be noted that the lower transcript abundance during stationary phase when grown with added 280 mg NO_2^- -N/L is in direct contrast to an increase in nirK during stationary phase, when grown without added NO_2^- -N (Figure 3 B4-C4). The more gradual build-up of nitrite in the latter case could have allowed for adaptation, whereas the initial spike of 280 mg NO_2^- -N/L might have imposed a significant toxic stress that resulted in reduced growth and different transcriptional profiles. Indeed, the toxic stress was possibly too severe at 560 mg NO_2^- -N/L, which resulted in no growth whatsoever.

Additionally, the reduction in transcript abundance of amoA and hao in the presence of NO_2^- -N, did not parallel the relatively unchanged sOUR in the presence or absence of NO_2^- -N. Given that sOUR is a measure of the sum of AMO and HAO activities, these results also suggest uncoupling of the responses at the gene transcription and post-transcriptional or translational levels (Figure 4). Responses at the protein abundance and activity levels would be needed to substantiate and provide an explanation for such uncoupling.

It should be noted that the severe impacts of added nitrite were possibly related to the application of these high nitrite concentrations at the beginning of the batch growth assays. Had the nitrite concentrations been applied during periods of relatively higher cell concentrations (during exponential or stationary phase), the impacts might have been less severe, given that the cells were already producing and responding to the increasing NO_2^- -N levels in the culture medium. Thus, in a sense, the results reported herein represent the most extreme response of N. europaea cultures to nitrite exposure.

Conclusions

The responses of N. europaea to cope with DO limitation and NO_2^- toxicity were wide-ranging from the gene transcription through whole cell levels. The results refuted the initial hypothesis that low DO is one of the main pre-requisite conditions for the transcription of nirK and norB genes in N. europaea. On the other hand, these results indeed supported our other hypothesis that higher NO_2^- concentrations constitute the principal trigger for increased relative transcription related to autotrophic denitrification reactions. The distinct responses observed during the exponential and stationary phase to both DO limitation and nitrite toxicity highlight the need to understand the specific regulatory mechanisms employed by N. europaea to jointly counter substrate starvation and stress.

Methods

Cultivation of Batch N. europaea Cultures

N. europaea (ATCC 19718, Manassas, VA) batch cultures were cultivated in the dark in batch bioreactors (Bellco Glass, Vineland, NJ, working volume = 4 L, agitation speed = 200 rpm) in a growth medium containing 280 mg-N/L and in addition (per liter): 0.2 g of $MgSO_4 7H_2O$, 0.02 g of $CaCl_2 2H_2O$, 0.087 g of K_2HPO_4, 2.52 g EPPS (3- [4-(2-Hydroxyethyl)-1-piperazine] propanesulfonic acid), 1 mL of 13% EDTA-Fe_3^+, 1 mL of trace elements solution (10 mg of $Na_2MoO_4 2H_2O$, 172 mg of $MnCl_2 4H_2O$, 10 mg of $ZnSO_4 7H_2O$, 0.4 mg of $CoCl_2 6H_2O$, and 100 mL of distilled water), 0.5 mL of 0.5% phenol red, and 0.5 mL of 2 mM $CuSO_4 5H_2O$. Reactor pH was controlled in the range 6.8-7.4 by manual addition of pre-sterilized 40% potassium bicarbonate solution.

Batch growth experiments were conducted at three DO concentrations, 0.5 ± 0.05, 1.5 ± 0.05 and 3.0 ± 0.05 mg O_2/L. Batch reactor DO was measured and controlled with a fermentation DO probe and benchtop dissolved oxygen meter and controller system (Cole-Parmer, Vernon Hills, IL) using a combination of filter sterilized (0.2 μm pore size, Millipore®, Ann Arbor, MI) nitrogen gas or air. In select experiments conducted at DO = 1.5 ± 0.05 mg O_2/L, the feed medium additionally contained 280, or 560 mg NO_2^- -N/L before N. europaea inoculation, which enabled the determination of batch growth in the presence of these high NO_2^- -N concentrations. NH_3 (gas-sensing electrode, Corning, Corning, NY), NH_2OH [30], NO_2^- (diazotization, [31], cell concentration (direct counting) and gaseous NO (chemiluminescence, CLD-64, Ecophysics, Ann Arbor, MI) were measured once a day during the batch growth profile. All batch growth experiments were conducted in duplicate.

Detection of Intracellular and Extracellular Nitric Oxide

Intracellular NO presence was determined by staining with 4-amino-5-methylamino-2',7'-difluorofluorescein diacetate (Molecular Probes, Eugene, OR) for 30 min in the absence of light. Stained cells were washed twice with sterile NH_3-free medium and quantified immediately with epifluorescence microscopy (Nikon ECLIPSE 80 i) using a minimum of 10 randomly-chosen microscopic fields (each 0.30×0.22 mm^2). NO was specifically the focus of gaseous bulk phase and intracellular measurements since it is the direct product of nitrite reduction, the main focus of this study. Additionally, the presence of NO inside N. europaea cells strongly implicates its direct production by the cells themselves rather than by extracellular abiotic reactions. In contrast to NO, there is currently no method that allows detection of intracellular N_2O. Therefore, N_2O data was not included in bulk or intracellular measurements.

Respirometry-Based Biokinetic Monitoring

The 'potential' maximum biokinetic rates of NH_3 oxidation were determined using a short-term (lasting approximately 30 min) batch respirometric assay [32]. The term 'potential' describes non-limiting NH_3 (initial concentration of 50 mg-N/L) and oxygen concentrations (supersaturated initial concentration of approximately 40 mg O_2/L, shown previously to be non-inhibitory to NH_3 oxidation [33]). Maximum NH_3 oxidation activity per cell was expressed as the specific oxygen uptake rate, sOUR and was calculated by dividing the slope of the respirograms (DO vs time) by the cell concentration.

RNA Extraction and Purification

40 ml cell suspensions were collected and immediately centrifuged at 4°C and 5000*g for 10 min. The resulting cell-pellets were resuspended and lysed in 1 mL TRIzol® solution (Invitrogen, Carlsbad, CA). RNA was isolated from lysed cell pellets using the TRIzol® RNA isolation protocol (Invitrogen). Subsequent DNA removal and reverse transcription was performed using the QuantiTect® Reverse Transcriptase kit (Qiagen, Valencia, CA).

Functional Gene Transcription

Transcript abundance of amoA, hao, nirK and norB was quantified by real-time reverse-transcriptase polymerase chain reaction (q-RT-PCR) using previously documented and newly designed primer sets (Table 1). Additional primers for conventional end-point PCR were also designed for hao, nirK and norB and used for preparing standard curves for q-RT-PCR (Table 1). Transcription of functional genes was normalized to 16S rRNA concentrations quantified using primers EUBF and EUBR [34]. q-RT-PCR and endpoint PCR were performed in duplicate on an iCycler iQ™5 (Bio-Rad Laboratories, Hercules, CA). A no-template-control was included for each set of PCR and q-RT-PCR reactions. Standard curves for q-RT-PCR consisted of six decimal dilutions of the respective plasmid DNA (corresponding to the four functional genes), containing a given endpoint PCR product. Plasmid concentrations were quantified (Cary 50 UV-Vis spectrophotometer, Varian, Palo Alto, CA) and translated to copy number assuming 660 Da per base pair of double-stranded DNA [35]. Transcript abundance was determined from samples obtained during exponential phase. For exponential phase cultures, sampling time points were 70 hr, 45 hr, and 52 hr for DO concentrations of 0.5, 1.5 and 3 mg/L, respectively, and corresponded to similar cell densities (Figure 3, A4-C4). For stationary phase cultures, the sampling time points were 165 hr, 116 hr, and 119 hr for DO concentrations of 0.5, 1.5 and 3 mg/L, respectively (Figure 3, A4-C4). The sampling time points for exponential and stationary phase cultures, which were grown with addded 280 mg NO_2^--N/L were 95 hr, and 143 hr, respectively (Figure 4, D4).

Authors' Contributions

RY performed the experiments and drafted the manuscript. KC conceived of and developed the study, helped to analyze and interpret the results and draft the manuscript. Both authors have read and approved the final manuscript.

Acknowledgements

This study was co-supported by the National Fish and Wildlife Foundation and the Water Environment Research Foundation.

References

1. Wood PM: Nitrification as a bacterial energy source. In Nitrification, Special Publications of the Society for General Microbiology. Volume 20. Edited by: Prosser JI. Oxford: IRL Press; 1986:39–62.

2. Ahn JH, Yu R, Chandran K: Distinctive microbial ecology and biokinetics of autotrophic ammonia and nitrite oxidation in a partial nitrification bioreactor. Biotechnol Bioeng 2008, 100(6):1078–1087.

3. Arp DJ, Chain PSG, Klotz MG: The impact of genome analyses on our understanding of ammonia-oxidizing bacteria. Annu Rev Microbiol 2007, 61(1)

4. Watson SW, Bock E, Harms H, Koops H-P, Hooper AB: Nitrifying Bacteria. In Bergey's Manual of Systematic Bacteriology. Baltimore, MD: Williams & Wilkins; 1989.

5. Hooper AB, Vannelli T, Bergmann DJ, Arciero DM: Enzymology of the oxidation of ammonia to nitrite by bacteria. Antonie van Leeuwenhoek 1997, 71:59–67.

6. Poth M, Focht DD: 15N Kinetic analysis of N2O production by Nitrosomonas europaea: An examination of nitrifier denitrification. Appl Environ Microbiol 1985, 49(5):1134–1141.

7. Beaumont HJE, van Schooten B, Lens SI, Westerhoff HV, van Spanning RJM: Nitrosomonas europaea expresses a nitric oxide reductase during nitrification. J Bacteriol 2004, 186(13):4417–4421.

8. Schmidt I, Steenbakkers PJM, op den Camp HJM, Schmidt K, Jetten MSM: Physiologic and proteomic evidence for a role of nitric oxide in biofilm formation by Nitrosomonas europaea and other ammonia oxidizers. J Bacteriol 2004, 186:2781–2788.

9. Beaumont HJE, Lens SI, Reijinders WNM, Westerhoff HV, van Spanning RJM: Expression of nitrite reductase in Nitrosomonas europaea involves NsrR, a novel nitrite-sensitive transcription repressor. Mol Microbiol 2004, 54(1).

10. Bock E: Nitrogen loss caused by denitrifying Nitrosomonas cells using ammonium or hydrogen as electron donors and nitrite as electron acceptor. Arch Microbiol 1995, 163:16–20.

11. Kester RA, de Boer W, Laanbroek HJ: Production of NO and N2O by pure cultures of nitrifying and denitrifying bacteria during changes in aeration. Appl Environ Microbiol 1997, 63:3872–3877.

12. Stein LY, Arp DJ: Ammonium limitation results in the loss of ammonia-oxidizing activity in Nitrosomonas europaea. Appl Environ Microbiol 1998, 64(4):1514–1521.

13. Korner H, Zumft WG: Expression of denitrification enzymes in response to the dissolved oxygen level and respiratory substrate in continuous culture of Pseudomonas stutzeri. Appl Environ Microbiol 1989, 55:1670–1676.

14. Chandran K, Hu Z, Smets BF: A critical comparison of extant batch respirometric and substrate depletion assays for estimation of nitrification biokinetics. Biotechnol Bioeng 2008, 101(1):62–72.

15. Chandran K, Love NG: Physiological state, growth mode, and oxidative stress play a role in Cd(II)-mediated inhibition of Nitrosomonas europaea 19718. Appl Environ Microbiol 2008, 74(8):2447–2453.

16. Chain P, Lamerdin J, Larimer F, Regala W, Lao V, Land M, Hauser L, Hooper A, Klotz M, Norton J, et al.: Complete genome sequence of the ammonia-oxidizing bacterium and obligate chemolithoautotroph Nitrosomonas europaea. J Bacteriol 2003, 185(9):2759–2773.

17. Hommes NG, Sayavedra-Soto L, Arp DJ: Mutagenesis and expression of amo, which codes for ammonia monooxygenase in Nitrosomonas europaea. J Bacteriol 1998, 180(13):3353–3359.

18. Stein LY, Arp DJ: Loss of ammonia monooxygenase activity in Nitrosomonas europaea upon exposure to nitrite. Appl Environ Microbiol 1998, 64(10):4098–4102.

19. Hommes NG, Sayavedra-Soto L, Arp DJ: Transcript analysis of multiple copies of amo (encoding ammonia monooxygenase) and hao (encoding hydroxylamine oxidoreductase) in Nitrosomonas europaea. J Bacteriol 2001, 183(3):1096–1100.

20. Ensign SA, Hyman MR, Arp DJ: In vitro activation of ammonia monooxygenase from Nitrosomonas europaea by copper. J Bacteriol 1993, 175(7):1971–1980.

21. Stein LY, Sayavedra-Soto LA, Hommes NG, Arp DJ: Differential regulation of amoA and amoB gene copies in Nitrosomonas europaea. FEMS Microbiol Lett 2000, 192(2):163–168.

22. Sayavedra-Soto LA, Hommes NG, Russell SA, Arp DJ: Induction of ammonia monooxygenase and hydroxylamine oxidoreductase mRNAs by ammonium in Nitrosomonas europaea. Mol Microbiol 1996, 20(3):541–548.

23. Wei X, Yan T, Hommes NG, Liu X, Wu L, McAlvin C, Klotz MG, Sayavedra-Soto LA, Zhou J, Arp DJ: Transcript profiles of Nitrosomonas europaea during growth and upon deprivation of ammonia and carbonate. FEMS Microbiol Lett 2006, 257(1):76–83.

24. Grady CPLJ, Daigger GT, Lim HC: Biological Wastewater Treatment. 2nd edition. New York: Marcel Dekker; 1999.

25. Cantera J, Stein L: Role of nitrite reductase in the ammonia-oxidizing pathway of Nitrosomonas europaea. Arch Microbiol 2007, 188(4):349–354.

26. Beaumont HJE, Hommes NG, Sayavedra-Soto LA, Arp DJ, Arciero DM, Hooper AB, Westerhoff HV, van Spanning RJM: Nitrite reductase of Nitrosomonas europaea is not essential for production of gaseous nitrogen oxides and confers tolerance to nitrite. J Bacteriol 2002, 184(9):2557–2560.

27. Davidson EA, Matson PA, Vitousek PM, Riley R, Dunkin K, Garcia-Mendez G, Maass JM: Processes Regulating soil emissions of NO and N2O in a seasonally dry tropical forest. Ecology 1993, 74(1):130–139.

28. Wrage N, Velthof GL, Laanbroek HJ, Oenema O: Nitrous oxide production in grassland soils: assessing the contribution of nitrifier denitrification. Soil Biol Biochem 2004, 36(2):229–236.

29. Nejidat A, Shmuely H, Abeliovich A: Effect of ammonia starvation on hydroxylamine oxidoreductase activity of Nitrosomonas europaea. J Biochem (Tokyo) 1997, 121(5):957–960.

30. Frear DS, Burrell RC: Spectrophotometric method for determining hydroxylamine reductase activity in higher plants. Anal Chem 1955, 27:1664–1665.

31. Eaton AD, Clesceri LS, Greenberg AE, eds: Standard Methods for the Examination of Water and Wastewater. 21st edition. Washington DC: APHA, AWWA and WEF; 2005.

32. Chandran K, Smets BF: Optimizing experimental design to estimate ammonia and nitrite oxidation biokinetic parameters from batch respirograms. Wat Res 2005, 39(20):4969–4978.

33. Chandran K: Biokinetic characterization of ammonia and nitrite oxidation by a mixed nitrifying culture using extant respirometry. In Ph. D. Dissertation. Storrs: University of Connecticut; 1999.

34. Nadkarni MA, Martin FE, Jacques NA, Hunter N: Determination of bacterial load by real-time PCR using a broad-range (universal) probe and primers set. Microbiol 2002, 148(1):257–266.

35. Madigan MT, Martinko JM: Brock Biology of Microorganisms. 11th edition. Upper Saddle River, NJ: Prentice Hall; 2006.

36. Holmes AJ, Costello A, Lidstrom ME, Murrell JC: Evidence that particulate methane monooxygenase and ammonia monooxygenase may be evolutionarily related. FEMS Microbiol Lett 1995, 132(3):203–208.

37. Okano Y, Hristova KR, Leutenegger CM, Jackson LE, Denison RF, Gebreyesus B, Lebauer D, Scow KM: Application of real-time PCR to study effects of ammonium on population size of ammonia-oxidizing bacteria in soil. Appl Environ Microbiol 2004, 70(2):1008–1016.

38. Yu R, Kampschreur MJ, van Loosdrecht MCM, Chandran K: Molecular mechanisms and specific directionality in autotrophic nitrous oxide and nitric oxide production in response to transient anoxia. Environ Sci Technol 2010, 44(4):1313–1319.

39. Moyer CL, Dobbs FC, Karl DM: Estimation of diversity and community structure through restriction fragment length polymorphism distribution analysis of bacterial 16S rRNA genes from a microbial mat at an active, hydrothermal vent system, Loihi Seamount, Hawaii. Appl Environ Microbiol 1994, 60(3):871–879.

CITATION

Originally published under the Creative Commons Attribution License. Yu R, Chandran K. Strategies of Nitrosomonas europaea 19718 to Counter Low Dissolved Oxygen and High Nitrite Concentrations. BMC Microbiology 2010, 10:70 doi:10.1186/1471-2180-10-70.

A Novel Pathogenicity Gene is Required in the Rice Blast Fungus to Suppress the Basal Defenses of the Host

Myoung-Hwan Chi, Sook-Young Park, Soonok Kim
and Yong-Hwan Lee

ABSTRACT

For successful colonization and further reproduction in host plants, pathogens need to overcome the innate defenses of the plant. We demonstrate that a novel pathogenicity gene, DES1, in Magnaporthe oryzae regulates counter-defenses against host basal resistance. The DES1 gene was identified by screening for pathogenicity-defective mutants in a T-DNA insertional mutant library. Bioinformatic analysis revealed that this gene encodes a serine-rich protein that has unknown biochemical properties, and its homologs are strictly conserved in filamentous Ascomycetes. Targeted gene deletion of DES1 had no apparent effect on developmental morphogenesis, including vegetative growth, conidial germination, appressorium formation, and appressorium-mediated penetration. Conidial size of the mutant became smaller than that of the wild type, but the mutant displayed no defects on cell wall integrity. The Δdes1 mutant was hypersensitive to exogenous oxidative stress and the activity and transcription level of extracellular enzymes including peroxidases and laccases were severely decreased in the mutant. In addition, ferrous ion leakage was observed in the Δdes1 mutant. In the interaction with a susceptible rice cultivar, rice cells inoculated with the Δdes1 mutant exhibited strong defense responses accompanied by brown granules in primary infected cells, the accumulation of reactive oxygen species (ROS), the generation of autofluorescent materials, and PR gene induction in neighboring tissues. The Δdes1 mutant displayed a significant reduction in infectious hyphal extension, which caused a decrease in pathogenicity. Notably, the suppression of ROS generation by treatment with diphenyleneiodonium (DPI), an inhibitor of NADPH oxidases, resulted in a significant reduction in the defense responses in plant tissues challenged with the Δdes1 mutant. Furthermore, the Δdes1 mutant recovered its normal infectious growth in DPI-treated plant tissues. These results suggest that DES1 functions as a novel pathogenicity gene that regulates the activity of fungal proteins, compromising ROS-mediated plant defense.

AUTHOR SUMMARY

Coevolution of plants and microbial pathogens leads to interactions that resemble a molecular war. Pathogens generate effector molecules to infect their hosts, and plants produce defense molecules against pathogen attacks. Interactions between these molecules results in plant immunity or disease. Plant disease could be likened to a complex and delicate matter of balance, where a number of molecules are involved in the battlefield. Discovering and understanding the tipping points in the battle are vital for developing disease-free crops. In the interaction of rice and rice blast fungus, a microbe sensor on rice stimulates the generation of reactive oxygen species (ROS) at the site of infection. ROS is known as an antimicrobial material and a stimulator for defense signaling that is important for preparing reinforcement in neighboring tissues. This paper presents the counter-defense mechanism of the fungus against plant-driven ROS. We found that a pathogenicity factor from rice blast fungus, DES1 (Defense Suppressor 1), is involved in overcoming oxidative stress for the counter-defense mechanism, suggesting that this gene is required for fungal pathogenicity.

Introduction

Plants are generally immune to most pathogenic microbes due to their innate defense systems, but the exceptional combination of a susceptible host and a pathogen species (or race) can result in disease [1]. Plants have two types of defense mechanism against attack by pathogenic microbes: one against general microorganisms, and the other against specific pathogen races [2],[3]. The general defense mechanism is known as a pathogen-associated molecular pattern (PAMP) triggered immunity (PTI). PTI is initiated by extracellular surface receptors that recognize general features of microorganisms such as bacterial flagellin [4],[5], chitosans (the deacetylated product of chitin [6]), and N-acetylchitooligosaccharides (the backbone fragment of the fungal cell wall [7]). As a result of coevolution, plant pathogens have developed various strategies to overcome PTI. One of them is an effector-triggered susceptibility (ETS), which deploys PTI-suppressing pathogen effectors [3]. Many effectors have been identified, and their functions and delivery systems are well studied in Gram-negative bacteria [8]. However, only a few effectors have been reported in plant pathogenic fungi, and their functions in PTI suppression and secretion mechanisms are still unknown [2],[3]. The more specific defense mechanism against pathogen ETS is known as effector-triggered immunity (ETI), which is stimulated by plant surveillance proteins (R-proteins) that specifically recognize one of the pathogen's effector proteins (Avr proteins). ETI is an accelerated and magnified defense response compared to PTI: in bacterial and fungal pathosystems, the same defense genes are related to both defense mechanisms, but they display stronger and faster activation in ETI than in PTI [9],[10]. ETI is accompanied by the active cell death of infected cells, the hypersensitive response (HR), which is known as the ultimate defense mechanism of plants [11]. However, certain pathogens avoid ETI by altering a target effector to prevent the recognition of a particular surveillance protein and/or by deploying other effectors that directly suppress ETI [12],[13].

One of the major and earliest responses of plant PTI is the rapid accumulation of reactive oxygen species (ROS) at the site of infection [14]. ROS act as direct reactive substrates to kill pathogens, to synthesize lignin and other oxidized phenolic compounds that have antimicrobial activity, and to strengthen plant cell walls by oxidative cross-linking to obstruct further extension of the pathogen [15]–[17]. ROS also function as signal molecules for programmed cell death of the infected cell and as diffusible second messengers in the production of various pathogenesis-related (PR) proteins and phytoalexins in neighboring cells [18],[19]. In rice, a membrane OsRac1 GTPase complex, which is required for PTI, controls ROS production through the direct regulation of NADPH oxidase [20],[21]. It is plausible that plant pathogens have counter-defense mechanisms against plant ROS-mediated resistance; however, little is known about how pathogens incapacitate plant-driven ROS. Recently, a study of the AP-1-like transcription factor in the maize pathogen Ustilago maydis suggested that peroxidases detoxify host-driven ROS [22].

The Ascomycete Magnaporthe oryzae, which causes rice blast disease, is the most destructive pathogen of cultivated rice worldwide [23]. The rice blast pathosystem is a model for studying fungal pathogen–plant interactions not only due to the economic importance of this disease, but also due to the molecular and genetic tractability of both the fungus and the host [24]. Complete genome sequence information is available for both the host and the pathogen, and various molecular functional genomics approaches have been initiated [25]. To investigate pathogenicity genes on a genome scale, our research group has generated >20,000 insertional mutants using Agrobacterium tumefaciens-mediated transformation (ATMT) and has evaluated the characteristics of each mutant in the essential steps for disease development [26]. The disease cycle of this pathogen consists of several steps that are essential for successful disease development. Asexual conidia are generated from conidiophores that emerge from diseased lesions and are released into the air. Upon contacting host leaves, conidia become firmly attached by the conidial tip mucilage and germinate upon hydration. Through environmental cues emanating from the plant surface, appressoria, the dome-shaped pre-penetration structures, develop at the end of germ tubes and generate enormous mechanical force to penetrate the outer surface of the plant [27],[28]. After penetration, specialized bulbous biotrophic infectious hyphae (IH) develop before necrotic lesion formation [29],[30]. In this early infectious stage, various interactive reactions are assumed to occur between the fungus and the host, possibly considered as a molecular war, which ultimately determines the level of disease. Therefore, there is growing researches on plant and pathogen factors focusing on this stage [31]. To date, studies on effector proteins in M. oryzae have relied on a genetic approach to find avirulence (AVR) genes interacting with plant resistance (R) genes on the early infectious stage. Two avirulence proteins were characterized, AVR-Pita [12] and ACE1 [32], whose putative functions are metalloprotease and polyketide synthase, respectively, but their roles as a virulence factor are insignificant. Study on MgAPT2, a member of P-type ATPase, suggested that delivery of fungal effectors including avirulence gene products is essential for infectious growth and HR induction [33]. Mig1 and SSD1 are pathogenicity factors dealing with plant innate defense in the early infectious stage, but how they counteract against host defense system and how they contribute to fungal virulence are still unknown [34],[35]. Thus, more detailed study of pathogenicity genes working in the early infectious stage could provide insights into their nature in the plant–fungi interaction.

We identified a T-DNA mutant from the ATMT mutant library, which displayed reduced pathogenicity. Investigation of the mutant led to identification of a fungal-specific gene that is required for plant innate defense suppression: DES1. The loss of DES1 in the fungus leads to the failure of host colonization and induces strong plant defense responses. DES1 is responsible for compromising oxidative signaling, and its function is related to extracellular peroxidase. Our results suggest that DES1 serves as a pathogenicity factor that counters plant defenses by restraining the oxidative component of PTI.

Results

Identification of a T-DNA Mutant with Defects in Pathogenicity

A T-DNA insertion mutant (ATMT0144A2) showing reduced virulence was identified from the M. oryzae ATMT mutant library [26]. This mutant developed restricted resistant-type lesions on a susceptible rice cultivar, Nakdongbyeo, and the number of lesions was much less than in the wild-type strain 70-15 (Fig. 1A). In addition, the mutant produced broader ellipsoidal conidia that were uniform and easily detected under a microscope (Fig. 1B). The conidia of the mutant were on average ~4 µm shorter and ~3 µm wider than those of the wild type (Fig. 1C). The T-DNA insertion mutant was not significantly defective in any

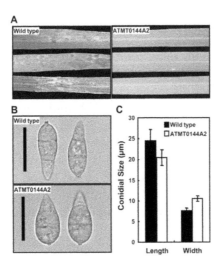

Figure 1. The Magnaporthe oryzae T-DNA mutant ATMT0144A2 has defects in lesion development and conidial morphology. (A) Rice seedlings (Nakdongbyeo) were inoculated with the wild-type strain 70-15 (left) and ATMT0144A2 (right). Diseased leaves were harvested 7 days after spray inoculation with conidial suspension (1×105 conidia/ml). (B) Light microscopy of conidia produced by 70-15 (top) and ATMT0144A2 (bottom). Bar = 20 µm. (C) Conidial size of the wild type and ATMT0144A2. Values are the mean±SD from >100 conidia of each strain, which were measured using the Axiovision image analyzer. Length is the distance from the base to apex of conidia. Width is the size of the longest septum.

other mycological phenotype tested, although the mycelial growth rate of ATM-T0144A2 was slightly faster than that of the wild type on agar medium (Table 1); the colony morphology of the mutant, however, was indistinguishable from that of the wild type. Despite the alteration in conidial morphology, conidia produced by the mutant had no defects in conidial adhesion, germination, and appressorium formation (Table 1). These phenotypes imply that the T-DNA insertion in ATMT0144A2 affects pathogenicity and conidial morphogenesis, but not other pre-penetration developmental stages.

Table 1. Comparison of mycological characteristics among strains.

Strain	Mycelial growth[a] (mm)	Conidiation[b] (10⁹ per ml)	Conidial adhesion[c] (%)	Conidial germination[d] (%)	Appressorium formation[e] (%)
70-15	63.5±1.0 B	102.3±3.8 B	90.9±5.7 A	100±0.0 A	99.1±1.5 A
DES1ᵀ⁻ᴰᴺᴬ	69.0±1.7 A	108.7±7.6 B	91.1±5.4 A	100±0.0 A	99.0±1.7 A
Δdes1	61.3±1.2 B	188.0±5.3 A	94.3±4.0 A	99.7±0.5 A	99.7±0.5 A

Within columns, means with different letters are significantly different, as estimated using Duncan's multiple range test (*P* = 0.05).
[a]Growth was measured as the diameter of the mycelium 12 days after inoculation.
[b]Conidiation was assayed by counting the number of conidia from the same culture plates used in growth measurements, flooded with 5 ml of sterile distilled water.
[c]Conidial adhesion ability was measured as the ratio of attached conidia after washing three times in distilled water to total conidia counted before washing.
[d]Germination ability was measured as the ratio of germinating conidia to total conidia.
[e]Appressorium formation was measured as the ratio of appressorium-forming conidia to germinating conidia on hydrophobic microscope coverslips.

ATMT0144A2 Phenotypes Are Caused by a Single T-DNA Insertion

Southern hybridization revealed that ATMT0144A2 has a single insertion of T-DNA in its genome (Fig. 2A). The presence of a single band of ~17 kb from BglII-digested DNA suggested the abnormal insertion of several copies of T-DNA at the same locus, since the band location was different from the expected (9 kb). The insertion locus was identified using thermal asymmetric interlaced polymerase chain reaction (TAIL-PCR) [36] with T-DNA border primers, as described in a previous study [37]. The PCR reactions using right border (RB) primers produced two distinct bands (RB-A and RB-B), but the left border (LB) primers produced no detectable band. Sequences from RB-A were matched to the supercontig 6.12 of the M. oryzae genome, and sequences from RB-B were matched to the pBHt2 vector region that is adjacent to RB. Both tandem and inverse repeats of T-DNA were detected by amplification with LB and RB primer combinations. From these results, a schematic diagram of the T-DNA integration in ATMT0144A2 was configured, and the insertion pattern was confirmed by PCR amplification with combinations of border primers and locus-specific primers, in which three to four copies of the T-DNA units were tandemly and inversely integrated (Fig. 2B). Junction sequences between the T-DNA and the M. oryzae genome revealed that the T-DNA had a typical RB border at one end, but had an abnormal RB read-through and 1-bp filler DNA at the other end (Fig. 2C). As a result of the T-DNA insertion, 6 bp of genomic DNA were deleted from the insertion site. During the in depth study of ATMT0144A2, it was found that the location of the T-DNA insertion was the same as that in another pathogenicity-defective mutant, AT-MT0144B3, which also produced broad ellipsoidal conidia [26].

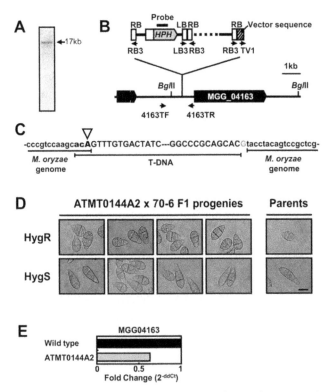

Figure 2. An abnormal T-DNA is integrated in the promoter region of MGG04163. (A) Southern hybridization with ATMT0144A2. Total genomic DNA was digested with BglII and probed with the HpaI-digested HPH fragment. (B) Schematic diagram of T-DNA in ATMT0144A2. Specific primers used for the confirmation of T-DNA insertion (small arrows), vector read-through (slashed box) and unknown regions (dashed line) are indicated. The T-DNA insertion point is −750 from MGG04163 start codon. (C) Sequences of the T-DNA junction sites. Sequences of both junctions between T-DNA (upper case letters) and the M. oryzae genome (lower case letters) are indicated. Typical right-border cleavage site (white arrowhead), micro-homology region (bold), and a filler DNA (gray) are denoted. (D) Co-segregation of conidial morphology and T-DNA in F1 progeny. Seven-day-old conidia produced by randomly selected F1 progeny from ATMT0144A2×70-6 crosses were observed under a light microscope, and they were examined on TB3 medium containing 200 ppm hygromycin B. Bar = 10 μm. (E) The transcriptional expression of MGG04163 in ATMT0144A2. The transcription level of MGG04163 was assayed by quantitative RT-PCR using mycelia of the wild type and ATMT0144A2 in 3-day-old liquid culture.

To confirm the single insertion and a correlation between the T-DNA insertion and ATMT0144A2 phenotypes, genetic analyses were performed with two different mating tester strains: 70-6 and 4091-5-8. The integrated T-DNA in ATMT0144A2 was stably segregated to F1 progeny in the genetic crosses. Of 102 F1 progeny from the ATMT0144A2×70-6 cross, 49 progeny were resistant to hygromycin B (HygR), whereas 53 progeny were susceptible (HygS). Chi-square analysis effectively supported 1:1 segregation ($X2 = 0.09$) at the 5% level of significance. All HygR F1 progeny from the cross produced broader ellipsoid conidia like the mutant parent, whereas all HygS F1 progeny produced normal shaped conidia like the wild-type parent (Fig. 2D). PCR amplification between 4163TF and RB3 revealed that all HygR F1 progeny had the T-DNA, whereas all HygS progeny had no T-DNA (data not shown). Thus, the T-DNA insertion is tightly linked to a locus that determines conidial morphology. In the genetic cross ATMT0144A2×4091-5-8, progeny from a single tetrad were the same. The T-DNA was located in a noncoding region between MGG04162.6 and MGG04163.6. MGG04163.6 was the nearest ORF from the T-DNA (750 bp upstream of the start codon), and quantitative RT-PCR showed that the expression level of MGG04163.6 in ATMT0144A2 was reduced to 60% of that in the wild type (Fig. 2E). Therefore, it seems that the integration of the T-DNA at the promoter region of the gene reduced the transcriptional expression and consequently affected pathogenicity and conidial morphogenesis. We named the gene MGG04163.6 (GenBank accession number: XP_361689) as DES1, derived from plant defense suppression. Because the T-DNA was present at the 3'-direction of MGG01462.6 beyond the stop codon with a distance of 2.5 kb, it is unlikely that the locus is responsible for the phenotypes of ATMT0144A2. It is further confirmed that the expression level of MGG04162.6 in ATMT0144A2 was not significantly different (1.15 fold) to that of the wild type when examined with quantitative RT-PCR (data not shown).

DES1 Encodes an Unknown Fungal-Specific Protein

The DES1 gene was located on chromosome IV in the genome of M. oryzae, and the predicted ORF was 3,864 bp long, encoding 1,287 amino acids, and there was no intron on the ORF. The sequence of the DES1 transcript was confirmed by sequencing the cDNA synthesized from mycelial mRNA using four pairs of primers spanning the ORF, and it was identical to the predicted ORF in the M. oryzae genome version 6 (data not shown). BLAST searches to find DES1 homologs resulted in only a few matches to hypothetical proteins of filamentous fungi. Additionally, the DES1 homologs were intensively analyzed in 59 recently released fungal genomes (including three Oomycetes) using the BLASTMatrix tool, which plots the BLAST results by taxonomic distribution (http://cfgp.snu. ac.kr [38]). Interestingly, DES1 homologs were found only in subphylum Pezizomycotina of Ascomycota, and each homolog was present as a single copy in each genome. Of 21 fungal species belonging to Pezizomycotina, the DES1 homologs were found in 20 species, the exception being Mycosphaerella graminicola. Sequence alignment of the DES1 homologs revealed that they are well conserved in length and amino acid composition, and they were grouped in distinct phylogenetic clades.

To predict the biochemical function of DES1, the amino acid sequences of DES1 and its homologs were analyzed using the bioinformatics tools Inter-ProScan [39], SignalP [40], and amino acid frequency analysis. Most of the DES1 homologs had no known functional domain when searched with InterProScan (v12.0). Exceptionally, Afu2g05410, the DES1 homolog in Aspergillus fumigatus, had the IPR002048 (calcium-binding EF-hand) domain consisting of 13 residues in the C-terminal region. Signal P (v3.0) predicted that all of the DES1 homologs had no signal peptide, indicating that they are likely non-secretory proteins. Amino acid frequency analysis revealed that DES1 and its homologs are serine-rich proteins: the average serine frequency of DES1 and its homologs was 13.25%, whereas that of whole ab initio annotations of M. oryzae was only 7.97%.

Targeted Gene Replacement of DES1 in M. Oryzae

Targeted gene deletion of DES1 confirmed that the gene is required for M. oryzae lesion development and conidial morphogenesis. A gene deletion vector was constructed by double joint PCR [41], in which the hygromycin resistance gene (HPH) cassette was combined with ~1-kb-long flanking regions. The gene deletion vector was introduced to wild-type protoplasts by PEG-mediated fungal transformation. After primary PCR screening of hygromycin-B-resistant transformants using a locus-specific primer (DES1KOSF) and an HPH gene primer (HPHF), a DES1 deletion mutant (Δdes1) and an ectopic transformant (E41) were confirmed by Southern hybridization. RT-PCR confirmed the null mutation of DES1 in which the Δdes1 mutant produced no DES1 transcript (data not shown). The morphology of conidia produced by the Δdes1 mutant was similar to that of conidia produced by ATMT0144A2 (designated as DES1T-DNA). Conidia produced by the Δdes1 mutant were significantly shorter in length than those of the wild type, although they did not become as wide as those of DES1$^{\text{T-DNA}}$. Conidia produced by an ectopic transformant exhibited the normal morphology of the wild type. The Δdes1 mutant was not defective in other mycological phenotypes, including growth rate and colony morphology on CM; conidial adhesion on hydrophobic surfaces; and development of germ tubes and appressorium formation (Table 1).

The DES1 Gene is Required for Successful Colonization of Host Tissues

In spray-inoculation tests, the Δdes1 mutant produced tiny and restricted lesions on a susceptible rice cultivar, Nakdongbyeo, whereas the wild type and the ectopic transformants caused susceptible-type spreading lesions. The level of virulence of DES1[T-DNA] was intermediate to those of the wild type and Δdes1 (Fig. 3A). Differences in disease severity were more dramatic when the diseased leaf area (%DLA) was measured. The %DLA of Δdes1 was 15±10%, which was less than one-quarter that of the wild type (68±20%) and the ectopic transformant (61±13%). The DES1T-DNA showed slightly higher levels of DLA (19±14%) compared to Δdes1 (Fig. 3B).

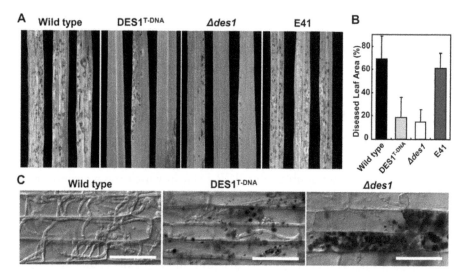

Figure 3. The loss of DES1 leads to reduced pathogenicity and a colonization defect in host tissues. (A) Pathogenicity assay. Five milliliters of conidial suspension (1×105 conidia/ml) of each strain were sprayed on rice seedlings (Nakdongbyeo). Diseased leaves were harvested 7 days after inoculation. (B) The disease severity of each strain was assessed from the percentage diseased leaf area as calculated using the Axiovision image analyzer. Values are the mean±SD from eight rice leaves inoculated by each strain. (C) Observation of infectious growth. Excised rice sheath from 5-week-old rice seedlings (Nakdongbyeo) was inoculated with conidial suspension (1×104 conidia/ml of each strain). Infectious growth was observed 96 h after inoculation. Bar = 50 μm.

Because no defect in appressorium development was observed in the Δdes1mutant, the development of infectious hyphae (IH) within the host cells was examined using an excised leaf sheath assay [42]. IH of the wild-type actively grew and occupied 10–20 cells neighboring the primary infected cells by 96 h after inoculation. However, IH of Δdes1 were mostly restricted to the primary infected cells, and there was an abundant accumulation of dark brown granules along IH of Δdes1 (Fig. 3C). Only a few IH of Δdes1 extended into neighboring cells (Fig. 3C). The DES1T-DNA displayed an intermediate phenotype, with poorly growing IH and scattered dark brown granules along the IH (Fig. 3C). The development of IH was further observed after destaining the brown granules with lactophenol and staining the IH with aniline blue. At 48 h after inoculation, bulbous IH of the wild type filled the primary infected cell, and the IH extended to neighboring cells. Contrary to the wild type, IH of the Δdes1 and DES1T-DNA mutants seemed to be broken down within primary infected cells, and very few slender IH were present in neighboring cells (Fig. 4A).

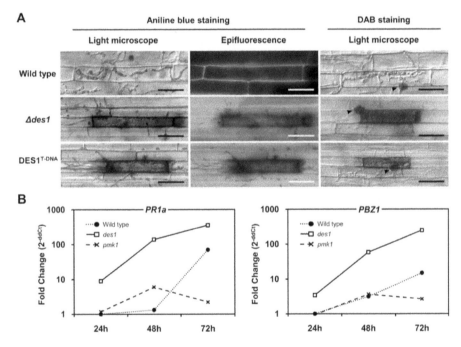

Figure 4. The deletion of DES1 caused the induction of strong plant defense responses. (A) DIC and fluorescence microscopy of infected rice sheaths (Nakdongbyeo) 48 h after inoculation. DIC images were captured using an 80-ms exposure time of transmission light with a DIC filter. Fluorescence images were captured using a 500-ms exposure for absorbed light using a GFP filter. Arrowheads on DAB staining panel indicate appressorium. Bar = 30 μm. (B) The expression of rice pathogenesis-related (PR) genes over time after inoculation. The transcriptional expression of PR1a and PBZ1 in the infected rice was analyzed using quantitative RT-PCR.

Plant Defense Responses were Induced by Challenge with the Δdes1 Mutant

Defense responses induced by the recognition of microbe-associated molecules are often associated with cell wall strengthening, the rapid production of ROS, and the transcriptional activation of PR genes [43]. Because rice cells infected by the Δdes1 mutant displayed brown granule generation and cell death, it is likely that plant defense responses might be involved in virulence attenuation of Δdes1. Thus, the defense responses against the wild type and the mutants were compared.

Autofluorescence at the site of infection indicates the accumulation of phenolic compounds and cell wall strengthening [44]. Under a fluorescence microscope, primary rice cells infected by the wild type emitted autofluorescence only in their cell walls (Fig. 4A). The fluorescence was severely diminished or absent in secondary and further infected rice cells and even in the cells that were occupied by actively growing IH (Fig. 4A). In contrast, strong autofluorescence was observed not only in rice cells directly infected by IH of the Δdes1 and DES1^{T-DNA} mutants, but also in neighboring cells that were not in contact with the fungus (Fig. 4A).

The accumulation of hydrogen peroxide (H_2O_2) at infection sites was also examined by staining with 3,3'-diaminobenzidine (DAB) 48 h after inoculation. Rice cells containing wild-type IH were not stained with DAB, whereas primary infected rice cells with Δdes1 and DES1^{T-DNA} were strongly stained with DAB, indicating high concentrations of H_2O_2 (Fig. 4A). Regardless of the plant responses, appressoria of both the wild type and mutants stained equally with DAB (Fig. 4A, arrowheads in DAB staining). The extent of defense responses seemed to be proportional to the level of DES1 expression because the levels of autofluorescence and H2O2 accumulation were lower in cells infected by the DES1^{T-DNA} mutant compared to those infected by the deletion mutant (data not shown).

To further investigate whether the plant defense genes were stimulated by infection with Δdes1, the expression patterns of two PR genes were analyzed by quantitative RT-PCR. A MAP-kinase mutant (pmk1), which is unable to infect plant tissue [45], was used as a negative control. Upon inoculation with wild-type conidia, the expression of PR1a and PBZ1 followed the typical pattern of compatible interaction, where induction of these genes was delayed to 72 hpi [9]. In contrast, expression of PR1a and PBZ1 was highly induced even in 24 and 48 hpi by inoculation with Δdes1 (Fig. 4B). Induction levels of PR1a and PBZ1 expression in Δdes1 challenged rice leaves were 104 and 19 folds at 48 hpi, respectively, compared to those in wild type challenged rice leaves (Fig. 4B). There was a little induction of PR1a or PBZ1 gene expression by inoculation with the pmk1 mutant, but the induction level was less or not much more than that in wild type challenged rice leaves (Fig. 4B). These results indicate that the induction of plant defense responses in Δdes1 challenged rice may contribute the retardation of the IH development.

Inhibition of Plant ROS Generation Restores IH Development of the Δdes1 Mutant

Plant NADPH oxidases generate ROS in response to pathogen attack [19]. To determine whether the virulence of the Δdes1 and DES1[T-DNA] mutants is affected by ROS generation in host plant tissues, diphenyleneiodonium (DPI), an inhibitor of NADPH oxidases [46], was applied to the rice sheath. Since treatment of high concentration (>25 μM) of DPI could affect conidial germination [47], we used 0.2–0.4 μM of DPI to prevent the effects on the fungal development. At these concentrations, conidial germination, appressorium-mediated penetration and IH development were not affected in the wild type (Fig. 5A). However, the attenuated virulence phenotypes of Δdes1 and DES1[T-DNA] were rescued in the rice sheath cells in which ROS generation was inhibited by DPI. Treatment with 0.2 μM DPI resulted in the reduction and fragmentation of the dark-brown granules around IH in the mutants (Fig. 5A). IH of the Δdes1 mutant were still restricted in the presence of 0.2 μM DPI, but IH of DES1[T-DNA] successfully occupied the primary infected cell and extended to neighboring cells (Fig. 5A). Treatment with 0.4 μM DPI completely prevented the production of brown granules, and both mutants could develop IH (Fig. 5A).

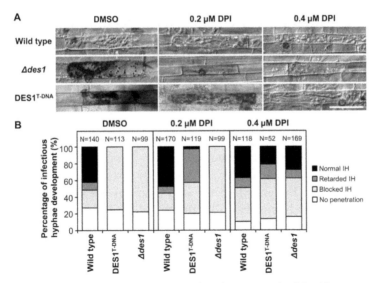

Figure 5. The inhibition of ROS generation recovers the infectious growth of the Δdes1 mutant. (A) The excised sheath of rice (Nakdongbyeo) was inoculated with conidial suspension (1×104 conidia/ml) of 70-15, Δdes1, or DES1[T-DNA] with or without diphenyleneiodonium (DPI) dissolved in DMSO. Samples were harvested and observed 48 h after inoculation. Bar = 50 μm. (B) Percentage of appressorium-mediated penetration and infectious hyphae development of 70-15, Δdes1, and DES1[T-DNA] in DPI treated onion epidermis. The total number of appressorium is indicated above each column. The level of IH development were scored after 72 h after inoculation (see Materials and Methods for details).

The level of plant defense response and the IH development were measured on onion epidermis. Under normal conditions (without DPI), ~70% of the appressoria of the wild type penetrated into onion epidermis, but finally ~40% of the appressoria successfully developed IH due to plant defense responses including callose deposition (Fig. 5B). In contrast, only a few penetrated appressoria of Δdes1 and DES1[T-DNA] developed IH, although the penetration rate was similar to that of the wild type (Fig. 5B). Treatment of DPI (0.4 μM) recovered the frequencies of IH development by Δdes1 and DES1[T-DNA] up to 28% and 21%, respectively (Fig. 5B). The shapes of the recovered IH of Δdes1 and DES1[T-DNA] were not distinguishable from that of the wild type (Fig. 5A). Similar to the results of the rice sheath test, the level of defense response of onion seemed to be related to DPI concentration and the expression level of the DES1 gene. These results suggest that the DES1 gene is related to either suppression of defense initiation in rice and onion (by similar mechanisms of DPI) or overcoming the defense responses by degrading brown granules and callose. Re-introduction of wild-type allele of DES1 gene into the Δdes1 mutant also recovers IH development in rice sheath and onion epidermis and the ability to suppress the plant basal defense. This result indicates that the deletion of DES1 gene is the very reason for the failure of infectious growth and PTI suppression.

The Δdes1 Mutant is Hypersensitive to Oxidative Stress

Because DPI is known to suppress plant ROS production, we investigated whether the DES1 gene is required to modulate either ROS or other diverse stress conditions that fungal pathogens may encounter in the plant cells. The Δdes1 and DES1[T-DNA] mutants did not show any differences in mycelial growth with high concentrations of osmolytes such as 1 M sorbitol or 0.5 M NaCl (data not shown). However, the mycelial growth of Δdes1 and DES1[T-DNA] was severely affected under oxidative stress conditions (Fig. 6A). The growth of mutants was altered at 2–3 mM H_2O_2, and the level of sensitivity was more significant in Δdes1 than in DES1[T-DNA]. The growth of the wild type was not significantly affected under these same conditions (Fig. 6B). These results indicate that the DES1 gene is related to ROS degradation.

Since the conidial morphology of Δdes1 and DES1[T-DNA] was altered, hypersensitivity to oxidative stress and reduced virulence may also be due to defects in cell wall composition, in spite of their insensitivities to osmotic stresses. To investigate this possibility, we added Nikkomycin Z, a chitin synthetase inhibitor, to germinating conidia. Treatment of Nikkomycin Z blocks conidial germination and induces protoplast-like swellings on cell wall-defective strains [48]. However, conidial germination of Δdes1 and DES1[T-DNA] was not inhibited in high concentrations of Nikkomycin Z (100 μM) and swellings on germ tubes were not distinguishable from the wild type. We also tested the sensitivity of these strains to a cell wall-degrading enzyme. Enzyme-treated mycelia of Δdes1and DES1T-DNA released no more or less protoplasts than the wild type when observed over a time course. We also tested mycelial growth on Calcofluor white (CFW) and Congo Red (CR) amended media, which inhibit fungal cell wall assembly by binding chitin and β-1,4-glucans, respectively [49],[50]. The mycelial growth of Δdes1on CFW media (100 ppm) was little reduced (88% of the wild type) when compared with normal CM (96% of the wild type), and it was more severely reduced (70% of the wild type) on CR media (100 ppm). However, since degradation halo was observed around the wild-type colony and no degradation halo was observed around the Δdes1 colony (Fig. 7A), the growth defect on CR media was assumed to be due to the absence of CR-degrading activity rather than defects in cell wall composition. The DES1[T-DNA] colonies showed intermediated levels of CR discoloration (Fig. 7A).

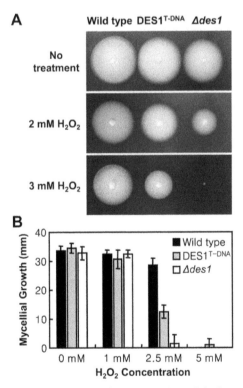

Figure 6. The Δdes1 mutant is hypersensitive to oxidative stress. (A) Mycelial colonies on complete agar medium with or without 2–3 mM H2O2 on day 4 after inoculation. (B) Mycelial growth on complete medium with or without H2O2 (1–5 mM) on day 7 after inoculation. The colony diameters of four replicates were measured. Error bars represent the standard deviation.

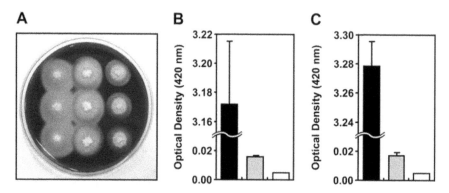

Figure 7. DES1 is related to activity of extracellular peroxidase and laccase. (A) The discoloration of Congo Red was tested on medium containing 100 ppm of the dye at final concentration. Strains were inoculated on CM agar medium containing Congo Red. Discoloration was observed on day 4. Left: wild type, middle: DES1T-DNA, right: Δdes1. (B) Peroxidase activity measured by ABTS oxidizing test under H2O2 supplemented condition (see Materials and Methods for details). Black column: wild type, grey column: DES1T-DNA, white column: Δdes1. (C) Laccase activity measured by ABTS oxidizing test without H2O2. The strain scheme is same with panel B. Error bars represent standard deviation.

The DES1 Gene is Related to the Activity of Extracellular Peroxidase

Since discolored halos were observed beyond the colony margins, extracellular enzymes were presumed to be responsible for CR degradation. Because the CR degradation reaction is known to be catalyzed by peroxidase, which requires H_2O_2 as a limiting substance [51],[52], we tested the influence of H_2O_2 on CR discoloration in M. oryzae. The sizes of the discolored halos increased considerably around the wild type when 1 mM H_2O_2 was added to CR medium. No color change was observed with the Δdes1 mutant, regardless of H_2O_2 treatment (data not shown). Considering the defective phenotypes of Δdes1 in scavenging H_2O_2 and discoloring CR, we reasoned that the DES1 gene is involved in extracellular peroxidase activity. Enzyme activity assay using 2, 2′-azino-di-3-ethylbenzthiazoline-6-sulphonate (ABTS, Sigma, A1888) [53] as substrate revealed that the Δdes1 mutant totally lost its peroxidase activity in extracellular culture filtrate (Fig. 7B). The culture filtrate of DES1[T-DNA] showed very low level of peroxidase activity (Fig. 7B). In addition, the ABTS oxidation test without H2O2 revealed that laccase activity [54],[55] was also diminished in the culture filtrates of Δdes1 and DES1[T-DNA] (Fig. 7C). We also compared the activity of another extracellular enzyme, xylosidase, in the culture filtrate of Δdes1, DES1T-DNA, and the wild type. However, xylosidase activities of the mutant strains were not different from that of the wild type (data not shown).

Deletion of DES1 Affects the Expression of Several Groups of Peroxidase Genes

We examined the transcriptional regulation of genes encoding peroxidases. Putative peroxidase-encoding genes were identified from the annotated M. oryzae genome database. There were 19 such genes that had peroxidase-related InterPro domains, including IPR010255 (haem peroxidase), IPR000889 (glutathione peroxidase), and IPR000028 (chloroperoxidase). Three of these 19 genes were excluded because their transcripts were not detected under the given experimental conditions. Sixteen putative peroxidase genes in M. oryzae could be classified into three clades by phylogenetic analysis, and most of them possessed a signal peptide when assessed using the SignalP program (Fig. 8). Differences in the transcriptional expression of the peroxidase genes between the wild type and the Δdes1 mutant under oxidative (1 mM H_2O_2) or normal (no H_2O_2) conditions were examined using quantitative RT-PCR, and fold changes were calculated using wild type under normal condition as a standard condition. The expression level of some peroxidase genes in clade 1 (plant ascorbate peroxidases: MGG04545, MGG10368, MGG08200, and MGG09398; fungal lignin peroxidase: MGG07790) was up-regulated under the oxidative condition in the wild type (Fig. 8). The transcription of MGG07790, MGG08200, and MGG09398 was down-regulated in the Δdes1 mutant, and the reduced transcription was not recovered by treatment with H2O2. The expression of MGG10368 and MGG04545 was also repressed in the Δdes1 mutant, but the reduced transcription was partially recovered by treatment of H_2O_2 (Fig. 8). The expression level of the other peroxidase genes in clade 1 (catalase peroxidases: MGG04337 and MGG09834; haem peroxidases: MGGG00461 and MGG10877) was not significantly altered by the deletion of the DES1 gene. Peroxidase genes in clade 2 (chloroperoxidases: MGG07574, MGG11849, MGG07871, and MGG07574) were responsive to H_2O_2 treatment, but their expression was not down-regulated in the Δdes1 mutant. Peroxidase genes in clade 3 (cytochrome P450 peroxidases: MGG10859 and MGG13239; glutathione peroxidase: MGG07460) were not responsive to H_2O_2 under the experimental conditions, but their expression was down-regulated in Δdes1 (Fig. 8). The expression of putative laccase-encoding genes was also examined in the Δdes1 mutant. Seventeen genes having two or three multicopper oxidase domains (IPR001117, IPR011706, or IPR011707) were identified from the M. oryzae genome database. We excluded one of them (MGG09102) from the analysis because the transcript was not detected in the given experimental conditions. Expression levels of laccase genes in the oxidative condition (1 mM H_2O_2) were rather reduced or similar in wild type. Transcription of all laccase genes except one (MGG07500) was also down-regulated in the Δdes1 mutant, even differences in the transcription level of the laccase genes between wild type and the Δdes1 mutant were more severe than those of the peroxidase genes.

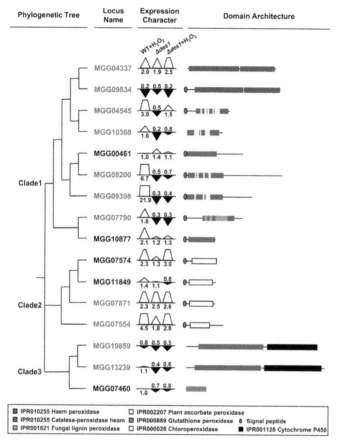

Figure 8. Expression Profiles of M. oryzae Peroxidases in the Δdes1 Mutant. A combination of the phylogenetic tree, expression characteristics, and domain architecture of 16 putative peroxidases in the M. oryzae genome were displayed. The phylogenetic tree was generated by ClustalW sequence alignment with 1000 bootstrappings and divided into three clades. The transcript levels of the the putative peroxidase encoding genes in the oxidative condition and/or in the Δdes1 mutant are indicated. Relative abundance of transcript compared with standard condition (wild type, normal condition) is displayed as a white triangle (up-regulated) or an inverted black triangle (down-regulated). Triangles indicating more than 2.0 (fold change) are displayed as trapezoids by cutting the top of the triangle. Fold changes of the standard condition (1.0) are not shown. Up-regulated genes in the Δdes1 mutant (more than 1.5 fold) were indicated in blue, and down-regulated genes in the Δdes1 mutant (less than 0.6 fold) were indicated in red. The InterPro terms and signal peptides are indicated (see legend).

The DES1 Gene is Required for Regulation of Ferrous Ions

Since the DES1 gene has no DNA-binding domain, assuming that DES1 acts as a direct transcriptional regulator of genes encoding peroxidase and laccase was difficult. So we investigated whether the Δdes1 mutant has defects in metal ion regulation, which may affect the expression and activity of enzymes with a metal core, including peroxidase and laccase. To investigate this hypothesis, bathophenanthroline sulfonate (BPS, Sigma, B1375), a chromogenic, and a specific chelator of the ferrous ion were used to detect extracellular ferrous ions [56]. Since complete media (CM) includes ~25 μM of ferrous ions, we used CM without trace elements as a negative reference for spectrophotometry. The BPS color reaction was stronger in the Δdes1 culture filtrate than in the wild type culture filtrate, and of an intermediate color in the DES1^{T-DNA} culture filtrate. This result suggests that the DES1 is related to either uptake or storage of ferrous ions.

Subcellular Localization of the DES1 Protein

To identify the cellular component to which the DES1 protein is targeted, a fluorescent reporter gene (eGFP) was fused to the C terminus of the DES1 gene. Since GFP observations using the native promoter (~1.2 kb) were not successful (data not shown), we used the Aspergillus nidulans TrpC promoter for constitutive expression. The ProTrpC-DES1-eGFP fusion construct and a plasmid containing the geneticin resistance gene (pII99) were introduced into wild-type protoplasts by co-transformation. DES1-eGFP fusion proteins were targeted to vacuoles in the conidia and growing mycelia of these transformants whereas eGFP without DES1 protein, which was expressed by the same TrpC promoter, was distributed to the cytosol (Fig. 9). Co-localization of the fluorescence signals with the vacuole-indicating dye 7-amino-4-chloromethylcoumarin (CMAC) confirmed the vacuolar localization of DES1-eGFP fusion proteins (Fig. 9).

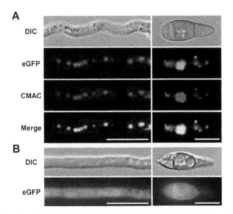

Figure 9. DES1-eGFP is localized in vacuoles. (A) Growing hyphae (left) and conidia (right) expressing DES1-eGFP in Czapek-Dox media. In the merged image, the original blue color from CMAC was changed to red for better visualization, so the co-localized spots were indicated as yellow. Bars = 10 μm. (B) Growing hyphae (left) and conidia (right) expressing eGFP without DES1 in Czapek-Dox media. Bars = 10 μm.

Expression Profiles of the DES1 Gene

The transcriptional expression of DES1 during fungal development was analyzed using quantitative RT-PCR. The cyclophilin-encoding gene (CYP1, MGG10447), which displayed stable expression during the developmental stages (Kim et al., unpublished data), was used as an endogenous control gene for normalization. Transcripts of the DES1 gene increased more than 2 folds during conidiation and infectious growth in planta. The level of expression did not change significantly under conditions that included conidial germination and appressorium formation. The expression of DES1 was 2.8-fold up-regulated by treatment with 1 mM H_2O_2.

Discussion

We described a novel pathogenicity gene of M. oryzae that we named DES1, which plays an essential role in colonization in planta. DES1 (involved in defense suppression of innate plant immunity) was originally identified as a pathogenicity-defective mutant generated by random insertional T-DNA mutagenesis of M. oryzae [26]. Subsequently, a gene deletion mutant was generated, and the inoculation of the null mutant on a susceptible rice cultivar resulted in a more significant reduction of pathogenicity than the T-DNA insertion mutant. Strong defense responses were observed along the infectious hyphae of both Δdes1 and DES1T-DNA, and the mutants had no apparent defects in hyphal growth, conidial germination, conidial adhesion, appressorium formation, appressorium-mediated penetration, and cell wall integrity. Moreover, treatment with a plant defense inhibitory chemical (DPI) recovered the IH development of Δdes1 and DES1T-DNA. The reduction in pathogenicity of the mutants seems to be due to strong plant defense responses, which resulted from failure of proper interactions between the host and the pathogen, rather than defects on IH development. Observation of PR1a and PBZ1 inductions in Δdes1-challenged rice tissue supports this hypothesis. The defense response against Δdes1 and DES1T-DNA seems to be stronger than the typical rice PTI, which has been reported as whole-plant specific resistance (WPSR) [42],[57]. However, it does not seem to be the ETI response, which is induced by an AVR–R gene interaction, since it occurred on both rice and onion—genetically different two species. Therefore, we assigned the plant response against Δdes1and DES1T-DNA as an intensified or non-suppressed version of PTI.

PTI in rice is triggered by cell wall derivatives of the fungus [58]. It is initiated via a signaling complex that includes a small GTPase OsRac1, which directly activates NADPH oxidases [20],[21],[59]. H_2O_2 produced by the NADPH oxidases is essential for the PTI response in plants, not only as a direct antimicrobial material but also a diffusible second messenger for defense gene induction [43]. It is presumed that virulent pathogens have developed ROS scavenging mechanisms to suppress the PTI defense responses of their hosts. The wild type M. oryzae has the ability to detoxify ~2 mM of H_2O_2 without disturbing hyphal growth. Since Δdes1and DES1^{T-DNA} showed growth defects under the same conditions, it is assumed that the H2O2-degrading ability in the mutants was lost or severely weakened. Note that the plant defense responses against Δdes1and DES1^{T-DNA} seemed to be regulated by the ROS level because not only did treatment with a ROS inhibitor (DPI) resulted in a dramatic reduction of the plant defense response on both rice and onion tissues but also the degree of the defense response was quantitatively affected by both the DPI and DES1 expression level. Therefore, we hypothesized that ROS scavenging ability controlled by DES1 is essential to the suppression of oxidative signaling, which is important in the induction of PTI. However, we cannot completely exclude other possibilities that unknown factors may be affected by the deletion of DES1.

We showed that the Δdes1 mutant lost H_2O_2-degrading ability. H_2O_2 is known to be scavenged by catalase and various peroxidases, including ascorbate and glutathione peroxidases [60]. In the Δdes1 mutant, the expression of putative peroxidase genes was down-regulated, and CR discoloration, which is known to be catalyzed by peroxidases, was completely abolished. It is thus suggested that some of the secreted peroxidase might be involved in extracellular H_2O_2 detoxification in M. oryzae. Molina and Kahmann [22] also reported that a transcription factor, YAP1, controls the expression of peroxidase genes (um01947 and um10672) in the maize pathogen U. maydis and is responsible for the scavenging of host-derived ROS in the fungus–plant interaction. We also found that the expression of a gene homologous to um01947 (MGG10368) was down-regulated in the Δdes1 mutant. This suggests that fungal peroxidases might play a role as common PTI suppressing effectors in rice and maize pathogens, although the regulatory mechanism is not likely to be identical (a DES1 gene homolog is absent in U. maydis). The matching homologous gene to um10672 was not found in the M. oryzae genome database. Catalase is also known to scavenge H_2O_2 [60]. However, CATB in M. oryzae seemed to be required for the strengthening of fungal cell walls, rather than the scavenging of host-driven H_2O_2 [54], and the expression of CATA and CATB was not affected by deletion of the DES1 gene. Therefore, we suggest that the extracellular peroxidases are a more likely candidate for host-driven H_2O_2 in rice–M. oryzae interactions.

The Δdes1 mutant had higher levels of ferrous ions in culture media than the wild type, which suggests defects in either uptake or storage of ferrous ions. Furthermore, we found that the DES1 protein is targeted to the vacuole. Since fungal vacuoles are considered to be areas for the storage of metal ions and the regulation of their homeostasis in the cell [61],[62], these results may provide a possible explanation that the DES1 may function on metal ion homeostasis, which could affect the activity of enzymes with a metal core [63],[64]. Considering that both peroxidase and laccase need a metal cofactor (iron and copper, respectively) and that the activity of laccase in M. oryzae is inhibited under copper-depleted conditions [65], these could explain why both activities of peroxidase and laccase were affected by the deletion of DES1. Since the Δdes1 mutant showed defects in regulation of ferrous ions, the DES1 function might be related to that of siderophores, which are high-affinity iron chelators used for iron uptake and storage in many fungal species [66]. In the recent study of a ferrichrome-type siderophore synthetase (SSM1) [67], however, expression of SSM1 was not correlated to the level of FeCl3 concentration, and the Δssm1 mutant was not sensitive to oxidative stress, unlike the Δdes1 mutant. These results suggest that there is no direct connection between DES1 and SSM1 in ferrous ion regulation in this fungus. An alteration of intracellular ROS has also been reported to affect appressorium-mediated penetration by M. oryzae [47],[68]. However, the cellular effects of ROS regulation by DES1 seem to be limited when compared to that by ABC3 (a multidrug resistance transporter gene) [68]. Although the Δdes1 mutant displayed high sensitivity to oxidative stress, unlike abc3Δ, Δdes1 was not lethal on 2 mM H_2O_2 and the appressoria of Δdes1 were fully functional without any treatment with antioxidants. Furthermore, the Δdes1 mutants generated the same level of DAB-positive material in their appressoria as the wild type. These indicate that deletion of DES1 does not lead to alteration of intracellular ROS, which are generated by the NADPH oxidases of M. oryzae [47],[68]. These observations suggest that DES1 might be related to the regulation of a limited range of ROS, or only extracellular ROS.

In conclusion, we identified and characterized a novel pathogenicity gene that is required for host colonization using two independent mutants with different alleles: DES1T-DNA and Δdes1. DES1 is responsible for scavenging extracellular ROS within host cells, which in turn results in a counter-defense of the pathogen against the plant innate defense responses. The discovery and functional assignment of more pathogenicity factors affecting plant defense systems may help to understand the nature of plant disease.

Materials and Methods

Fungal Strains and Culture Conditions

Magnaporthe oryzae strain 70-15 (Mat1-1) and 70-6 (Mat1-2) were obtained from A.H. Ellingboe (University of Wisconsin-Madison, USA), and 70-15 was used as wild-type strain in this study. Strain 4091-5-8 (Mat1-2) was obtained from B. Valent (Kansas State University, USA). All fungal strains are stored in the Center for Fungal Genetic Resources (Seoul National University, Seoul, Korea; http:// genebank.snu.ac.kr). Strains were normally maintained on oatmeal agar medium (OMA, 5% oatmeal and 2.5% agar powder (w/v)) and grown at 25°C under constant fluorescent light to promote conidiation. The strains were cultured for 3 to 12 days on complete agar media [69] to assess the growth and colony characteristics. Hygromycin B resistant transformants generated by fungal transformation were selected on solid TB3 agar media (0.3% yeast extract, 0.3% casamino acids, 1% glucose, 20% sucrose (w/v), and 0.8% agar powder) supplemented with 200 ppm hygromycin B. Mycelia used for nucleic acid extraction were prepared by growing the relevant strains in liquid CM for 3 days at 25°C with agitation (150 rpm), or directly obtained from the TB3 agar media for the quick DNA extraction method described previously [70]. Genetic crosses and progeny analysis (tetrads or random ascospore analysis) were performed as previously described [71]. To observe the vegetative growth or RNA extraction under stress conditions, the fungal strains were treated as follows; Oxidative stress was applied by amending solid or liquid CM with the proper volume of H_2O_2 solution (Aldrich, 323381, 3 wt. %). Three-day-old mycelia in liquid CM were treated with or without 1 mM of H_2O_2 for 30 min before harvesting for RNA extraction. Stress conditions for cell wall biogenesis was performed by supplementation of Congo Red (CR, Aldrich, 860956) and Calcofluor White (CW, Sigma, F3543) in 100 ppm final concentration in CM agar media, both of which are known to interfere with the assembly of fungal cell walls [72]. For the osmotic stress conditions, CM agar media was amended with 500 mM NaCl and 1 M sorbitol in final concentration.

Developmental Phenotypes Assays

Radial colony growth rate was measured on CM agar plates 12 days after inoculation with triplicate. Colony color and morphology were also observed in the condition above. Conidiation was assayed with the 12-day-old colonies grown on OMA. Conidia were collected in 5 ml of distilled water by scraping and counted with a haemacytometer under a microscope. Conidial germination and appressoria formation were measured on hydrophobic microscope coverslip (Marienfeld, Landa-Königshofen, Germany). Conidia harvested from 12-day-old OMA culture were diluted into 2×104 conidia per milliliter in sterile distilled water. Drops of conidial suspension (40 μl) were placed on the coverslips with three replicates, then placed in a moistened box and incubated at 25°C. After 8 hr incubation, the percentage of conidial germination and appressorium formation was determined by microscopic examination of at least 100 conidia per replicate in at least three independent experiments.

Fluorescence Microcopy

Fluorescence and DIC imaging was done using a Zeiss Axio Imager A1 fluorescence microscope (Carl Zeiss, Oberkochen, Germany). A filter set with excitation at 470/40 nm and emission at 525/50 nm was used for enhanced green fluorescence protein (eGFP) observation, another filter set with excitation at 365 nm and emission at 445/50 nm was used for 7-amino-4-chloromethylcoumarin (CellTracker™ Blue CMAC, Invitrogen, Carlsbad, CA, USA) and Aniline blue fluorochrome observation. The staining of conidia and growing mycelia with CMAC was performed as previously described [73].

Pathogenicity Assays and Infectious Growth Visualization

For spray inoculation, conidial suspension (10 ml) containing Tween 20 (250 ppm) and conidia harvested from 12-day-old cultures on OMA plate ($1–5×10^5$ conidia/ml) was sprayed onto four-weeks old susceptible rice seedlings (Oryza sativa cv. Nakdongbyeo). Inoculated plants were placed in a dew chamber at 25°C for 24 hours in the dark, and then transferred back to the growth chamber with a photoperiod of 16 hours using fluorescent lights [74]. Disease severity was assessed at seven days after inoculation. The %DLA was recorded to permit more accurate evaluation of the virulence of the mutants. Photographs of diseased rice leaves including eight centimeter long leaf blades were taken. The number of pixels under lesion areas and healthy areas of diseased leaves was calculated by Axiovision image analyzer with the photographs. For microscopic observation of penetration and infectious growth on rice tissue, excised rice leaf sheath of Nakdongbyeo were prepared as previously described [30],[42] and inoculated by conidia suspension ($1×104$ conidia/ml) on the adaxial surface. After 24, 48 and 96 hours incubation in a moistened box, the sheaths were trimmed to remove chlorophyll enriched plant parts. Remaining epidermal layer of mid vein (three to four cell layers thick) were utilized for microscopic observations. Inoculation on onion epidermis was performed as previously described [75]. Fixation and aniline blue staining of rice sheath and onion epidermis were performed as previously described [75]. Samples were incubated in lactophenol at room temperature for 1hour and directly mounted with 70% glycerin or transferred into 0.01% aniline blue for 1hour and destained with lactophenol. For 3, 3'-diaminobenzidine (DAB, Sigma, D-8001) staining, samples were incubated in 1mg/ml DAB solution (pH 3.8) at room temperature for 8 hours and destained with clearing solution (ethanol:acetic acid = 94:4, v/v) for 1 hour. For observation and scoring penetration rate and IH development, conidia suspension were dropped on onion epidermis and incubated for 72 hours in moistened culture plate. Samples were fixed and stained as rice sheath described above. Extensive IH from single appressoria with no (or scatterd) callose were scored as normal IH, relative short and attenuated IH with accumulated callose were scored as retarded IH, appressorium developing very short IH or penetration peg with strong callose were scored as blocked IH, and appressorium without IH and callose deposition were scored no penetration.

Nucleic Acid Manipulation and Polymerase Chain Reaction

For Southern hybridization analysis, genomic DNA was isolated according to the method described [76] with slight modification. Restriction enzyme digestion, agarose gel separation, and cloning were performed following standard procedures [77]. Southern hybridization analysis was carried out as described previously [75]. HpaI fragment (1.4 kb) including hygromycin B phosphotransferase gene (HPH), and pCB1004 were used as the hybridization probe. Genomic DNA of transformants for PCR screening was isolated by the quick extraction procedure [70]. About 50 ng of genomic DNA (2 μl) was used for PCR reactions with 1μl of 100 nM of each primer and 5 ul of 2× PCR mixture containing dNTP, PCR buffer, 1 unit of Taq polymerase and loading dye (Enzynomics™, Daejeon, Korea). Perkin-Elmer 9600 DNA Thermal Cycler was employed for PCR. Plasmid DNA was prepared by standard methods [77]. Total RNA was isolated from the frozen fungal and plant tissues with Easy-spin™ total RNA extraction kit (iN-tRON Biotechnology, Seoul, Korea) according to the manufacturer's instruction. To quantify levels of transcript, quantitative RT-PCR was performed as described [78]. Briefly, 5 μg of total RNA was reverse transcribed into first-strand cDNA with oligo (dT) primer using SuperScript™ First-Strand Synthesis System (Invitrogen™ Life Technologies, Carlsbad, CA, USA) according to the manufacturer's instruction. Reactions were performed in a 25 μl volume containing 100 nM of each primer, 2 μl of cDNA (25 ng of input RNA) and 12.5 μl of 2× Power SYBR® Green PCR Master Mix (Applied Biosystems, Warrington, UK). Real-time PCR was run on the Applied Biosystems 7500 Real Time PCR System (Applied Biosystems, Foster City, CA). After each run, amplification specificity was checked with a dissociation curve acquired by heating the samples from 60 to 95°C. Normalization and comparison of mean Ct values were performed as described [79]. To compare relative abundance of transcripts of target genes, the mean threshold cycle (Ct) of triplicate reactions was normalized by that of M. oryzae cyclophilin gene (CYP1, MGG10447), which displayed stable expressions in the developmental stages (unpublished data) and used previously [78],[80],[81], or by that of O. sativa elongation factor 1α gene (Os03g08020, [82]). Fold changes were compared among treatments or conditions with standard condition. Quantitative RT-PCR was conducted at least twice with three replicates from independent biological experiments. Genomic DNA adjacent to the T-DNA insertion of DES1T-

DNA was isolated by TAIL-PCR, which was performed as described previously [37]. The insertion was confirmed by PCR amplification with pairs of locus specific primers (4163TF and 4163 TR, see primer list), and T-DNA specific primers (RB3 and TV1). The amplified PCR fragments were sequenced twice to analyze insertion characteristics.

Targeted Disruption and Complementation of DES1 in M. Oryzae

The targeted gene disruption vector was designed using modified double-joint PCR [41]. The target region was a ~4.5 kb size fragment including DES1 ORF and short putative UTR (5'-70 bp and 3'-500 bp) sequences. An 1188 bp long 5' flanking region was amplified with a primer pair, DES1KOSF and DES1KO5R. A 942 bp long 3' flanking region was amplified with a primer pair, DES1KO3f and DES1KOSR. Both 5' and 3' flanking regions of target sequences were fused to HPH cassette using a specific primer pair with 23 bp tail sequence in DES1KO5R and DES1KO3f. After fusion with HPH cassette, a nested primer pair (DES1KO5F and DES1KO3R) was used for amplification of the final construct. Fungal protoplasts of the wild-type 70-15 were directly transformed with double-joint PCR product after purification. Protoplast generation and subsequent transformation were conducted by following the established procedures with slight modification. Initial identification of the gene disruption mutants was performed by PCR with the primer par of DES1KOSF and HPHF. Genomic DNA was isolated by the quick procedure described previously [70]. Candidates of gene disruption mutant were genetically purified by single conidia isolation. To confirm the disruption mutant, the genomic DNA of candidate strains was digested with BamHI and Southern hybridization analysis was performed with 855 bp long 3' flanking fragment (amplified with the primer pair of DES1KO3F and DES1KO3R) as a probe. For complementation of Δdes1, a 6 kb fragment carrying the DES1 ORF and 1.4 kb of 5' region (putative promoter and UTR) was amplified from wild-type genomic DNA using a primer pair of DES1_1400pF and DES1KO3R, and it cloned into pCR®-TOPO® 2.1 vector (Invitrogen, Carlsbad, CA, USA). The resulting plasmid was used for co-transforming Δdes1 protoplasts with pII99 vector. The transformants were selected on TB3 agar medium amended with 800 ppm geneticin. After genetic purification by single conidium isolation, the presence of DES1 ORF was checked by PCR amplification using a primer pair of DES1_QF and DES1_QR.

Construction of DES1-eGFP Vector

For fusion constructs, Aspergillus nidulans TrpC promoter (0.3 kb ClaI-HindIII fragment), eGFP coding sequence (0.7 kb HindIII-XbaI fragment) and DES1

ORF (4 kb SpeI or BamHI fragment) were amplified by PCR from pSK1093, pSK2702 and genomic DNA from 70-15 using primers. Each primer contains a restriction enzyme site at its 5′ end to facilitate subsequent cloning. All three PCR products were isolated from gels using QIAquick spin columns and were cloned in pGEM-T Easy. All clones were verified by sequencing. Subsequently, the ProTrpC-DES1-eGFP construct was inserted between the ClaI-XbaI sites of pGEM-3Zf (Promega, Madison, WI). The fusion constructs were co-transformed into wild-type 70-15 with pII99 that containing the geneticin resistant gene [83]. Three transformants per construct were selected and observed.

Cell Wall Integrity Test

For Nikkomycin Z sensitivity assays, conidia were incubated on slide glass with 100 μM of the drug, and germ tubes were observed after 2 hours. Protoplast production assay were performed as previously described [84] using 3-day-old mycelium (0.5 g) from CM liquid culture.

Measurement of Enzyme Activity and Detection of Ferrous Ion in Extracellular Culture Filtrate

Enzyme activity was assayed using culture filtrate from 3-day-old CM liquid culture. Mycelia were completely removed by filtration and centrifugation (5,000 g at 4°C). For measurement of peroxidase and laccase activity, a reaction mixture (1 ml) containing 50 mM acetate buffer (pH 5.0) and 10 mM ABTS was mixed with the culture filtrate (200 μl) and incubated at 25°C for 5 minutes with or without 3mM of H_2O_2. Absorbance was evaluated at 420 nm.

Ferrous ion in the culture filtrate was measured with Bathophenanthroline disulfonate (BPS) color reaction. BPS was added to the culture filtrate (final concentration 1 mM), and the mixture was incubated for 3 hours at room temperature. Concentration of ferrous ion was monitored spectrophotometrically at 535 nm as BPS-Fe(II) complex formation. A standard curve was generated between color intensity and ferrous ion concentration by using standard solutions of varying concentrations (0–30 μM) of ferrous ion (OD535 = $0.213[Fe^{2+}]$, R^2 = 0.9995), and the absorbance at 535 nm was converted into ferrous ion concentration according to the standard curve.

Bioinformatics

All sequence information used in this study was obtained from the online database CFGP ([38], http://cfgp.snu.ac.kr) which containing the latest annotated genome information of 59 fungi including M. oryzae. To identify DES1 homologs, GeneBank (http://www.ncbi.nlm.nih.gov/BLAST) and CFGP database were searched using the BLAST algorithm [85]. Gene distribution analysis after DES1 homolog search was performed automatically by the BLAST matrix program incorporated in CFGP. Sequence alignment using the ClustalW algorithm [86] and generation of bootstrapped phylogenetic trees were performed in CFGP. Results of InterPro Scan v12.0 [39], domain architecture visualization, SignalP v3.0 [40] and amino acid frequency analysis also were automatically provided from CFGP. Primers used in this study were designed using Primer Select™ program (DNASTAR Inc., Madison, USA) and commercially synthesized (BIONEER Corp., Daejeon, Korea).

Competing Interests

The authors have declared that no competing interests exist.

Acknowledgements

We thank Dr. Barbara Valent (Kansas State University) for mating tester strains, Dr. Seogchan Kang for providing eGFP plasmid, and Dr. Nicole Donofrio (University of Delaware) for critical reading of the manuscript. We also thank Dr. In-Suk Oh, Dr. Jae-Hwan Rho, and Mr. Myoung Gil Cho (National Institute of Crop Science) for providing plants.

Authors' Contributions

Conceived and designed the experiments: MHC YHL. Performed the experiments: MHC SYP SK. Analyzed the data: MHC SYP SK. Wrote the paper: MHC YHL.

References

1. Heath MC (2000) Nonhost resistance and nonspecific plant defenses. Curr Opin Plant Biol 3: 315–319.

2. Chisholm ST, Coaker G, Day B, Staskawicz BJ (2006) Host-microbe interactions: shaping the evolution of the plant immune response. Cell 124: 803–814.

3. Jones JD, Dangl JL (2006) The plant immune system. Nature 444: 323–329.

4. Gómez-Gómez L, Boller T (2000) FLS2: an LRR receptor-like kinase involved in the perception of the bacterial elicitor flagellin in Arabidopsis. Mol Cell 5: 1003–1011.

5. Felix G, Duran JD, Volko S, Boller T (1999) Plants have a sensitive perception system for the most conserved domain of bacterial flagellin. Plant J 18: 265–276.

6. Lin WL, Hu XY, Zhang WQ, Rogers WJ, Cai WM (2005) Hydrogen peroxide mediates defence responses induced by chitosans of different molecular weights in rice. J Plant Physiol 162: 937–944.

7. Kuchitsu K, Kosaka H, Shiga T, Shibuya N (1995) EPR evidence for generation of hydroxyl radical triggered by N-acetylchitooligosaccharide elicitor and a protein phosphatase inhibitor in suspension-cultured rice cells. Protoplasma 188: 138–142.

8. Hauck P, Thilmony R, He SY (2003) A Pseudomonas syringae type III effector suppresses cell wall-based extracellular defense in susceptible Arabidopsis plants. Proc Natl Acad Sci USA 100: 8577–8582.

9. Ahn IP, Kim S, Kang S, Suh SC, Lee YH (2005) Rice defense mechanisms against Cochliobolus miyabeanus and Magnaporthe grisea are distinct. Phytopathology 95: 1248–1255.

10. Tao Y, Xie Z, Chen W, Glazebrook J, Chang HS, et al. (2003) Quantitative nature of Arabidopsis responses during compatible and incompatible interactions with the bacterial pathogen Pseudomonas syringae. Plant Cell 15: 317–330.

11. Bowles DJ (1990) Defense-related proteins in higher plants. Annu Rev Biochem 59: 873–907.

12. Orbach MJ, Farrall L, Sweigard JA, Chumley FG, Valent B (2000) A telomeric avirulence gene determines efficacy for the rice blast resistance gene Pi-ta. Plant Cell 12: 2019–2032.

13. Abramovitch RB, Kim YJ, Chen S, Dickman MB, Martin GB (2003) Pseudomonas type III effector AvrPtoB induces plant disease susceptibility by inhibition of host programmed cell death. Embo J 22: 60–69.

14. Apostol I, Heinstein PF, Low PS (1989) Rapid stimulation of an oxidative burst during elicitation of cultured plant cells: role in defense and signal transduction. Plant Physiol 90: 109–116.

15. Bradley DJ, Kjellbom P, Lamb CJ (1992) Elicitor- and wound-induced oxidative cross-linking of a proline-rich plant cell wall protein: a novel, rapid defense response. Cell 70: 21–30.

16. Chen SX, Schopfer P (1999) Hydroxyl-radical production in physiological reactions. A novel function of peroxidase. Eur J Biochem 260: 726–735.

17. Levine A, Tenhaken R, Dixon R, Lamb C (1994) H2O2 from the oxidative burst orchestrates the plant hypersensitive disease resistance response. Cell 79: 583–593.

18. Tanaka N, Che FS, Watanabe N, Fujiwara S, Takayama S, et al. (2003) Flagellin from an incompatible strain of Acidovorax avenae mediates H2O2 generation accompanying hypersensitive cell death and expression of PAL, Cht-1, and PBZ1, but not of Lox in rice. Mol Plant Microbe Interact 16: 422–428.

19. Torres MA, Jones JD, Dangl JL (2005) Pathogen-induced, NADPH oxidase-derived reactive oxygen intermediates suppress spread of cell death in Arabidopsis thaliana. Nat Genet 37: 1130–1134.

20. Ono E, Wong HL, Kawasaki T, Hasegawa M, Kodama O, et al. (2001) Essential role of the small GTPase Rac in disease resistance of rice. Proc Natl Acad Sci USA 98: 759–764.

21. Wong HL, Pinontoan R, Hayashi K, Tabata R, Yaeno T, et al. (2007) Regulation of rice NADPH oxidase by binding of Rac GTPase to its N-terminal extension. Plant Cell 19: 4022–4034.

22. Molina L, Kahmann R (2007) An Ustilago maydis gene involved in H2O2 detoxification is required for virulence. Plant Cell 19: 2293–2309.

23. Ou SH (1985) Rice Diseases. Kew, England: Commonwealth Mycological Institute.

24. Talbot NJ (2003) On the trail of a cereal killer: Exploring the biology of Magnaporthe grisea. Annu Rev Microbiol 57: 177–202.

25. Dean RA, Talbot NJ, Ebbole DJ, Farman ML, Mitchell TK, et al. (2005) The genome sequence of the rice blast fungus Magnaporthe grisea. Nature 434: 980–986.

26. Jeon J, Park SY, Chi MH, Choi J, Park J, et al. (2007) Genome-wide functional analysis of pathogenicity genes in the rice blast fungus. Nat Genet 39: 561–565.

27. Howard RJ, Ferrari MA, Roach DH, Money NP (1991) Penetration of hard substrates by a fungus employing enormous turgor pressures. Proc Natl Acad Sci USA 88: 11281–11284.

28. Howard RJ, Valent B (1996) Breaking and entering: host penetration by the fungal rice blast pathogen Magnaporthe grisea. Annu Rev Microbiol 50: 491–512.

29. Heath MC, Valent B, Howard RJ, Chumley FG (1990) Correlation between cytologically detected plant-fungal interactions and pathogenicity of Magnaporthe grisea toward weeping lovegrass. Phytopathology 80: 1382–1386.

30. Kankanala P, Czymmek K, Valent B (2007) Roles for rice membrane dynamics and plasmodesmata during biotrophic invasion by the blast fungus. Plant Cell 19: 706–724.

31. Caracuel-Rios Z, Talbot NJ (2007) Cellular differentiation and host invasion by the rice blast fungus Magnaporthe grisea. Curr Opin Microbiol 10: 339–345.

32. Bohnert HU, Fudal I, Dioh W, Tharreau D, Notteghem JL, et al. (2004) A putative polyketide synthase/peptide synthetase from Magnaporthe grisea signals pathogen attack to resistant rice. Plant Cell 16: 2499–2513.

33. Gilbert MJ, Thornton CR, Wakley GE, Talbot NJ (2006) A P-type ATPase required for rice blast disease and induction of host resistance. Nature 440: 535–539.

34. Tanaka S, Yamada K, Yabumoto K, Fujii S, Huser A, et al. (2007) Saccharomyces cerevisiae SSD1 orthologues are essential for host infection by the ascomycete plant pathogens Colletotrichum lagenarium and Magnaporthe grisea. Mol Microbiol 64: 1332–1349.

35. Mehrabi R, Ding S, Xu JR (2008) MADS-box transcription factor mig1 is required for infectious growth in Magnaporthe grisea. Eukaryot Cell 7: 791–799.

36. Liu YG, Whittier RF (1995) Thermal asymmetric interlaced PCR: automatable amplification and sequencing of insert end fragments from P1 and YAC clones for chromosome walking. Genomics 25: 674–681.

37. Choi J, Park J, Jeon J, Chi MH, Goh J, et al. (2007) Genome-wide analysis of T-DNA integration into the chromosomes of Magnaporthe oryzae. Mol Microbiol 66: 371–382.

38. Park J, Park B, Jung K, Jang S, Yu K, et al. (2008) CFGP: a web-based, comparative fungal genomics platform. Nucleic Acids Res 36: D562–571.

39. Mulder NJ, Apweiler R, Attwood TK, Bairoch A, Bateman A, et al. (2005) InterPro, progress and status in 2005. Nucleic Acids Res 33: D201–205.

40. Bendtsen JD, Nielsen H, von Heijne G, Brunak S (2004) Improved prediction of signal peptides: SignalP 3.0. J Mol Biol 340: 783–795.

41. Yu JH, Hamari Z, Han KH, Seo JA, Reyes-Dominguez Y, et al. (2004) Double-joint PCR: a PCR-based molecular tool for gene manipulations in filamentous fungi. Fungal Genet Biol 41: 973–981.

42. Koga H, Dohi K, Nakayachi O, Mori M (2004) A novel inoculation method of Magnaporthe grisea for cytological observation of the infection process using intact leaf sheaths of rice plants. Physiol Mol Plant Path 64: 67–72.

43. Nürnberger T, Brunner F, Kemmerling B, Piater L (2004) Innate immunity in plants and animals: striking similarities and obvious differences. Immunol Rev 198: 249–266.

44. Nicholson RL, Hammerschmidt R (1992) Phenolic-compounds and their role in disease resistance. Annu Rev Phytopathol 30: 369–389.

45. Xu JR, Hamer JE (1996) MAP kinase and cAMP signaling regulate infection structure formation and pathogenic growth in the rice blast fungus Magnaporthe grisea. Gene Dev 10: 2696–2706.

46. Morré DJ (2002) Preferential inhibition of the plasma membrane NADH oxidase (NOX) activity by diphenyleneiodonium chloride with NADPH as donor. Antioxid Redox Signal 4: 207–212.

47. Egan MJ, Wang ZY, Jones MA, Smirnoff N, Talbot NJ (2007) Generation of reactive oxygen species by fungal NADPH oxidases is required for rice blast disease. Proc Natl Acad Sci USA 104: 11772–11777.

48. Odenbach D, Breth B, Thines E, Weber RW, Anke H, et al. (2007) The transcription factor Con7p is a central regulator of infection-related morphogenesis in the rice blast fungus Magnaporthe grisea. Mol Microbiol 64: 293–307.

49. Ram AF, Wolters A, Ten Hoopen R, Klis FM (1994) A new approach for isolating cell wall mutants in Saccharomyces cerevisiae by screening for hypersensitivity to calcofluor white. Yeast 10: 1019–1030.

50. Wood PJ, Fulcher RG (1983) Dye interactions. A basis for specific detection and histochemistry of polysaccharides. J Histochem Cytochem 31: 823–826.

51. Cripps C, Bumpus JA, Aust SD (1990) Biodegradation of azo and heterocyclic dyes by Phanerochaete chrysosporium. Appl Environ Microbiol 56: 1114–1118.

52. Woo SW, Cho JS, Hur BK, Shin DH, Ryu KG, et al. (2003) Hydrogen peroxide, its measurement and effect during enzymatic decoloring of Congo red. J Microbiol Biotechn 13: 773–777.

53. Shindler JS, Childs RE, Bardsley WG (1976) Peroxidase from human cervical mucus. The isolation and characterisation. Eur J Biochem 65: 325–331.

54. Skamnioti P, Henderson C, Zhang Z, Robinson Z, Gurr SJ (2007) A novel role for catalase B in the maintenance of fungal cell-wall integrity during host invasion in the rice blast fungus Magnaporthe grisea. Mol Plant Microbe Interact 20: 568–580.

55. Wolfenden BS, Willson RL (1982) Radical-cations as reference chromogens in kinetic-studies of one-electron transfer-reactions: pulse radiolysis studies of 2,2′-azinobis-(3-ethylbenzthiazoline-6-sulphonate). J Chem Soc, Perkin Trans 2: 805–812.

56. Nilsson UA, Bassen M, Savman K, Kjellmer I (2002) A simple and rapid method for the determination of "free" iron in biological fluids. Free Radic Res 36: 677–684.

57. Koga H, Dohi K, Mori M (2004) Abscisic acid and low temperatures suppress the whole plant-specific resistance reaction of rice plants to the infection of Magnaporthe grisea. Physiol Mol Plant Path 65: 3–9.

58. Kaku H, Nishizawa Y, Ishii-Minami N, Akimoto-Tomiyama C, Dohmae N, et al. (2006) Plant cells recognize chitin fragments for defense signaling through a plasma membrane receptor. Proc Natl Acad Sci USA 103: 11086–11091.

59. Thao NP, Chen L, Nakashima A, Hara SI, Umemura K, et al. (2007) RAR1 and HSP90 form a complex with Rac/Rop GTPase and function in innate-immune responses in rice. Plant Cell.

60. Lamb C, Dixon RA (1997) The oxidative burst in plant disease resistance. Annu Rev Plant Physiol Plant Mol Biol 48: 251–275.

61. De Freitas J, Wintz H, Kim JH, Poynton H, Fox T, et al. (2003) Yeast, a model organism for iron and copper metabolism studies. Biometals 16: 185–197.

62. Klionsky DJ, Herman PK, Emr SD (1990) The fungal vacuole: composition, function, and biogenesis. Microbiol Rev 54: 266–292.

63. Bellemare DR, Shaner L, Morano KA, Beaudoin J, Langlois R, et al. (2002) Ctr6, a vacuolar membrane copper transporter in Schizosaccharomyces pombe. J Biol Chem 277: 46676–46686.

64. Luk EE, Culotta VC (2001) Manganese superoxide dismutase in Saccharomyces cerevisiae acquires its metal co-factor through a pathway involving the Nramp metal transporter, Smf2p. J Biol Chem 276: 47556–47562.

65. Iyer G, Chattoo BB (2003) Purification and characterization of laccase from the rice blast fungus, Magnaporthe grisea. FEMS Microbiol Lett 227: 121–126.

66. Haas H (2003) Molecular genetics of fungal siderophore biosynthesis and uptake: the role of siderophores in iron uptake and storage. Appl Microbiol Biotechnol 62: 316–330.

67. Hof C, Eisfeld K, Welzel K, Antelo L, Foster AJ, et al. (2007) Ferricrocin synthesis in Magnaporthe grisea and its role in pathogenicity in rice. Mol Plant Pathol 8: 163–172.

68. Sun CB, Suresh A, Deng YZ, Naqvi NI (2006) A multidrug resistance transporter in Magnaporthe is required for host penetration and for survival during oxidative stress. Plant Cell 18: 3686–3705.

69. Talbot NJ, Ebbole DJ, Hamer JE (1993) Identification and characterization of MPG1, a gene involved in pathogenicity from the rice blast fungus Magnaporthe grisea. Plant Cell 5: 1575–1590.

70. Chi MH, Park SY, Kim S, Lee YH (2009) A quick and safe method for fungal DNA extraction. Plant Pathol J 25: 108–111.

71. Valent B, Farrall L, Chumley FG (1991) Magnaporthe grisea genes for pathogenicity and virulence identified through a series of backcrosses. Genetics 127: 87–101.

72. Ram AF, Klis FM (2006) Identification of fungal cell wall mutants using susceptibility assays based on Calcofluor white and Congo red. Nat Protoc 1: 2253–2256.

73. Ohneda M, Arioka M, Nakajima H, Kitamoto K (2002) Visualization of vacuoles in Aspergillus oryzae by expression of CPY-EGFP. Fungal Genet Biol 37: 29–38.

74. Choi WB, Chun SJ, Lee YH (1996) Host range of Korean isolates of Magnaporthe grisea. Kor J Plant Pathol 12: 453–454.

75. Kim S, Ahn IP, Rho HS, Lee YH (2005) MHP1, a Magnaporthe grisea hydrophobin gene, is required for fungal development and plant colonization. Mol Microbiol 57: 1224–1237.

76. Rogers SO, Bendich AJ (1985) Extraction of DNA from milligram amounts of fresh, herbarium and mummified plant-tissues. Plant Mol Biol 5: 69–76.

77. Sambrook J, Fritsch EF, Maniatis T (1989) Molecular Cloning: A Laboratory Manual. Cold Spring Harbor, NY, USA: Cold Spring Harbor Laboratory Press.

78. Yi M, Chi MH, Khang CH, Park SY, Kang S, et al. (2009) The ER chaperone LHS1 is involved in asexual development and rice infection by the blast fungus Magnaporthe oryzae. Plant Cell. 10.1105/tpc.1107.055988.

79. Livak KJ, Schmittgen TD (2001) Analysis of relative gene expression data using real-time quantitative PCR and the 2−ΔΔCt method. Methods 25: 402–408.

80. Yi M, Lee YH (2008) Identification of genes encoding heat shock protein 40 family and the functional characterization of two Hsp40s, MHF16 and MHF21, in Magnaporthe oryzae. Plant Pathol J 24: 131–142.

81. Yi M, Park JH, Ahn JH, Lee YH (2008) MoSNF1 regulates sporulation and pathogenicity in the rice blast fungus Magnaporthe oryzae. Fungal Genet Biol 45: 1172–1181.

82. Caldana C, Scheible WR, Mueller-Roeber B, Ruzicic S (2007) A quantitative RT-PCR platform for high-throughput expression profiling of 2500 rice transcription factors. Plant Methods 3: 7.

83. Lee J, Lee T, Lee YW, Yun SH, Turgeon BG (2003) Shifting fungal reproductive mode by manipulation of mating type genes: obligatory heterothallism of Gibberella zeae. Mol Microbiol 50: 145–152.

84. Jeon J, Goh J, Yoo S, Chi MH, Choi J, et al. (2008) A putative MAP kinase kinase kinase, MCK1, is required for cell wall integrity and pathogenicity of the rice blast fungus, Magnaporthe oryzae. Mol Plant Microbe Interact 21: 525–534.

85. McGinnis S, Madden TL (2004) BLAST: at the core of a powerful and diverse set of sequence analysis tools. Nucleic Acids Res 32: W20–25.

86. Thompson JD, Higgins DG, Gibson TJ (1994) Clustal W: improving the sensitivity of progressive multiple sequence alignment through sequence weighting, position-specific gap penalties and weight matrix choice. Nucleic Acid Res 22: 4673–4680.

CITATION

Originally published under the Creative Commons Attribution License. Chi M-H, Park S-Y, Kim S, Lee Y-H (2009). A Novel Pathogenicity Gene Is Required in the Rice Blast Fungus to Suppress the Basal Defenses of the Host. PLoS Pathog 5(4): e1000401. doi:10.1371/journal.ppat.1000401.

Differential Gene Expression in Incompatible Interaction between Wheat and Stripe Rust Fungus Revealed by Cdna-AFLP and Comparison to Compatible Interaction

Xiaojie Wang, Wei Liu, Xianming Chen, Chunlei Tang, Yanling Dong, Jinbiao Ma, Xueling Huang, Guorong Wei, Qingmei Han, Lili Huang and Zhensheng Kang

ABSTRACT

Background

Stripe rust of wheat, caused by Puccinia striiformis f. sp. tritici (Pst), is one of the most important diseases of wheat worldwide. Due to special features of

hexaploid wheat with large and complex genome and difficulties for transformation, and of Pst without sexual reproduction and hard to culture on media, the use of most genetic and molecular techniques in studying genes involved in the wheat-Pst interactions has been largely limited. The objective of this study was to identify transcriptionally regulated genes during an incompatible interaction between wheat and Pst using cDNA-AFLP technique

Results

A total of 52,992 transcript derived fragments (TDFs) were generated with 64 primer pairs and 2,437 (4.6%) of them displayed altered expression patterns after inoculation with 1,787 up-regulated and 650 down-regulated. We obtained reliable sequences (>100 bp) for 255 selected TDFs, of which 113 (44.3%) had putative functions identified. A large group (17.6%) of these genes shared high homology with genes involved in metabolism and photosynthesis; 13.8% to genes with functions related to disease defense and signal transduction; and those in the remaining groups (12.9%) to genes involved in transcription, transport processes, protein metabolism, and cell structure, respectively. Through comparing TDFs identified in the present study for incompatible interaction and those identified in the previous study for compatible interactions, 161 TDFs were shared by both interactions, 94 were expressed specifically in the incompatible interaction, of which the specificity of 43 selected transcripts were determined using quantitative real-time polymerase chain reaction (qRT-PCR). Based on the analyses of homology to genes known to play a role in defense, signal transduction and protein metabolism, 20 TDFs were chosen and their expression patterns revealed by the cDNA-AFLP technique were confirmed using the qRT-PCR analysis.

Conclusion

We uncovered a number of new candidate genes possibly involved in the interactions of wheat and Pst, of which 11 TDFs expressed specifically in the incompatible interaction. Resistance to stripe rust in wheat cv. Suwon11 is executed after penetration has occurred. Moreover, we also found that plant responses in compatible and incompatible interactions are qualitatively similar but quantitatively different soon after stripe rust fungus infection.

Background

Plant disease resistance and susceptibility are governed by the combined genotypes of host and pathogen, and depend on a complex exchange of signals and responses occurring under given environmental conditions. During the long processes of host-pathogen co-evolution, plants have developed various elaborate mechanisms to ward off pathogen attack [1]. A key difference between resistant and susceptible plants is the timely recognition of the invading pathogen, and the rapid and effective activation of host defense mechanisms. The activation of defense responses in plants is initiated by host recognition of pathogen-encoded molecules called elicitors [2]. The interaction of pathogen elicitors with host receptors likely activates a signal transduction cascade that may involve protein phosphorylation, ion fluxes, reactive oxygen species (ROS), and other signaling events [3,4]. Subsequent transcriptional and/or posttranslational activation of transcription factors eventually leads to the induction of plant defense related genes [5]. In addition to eliciting primary defense responses, pathogen signals may be amplified through the generation of secondary plant signal molecules such as salicylic acid [6]. Both primary pathogen elicitors and secondary endogenous signals may trigger a diverse array of plant defense related genes, encoding glutathione S-transferases (GST), peroxidases, cell wall proteins, proteinase inhibitors, hydrolytic enzymes, pathogenesis-related (PR) proteins and phytoalexin biosynthetic enzymes [7].

At the macroscopic level, induced defense responses are frequently manifested in part as a hypersensitive response (HR), which is characterized by necrotic lesions resulting from localized host cell death at the site of infection [8]. Plant cell death occurring during the HR plays an important role in preventing the growth and spread of biotrophic pathogen into healthy tissues [9,10]. In addition to the localized HR, plants may respond to pathogen infection by activating defense responses in uninfected parts of the plant, expressing so called systemic acquired resistance (SAR) that can be long-lasting and often confers broad-based resistance to a variety of different pathogens [11,12].

Stripe rust (or yellow rust), caused by Puccinia striiformis Westend. f. sp. tritici Eriks. (Pst), is one of the most important diseases of wheat (Triticum aestivum L.) worldwide. Severe yield losses can result as a consequence of the rapid development and large-scale spread of the disease epidemic under optimal environmental conditions. Furthermore, the ability of Pst to form new races that can attack previously resistant cultivars, along with the capacity of fungal spores to travel long distances, can make control of stripe rust difficult. Over the last few years, epidemiological [13,14], genetic [15-18], histological [19] and molecular [20-23] studies on the disease and pathogen have been reported. The wheat-Pst interactions have been studied at the molecular genetics and ultrastructural levels [24,25]. Due to special features of hexaploid wheat with large and complex genome and difficulties for transformation, and of Pst without sexual reproduction and hard to culture on media [26], the use of most genetic and molecular techniques in studying genes involved in the wheat-Pst interactions has been largely limited. Thus, a global gene expression approach should be useful for elucidating the molecular mechanisms of the wheat-Pst interactions.

Significant progresses have been made for understanding the signaling processes involved in several plant-pathogen interactions [27]. A few studies on the wheat-rust fungus interaction have been carried out using the Wheat GeneChip® [21,22,28,29]. The use of the Wheat GeneChip® technique is often conditioned by known gene sequences arrayed on the chip, with limited ESTs unspecific to different wheat materials. In contrast, cDNA-amplified fragment length polymorphism (cDNA-AFLP) does not require prior sequence information and is universal for any organisms or interactions, and is, therefore, a powerful tool for identifying novel genes in non-model organisms [30,31], such as wheat [32]. As described by Bachem et al., cDNA-AFLP is an efficient, sensitive, and reproducible technique to detect differentially expressed genes dynamically [33].

In our previous study, we identified 186 genes likely involved in a compatible interaction between wheat (cv. Suwon 11) and Pst (pathotype CYR31) using the cDNA-AFLP technique [34]. A parallel study for an incompatible interaction should allow us to compare genes involved in the compatible and incompatible interactions, which should provides insights to molecular mechanisms of the different interactions of the important wheat-Pst pathosystem. The objective of this study was to determine wheat genes that are transcriptionally regulated in response to Pst infection using the cDNA-AFLP technique in a whole-genome scale. The quantitative real-time polymerase chain reaction (qRT-PCR) analysis was also used to validate the expression patterns of some important genes. Here, we report a number of transcript derived fragments (TDFs) that were found to be activated or suppressed during the incompatible interaction between wheat and Pst. In particular, a large number of genes encoding signal molecules were identified as early pathogen responsive genes and potential defense-related genes. Genes specifically expressed during the incompatible interaction were identified through comparing the transcription profiling in the present study with that of our previous study on compatible interaction [34].

Results

Infection Process of Stripe Rust Fungus and HR

In the incompatible interaction, at 18 hpi when haustorial mother cells were in contact with mesophyll cells, HR was observed at the infection sites (Figure 1A). At 24 hpi, the host cells undergoing HR still looked intact (Figure 1B). With advancing incubation time, an increasing number of host cells took up HR and started to lose their original shape. Necrotic host cells could be observed at almost every infection site by 72 hpi, (Figure 1C). Up to 120 hpi, large number of host cell deceased and fungal spread were inhibited at infection sites (Figure 1D).

However, in the compatible interaction, there was no indication of cell death at infection sites (data not shown). The results showed that these samples were suitable for further experiments and analyses.

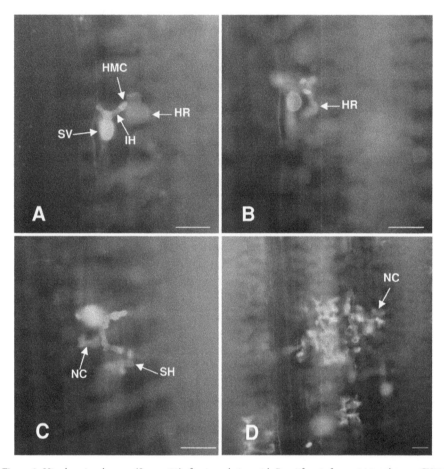

Figure 1. Histology in wheat cv. 'Suwon 11' after inoculation with P. striiformis f. sp. tritici pathotype CY23 (incompatible interaction). (A) hypersensitive cell death (HR) was only observed in the mesophyll cells in contact with the haustorial mother cell at few infection sites, 18 hpi. (B) HR could be observed in most of infection sites, 24 hpi. (C) Second hyphae were formed and mesophll cells which around intercellular hyphae showed cell death, 72 hpi. (D) Host cell death and fungal spread inhibit at infection sites, 120 hpi. Bars = 50 μm, SV, substomatal vesicle; IH, infection hypha; HMC, haustorial mother cell; SH, secondary hyphae; NC, necrotic cell.

Isolation of Differentially Expressed Genes

Transcript derived fragments displayed by cDNA-AFLP analysis ranged in size from 50 to 750 bp, depending on primer combinations and time points. For each of the 64 primer combinations, 55–83 bands were observed. Figure 2 showed an example of the expression patterns of the genes revealed using cDNA-AFLP with the primer pair MTT+TAC. A total of 52,992 fragments were obtained with 64 primer pairs. The cDNA-AFLP fragments were highly reproducible as the band intensities were similar from the three biological replications for each time-point. Altered expression patterns after inoculation were detected for 2,437 TDFs compared to the near 0 hpi mock-inoculation control and among the different time points, accounting for 4.6% of displayed fragments. Of the 2,437 TDFs, 1,787 were up-regulated and 650 down-regulated. A total of 300 TDFs were selected based on their intensity differences at various time points for attempted further analysis, of which 255 were recovered from gels, re-amplified, cloned and sequenced.

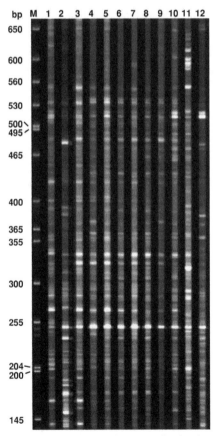

Figure 2. Expression of wheat (cv. Suwon 11) genes in leaves inoculated with Puccinia striiformis f. sp. tritici (pathotype CY23) transcripts displayed by cDNA-AFLP. An example showing selective amplification with primers MTT+TAC. Lanes 1–11 are: 6, 12, 18, 24, 36, 48, 72, 96, 120, 144 and 168 hpi, respectively; lane 12: 0 h (control plants of mock inoculation with sterile water near 0 hpi; M = molecular weight marker.

Gene Sequence Analysis

The 255 TDFs produced reliable (>100 bp) sequences. To verify the sequences for the respective bands in the cDNA-AFLP analysis, at least three clones were sequence for each re-amplified TDF and they produced an identical sequence. The sequence of each TDF was identified by similarity search using the BLASTX program against the GenBank non-redundant public sequence database. The TDF sequences were classified into functional categories based on their homology to known proteins according to Bevan's method [35]. No function could be assigned to 142 (55.7%) of the TDFs as they showed no or low sequence similarities in the database search. A group of 45 sequences (17.6%) were identified to be involved in metabolism and photosynthesis. Eighteen (7.1%) sequences shared high similarities to genes with functions in disease defense, and 17 (6.7%) sequences were found to be involved in signal transduction. The remaining 33 (12.9%) sequences were classified into groups of genes involved in transcription/transport process, protein metabolism and cell structure (Figure 3). The number of TDFs that were up- or down-regulated in each function category is summarized in Table 1. All of the 255 unigenes were submitted to the NCBI database with accession numbers assigned.

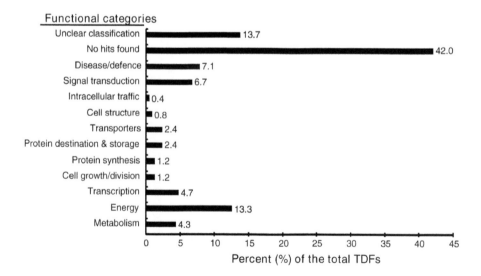

Figure 3. Classification of differentially accumulated transcript derived fragments (TDFs) after inoculation with Puccinia striiformis f. sp. tritici. A total of 255 TDFs were classified based on the BLASTX homology search.

Table 1. The numbers of transcript derived fragments (TDFs) in some of the biological function categories showing up- or down- regulated expression pattern in wheat (cv. Suwon 11) leaves inoculated with Puccinia striiformis f. sp. tritici (pathotype CYR23)

Function group	Number	Percentage (%)	Expression pattern (% in category)	
			Up (%)	Down (%)
1. Sequenced	255	10.5	187 (73.3)	68 (26.7)
Metabolism	11	4.3	8 (72.7)	3 (27.3)
Energy	34	13.3	26 (76.5)	8 (23.5)
Cell growth/division	3	1.2	2 (66.7)	1 (33.3)
Transcription	12	4.7	11 (91.7)	1 (8.3)
Protein synthesis	3	1.2	3	0
Protein destination and storage	6	2.4	3 (50.0)	3 (50.0)
Transporters	6	2.4	5 (83.3)	1 (16.7)
Intracellular traffic	1	0.4	1	0
Cell structure	2	0.8	2	0
Signal transduction	17	6.7	14 (82.4)	3 (17.6)
Disease/defense	18	7.1	15 (83.3)	3 (16.7)
Unclassified	107	42.0	72 (67.3)	35 (32.7)
Unclear classification	35	13.7	25 (71.4)	10 (28.6)
2. Unsequenced	2,182	89.5	1,600 (73.3)	582 (26.7)
Total	2,437	100.0	1,787 (73.3)	650 (26.7)

The BLASTN searching of the P. graminis f. sp. tritici genome database for the 255 TDFs indicated that 19 (7.4%) of the sequenced TDFs were likely from Pst and 129 (50.6%) were likely from wheat, and remaining 107 (42%) were unclear about their origin because they had no hit. Of the 19 TDFs, putatively encoding ATP synthase, glycine dehydrogenase, fructose-1,6-bisphosphatase, ATP-dependent RNA helicase, T-complex protein, enolase and conserved hypothetical proteins, 11 had significant homologies (e < 1 × 10-20) to P. graminis f. sp. tritici genes. However, all of the 19 TDFs were finally determined to be from wheat through PCR amplification of the genomic DNA of the Suwon 11 wheat and CYR23 Pst pathotype, as 17 fragments were amplified only from the wheat, and 2 from both wheat and stripe rust pathogen but the wheat sequences were identical to those of the TDFs.

The comparative analysis between the incompatible and compatible interactions by the TBLASTN searching showed that 161 of the 255 TDFs identified in the incompatible interaction were also induced in the compatible interaction [34], indicating that the 161 TDFs are involved in the basal defense. The remaining 94 TDFs were considered to be expressed more specifically in the incompatible interaction, 43 of 94 TDFs were further analyzed by qRT-PCR.

Validation of Expression Patterns by qRT- PCR Analysis

To validate the results of the cDNA-AFLP and comparative analyses, twenty TDFs were studied for verifying the expression patterns identified in the cDNA-AFLP study using 7 time-points (0, 12, 18, 24, 48, 72 and 120 hpi) (Figure 4, Figure 5), forty three TDFs were analyzed mainly for comparing their expressions in the incompatible and compatible interactions, and therefore, only 4 time-points (0, 12, 24 and 48 hpi) were chosen based on their major differences in the cDNA-AFLP study (Table 2). The results showed that 27 of 43 TDFs were induced in both incompatible and compatible interactions. Although in similar expression patterns, the 27 genes expressed earlier and at higher levels in the incompatible interaction than in the compatible interaction, of which 8 TDFs had expression levels great than 10 folds of the mock inoculated controls (Table 2, Figure 4, Figure 5). The transcriptional products of 5 TDFs were decreased slightly in both interactions, compared with those in the controls (Table 2). The remaining 11 TDFS were up-regulated in the incompatible interaction, but did not have significant changes or were down-regulated in the compatible interaction (Table 2).

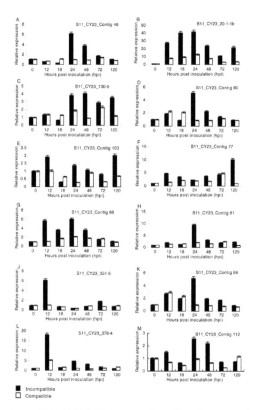

Figure 4. Quantitative real-time PCR (qRT-PCR) analyses of 12 selected genes. Leaf tissues were sampled for both inoculated and mock-inoculated plants at 12, 18, 24, 48, 72 and 120 hpi, as well as mock-inoculated near 0 hpi. Three independent biological replications were performed. Relative gene quantification was calculated by comparative $\Delta\Delta$CT method. All data were normalized to the 18S rRNA expression level. The mean expression value was calculated for every transcript derived fragment (TDF) with three replications.

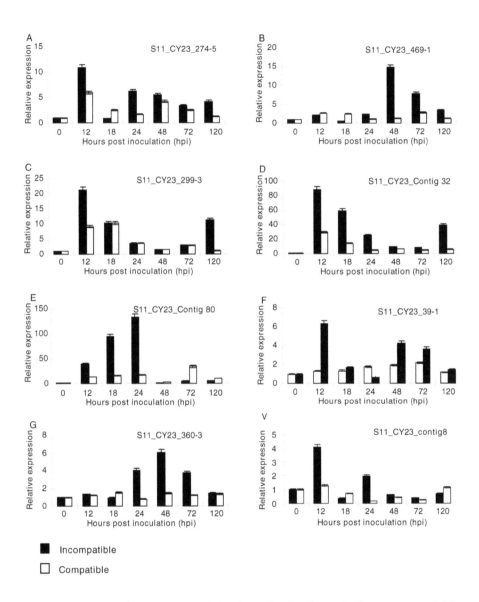

Figure 5. Quantitative real-time PCR (qRT-PCR) analyses of 8 selected genes. Leaf tissues were sampled for both inoculated and mock-inoculated plants at 12, 18, 24, 48, 72 and 120 hpi, as well as mock-inoculated near 0 hpi. Three independent biological replications were performed. Relative gene quantification was calculated by comparative ΔΔCT method. All data were normalized to the 18S rRNA expression level. The mean expression value was calculated for every transcript derived fragment (TDF) with three replications.

Table 2. Transcript derived fragments (TDFs), which were specifically expressed in incompatible interaction by BLASTN analyzing, commonly or differentially expressed in the incompatible and compatible interactions between wheat cultivar Suwon 11 and Puccinia striiformis f. sp. tritici pathotypes CYR23 and CYR31, respectively.

TDF	Accession No.	12 hpi		24 hpi		48 hpi	
		I	C	I	C	I	C
Up-regulated in both interactions							
S11_CY23_129-1	FG618817	+	+	-	-	+	NC
S11_CY23_Contig39	FG618708	NC	NC	+	-	NC	NC
S11_CY23_42b-2	FG618898	++	+	+	+	+	-
S11_CY23_253-1	FG618863	+	+	+	NC	+	+
S11_CY23_130-5	FG618819	NC	NC	+	+	+	NC
S11_CY23_Contig66	FG618735	++	NC	++	+	+	NC
S11_CY23_274-5*	FG618919	+++	++	++	NC	+	+
S11_CY23_299-3*	FG618916	+++	++	+	+	NC	NC
S11_CY23_90-6	FG618918	+	+	+	NC	NC	NC
S11_CY23_Contig32*	FG618701	+++	++	++	NC	+	NC
S11_CY23_Contig69	FG618738	+	+	++	NC	+	+
S11_CY23_39-1*	FG618892	++	NC	+	NC	+	+
S11_CY23_181-4	FG618829	+	NC	+	+	NC	NC
S11_CY23_Contig7	FG618676	+	NC	+	-	NC	NC
S11_CY23_Contig58	FG618727	+	+	NC	NC	+	-
S11_CY23_Contig93	FG618762	+	NC	-	NC	-	-
S11_CY23_Contig98	FG618767	+	+	+	NC	+	NC
S11_CY23_Contig130	FG618798	+	+	+	+	+	+
S11_CY23_349-2	FG618877	+	NC	+	+	+	NC
S11_CY23_Contig134	FG618802	-	-	+	+	+	NC
S11_CY23_209-5	FG618844	+	+	+	NC	NC	NC
S11_CY23_257-1-3*	FG618866	+	+	+++	++	+	NC
S11_CY23_257-3	FG618869	++	+	NC	+	+	+
S11_CY23_397p-5	FG618896	NC	NC	NC	NC	+	NC
S11_CY23_376-4*	FG618889	+++	++	++	NC	+	NC
S11_CY23_257-1-5*	FG618868	+++	+	NC	NC	+	NC
S11_CY23_360-3*	FG618886	NC	NC	+	NC	++	+
Down-regulated in both interactions							
S11_CY23_Contig4	FG618673	NC	NC	NC	-	-	NC
S11_CY23_236-3	FG618859	-	-	NC	-	NC	NC
S11_CY23_231-1	FG618855	-	-	NC	-	NC	NC
S11_CY23_Contig5	FG618674	-	-	NC	NC	-	-

Table 2. (*Continued*)

Gene	Accession					
S11_CV23_Contig126	FG618794	-		NC	NC	NC
Up in incompatible, but did not change in compatible						
S11_CV23_193-4	FG618834	+	+++	NC	+	NC
S11_CV23_128-2	FG618814	-	+	NC	+	NC
S11_CV23_351-6	FG618917	++	NC	NC	NC	NC
S11_CV23_469-1	FG618920	+	+	NC	+++	NC
S11_CV23_397p-3	FG618895	+	NC	NC	NC	NC
S11_CV23_Contig99	FG618768	NC	NC	NC	+	NC
S11_CV23_148-5	FG618824	NC	+	NC	NC	NC
S11_CV23_181-5	FG618830	+	+	NC	NC	NC
Up in incompatible and down in compatible						
S11_CV23_Contig73	FG618742	NC	+	·	NC	-
S11_CV23_Contig105	FG618774	NC	+	·	+	NC
S11_CV23_12-I-3	FG618811	+	+	-	NC	-

* Strongly induced in incompatible interaction
NC No changes in expression profiling
+ Transcripts >2 folds
++ Transcripts > 2 folds and < 5 folds
+++ Transcripts >10 folds
- Transcripts were repressed

The 20 genes selected to verify the cDNA-AFLP results included 8 genes (S11_CY23_contig46, S11_CY23_contig8, S11_CY23_contig69, S11_CY23_contig112, S11_CY23_contig103, S11_CY23_contig32, S11_CY23_contig80 and S11_CY23_360-3) putatively involved in disease defense, 6 genes (S11_CY23_contig66, S11_CY23_351-6, S11_CY23_contig90, S11_CY23_274-5, S11_CY23_299-3, and S11_CY23_469-1) putatively involved in signal transduction, 2 gene (S11_CY23_376-4, S11_CY23_39-1) in the "no-hit" group, and 4 genes (S11_CY23_20-1-1b, S11_CY23_130-5, S11_CY23_contig77, and S11_CY23_contig81) in other categories. All of the genes were up-regulated and their transcripts increased as early as 12 hpi, except for TDFs S11_CY23_contig46 and S11_CY23_130-5, whose expression did not increased until 24 hpi and for S11_CY23_469-1 until 48 hpi. The accumulation of transcripts of eight genes (S11_CY23_contig8, S11_CY23_39-1, S11_CY23_376-4, S11_CY23_contig103, S11_CY23_contig32, S11_CY23_351-6, S11_CY23_274-5, and S11_CY23_ 299-3) peaked at 12 hpi with Pst, and the others peaked at 24 hpi, except that the maximum induction of S11_CY23_469-1 and S11_CY23_130-5 transcripts occurred at 48 hpi and then steadily decreased to the original levels. TDF S11_CY23_contig77 was activated as early as 12 hpi, and followed by a slight decrease, this gene reached their maximum accumulation of transcripts at 120 hpi. For all of the 20 genes, the expression patterns of the qRT-PCR were similar to those observed in the cDNA-AFLP tests. The results showed that the cDNA-AFLP technique was more efficient in identified expressed genes and also indicated that all of the studied genes were induced by the Pst infection

Discussion

Transcriptomics is a powerful approach for the global analysis of plant-pathogen interactions. Using the cDNA-AFLP technique, we observed widespread modulation of transcriptional activity, with 4.6% of all transcripts showing some form of differential expression. The gene expression patterns revealed by the cDNA-AFLP and qRT-PCR analyses were largely consistent with the physiological and biochemistry changes corresponding to the Pst infection events in the wheat leaf tissue.

About 73% (187) of the 255 differentially expressed genes in wheat were up-regulated during the infection process. Most of these genes peaked at 12-24 hpi, possibly reflecting the exploitation of cellular resources and/or the activation of defense responses [36,31]. The up-regulated genes were similar to that of a recent study reported by Coram et al. [22]. They reported that 64 genes specifically involved in the incompatible reaction between the Yr5 single gene line with a US Pst pathotype, PST-78, and these genes were up-regulated and peaked at 12-24 hpi. In these study, we identified 94 genes preferably induced in the incompatible interaction, also around 12-24 hpi. The most of the genes identified in both studies were characterized in the same functional categories. In contrast, Coram et al. identified only one gene down-regulated in the incompatible interaction [22]. Comparatively, 68 TDFs of 255 differentially expressed genes showed down-regulation in the present study. Such difference might be due to the fact that the probes of Affymetrix GeneChip were designed based on known wheat genes. Meantime, the transcriptional profiling obtained by cDNA-AFLP technique largely covered overall wheat transcriptome. Moreover, different genotypes of wheat or Pst pathotype, as well as different temperatures used in the tests, might also contribute to the difference. Given that genetic manipulation for Pst is unavailable, along with unstable wheat transformation system, the putative functions of a large number of genes identified in this study have only been predicted by bioinformatical approaches combined with altered expression patterns. Of the 255 sequenced TDFs, 113 had relatively clear functions in various categories when searching the non-redundant protein database. Thus, these genes can be valuable resources for understanding molecular changes in the incompatible interaction.

A fascinating discovery in this study is the quenching of divergent expression of Pst-regulated genes in both incompatible and compatible interactions in the middle stages of Pst infection. Similar to the results of our previous study [34], the expression of nearly all wheat genes that were differentially regulated at the early time frame returned to the levels of the mock-inoculted plants by 48 hpi. The low level of expression remained up to 120 hpi. The lack of differential gene expression at the period from 72 to 120 hpi could be because Pst might initiatively inhibit the early host responses in both incompatible and compatible interactions. Haustorium-forming fungi and oomycetes secrete many proteins from the haustoria into the extrahaustorial matrix during the parasitic stage of host infection, subsequently, a subset of proteins are further transported into the host cell [37]. Presumably, to establish infection, these proteins enable the pathogens to obtain nutrients, or to evade or manipulate host defenses [37,38]. It is thought that the oomycete Phytophthora also forms haustoria and secretes molecular signals that functions in the plant apoplast and cytoplasm to reprogram molecular host defenses [39]. The response of resistant plants at 48-72 hpi suggests that an avirulence gene is recognized by a resistance gene prior to this time, which appears to lead to a depression of host defenses. Our results may reflect that the resistance gene can be recognized by avirulence gene prior to 48 hpi in the interaction of wheat-Pst. Similar to our results, a study of the soybean response to Asian soybean rust (ASR) controlled by the Rpp2 gene also found many ASR-regulated genes responded early during the infection (6 to 36 hpi), followed by a period (24 to 72 hpi) in which expression levels returned to the mock levels and a new differentiation in gene expression during late infection (72 to 168 hpi) [40].

The development of Pst on resistant and susceptible wheat cv. Suwon 11 was found to be similar in urediniospore germination, appressorium formation, and penetration (foundation of substomatal vesicle, infection hypha, haustorial mother cell, and haustorium initial). However, after penetration (24 hpi), distinct differences in fungal spread between the compatible and incompatible interaction could be observed. In the compatible interaction, hyphae of CYR31 rapidly colonized host tissues intercellularly and numerous haustoria in the adjacent host cells were formed. In the incompatible interaction, the host cells showed hypersensitive cell death and the density of the intercellular hyphae and the number of haustoria were greatly reduced compared to the compatible interaction [24]. These results suggested that resistance to stripe rust in wheat cv. Suwon11 is executed after penetration has occurred. Yet, for host response to Pst, H_2O_2 accumulation was detected in host guard cells as early as 6-8 hpi [19], therefore, the perception of the Pst fungus by wheat and the ability of the pathogen to avoid or overcome the host's defense imply a complex, dynamic network of communication, a series of signal events should be operated before the resistance gene is expressed. Relative specifically expression of the 94 genes in the incompatible interaction and their diverse putative functions in various metabolisms support the hypothesis. However, how and when the signal is perceived by the host and transduced is still poorly understood. In this study, we focused on genes that accumulated preferentially in the incompatible reaction before 48 hpi. Dissection of these genes and their involved biochemical pathways in the future studies might provide answers to the questions.

Because the regulation of gene expression is a dynamic process, the expression profiling was presented over a time course by cDNA-AFLP, which allowed us to study the dynamic behavior of gene expression and characterize their changes over time. The induction and signal transduction of defense responses specific to the interaction require up- or down- regulation of many genes. We were primarily interested in genes whose expression might be used to distinguish incompatible from compatible interactions in wheat. A different analysis was conducted to achieve the goal by comparing gene expressions in Suwon 11 challenged with CYR23 (avirulent) or CYR31 (virulent). The comparison analysis of 255 TDFs in the incompatible interaction with those in the compatible interaction as previously described [34] showed that of the 255 transcripts induced during the incompatible interaction, 161 TDFs (63%) were also induced during the compatible interaction, and thus were classified as basal defense-related. 94 TDFs were expressed preferably in the incompatible interaction. The large proportion of TDFs were shared in both interactions, these results were similar to the reports of Coram et al. [22] with the same pathosystem. Coram et al. [22] reported 51 genes commonly induced in both incompatible and compatible interactions between wheat and Pst. Tao et al. [41] also found that plant responses in compatible and incompatible interactions are qualitatively similar but quantitatively different soon after infection. Another study of the barley response to powdery mildew controlled by the Mla6, Mla13 and Mla1 single resistance genes also provided evidence for a shared response between compatible and incompatible interactions up to the point of pathogen penetration [42].

Of our special interest is that 11 of 94 TDFs were up-regulated in the incompatible interaction, but did not change or were repressed in compatible interaction through the qRT-PCR validation. Of the 11 TDFs, 6 (S11_CY23_193-4, S11_CY23_128-2, S11_CY23_181-5, S11_CY23_Contig73, S11_CY23_Contig105 and S11_CY23_12-1-3) have unknown functions and 5 (S11_CY23_397p-3, S11_CY23_469-1, S11_CY23_351-6, S11_CY23_Contig99 and S11_CY23_148-5) encode Leucine Rich Repeat family protein, CBL-interacting protein kinase, Serine/threonine Kinase, ethylene-responsive RNA helicase and protein phosphatase type 2C, respectively. Protein phosphatase type 2C is a negative regulator of ABA responses. Gosti et al. [43] reported that suppressor mutants were more sensitive to applied ABA than the wild type and displayed increases in seed dormancy, whole-plant drought tolerance, and drought rhizogenesis intensity. However, ABA is required for plant defense. Adie et al. [44] measured ABA hormone levels in wild-type and JA/ET/SA/ABA-related mutants after Pythium irregulare infection to determine whether ABA is required for overall plant resistance. They found that ABA mutants showed an increased susceptibility to P. irregulare compared with the wild-type background, indicating that ABA is a positive signal involved in the activation of effective defenses against this pathogen. However, several reports showed that ABA increases susceptibility by counteracting SA-dependent defenses, and ABA-dependent priming of callose biosynthesis promotes enhanced resistance to some pathogens [45]. These results supported that ABA should have a negative effect on resistance. Our results also indicated that ABA should be expected to play a negative role in response to Pst. DEAD-box RNA helicases had been reported to play an important role during development and stress responses in various organisms [46-48]. Rice OsBIRH1 encoding DEAD box RNA helicase was shown to function in defense responses against pathogen and oxidative stresses [46]. STRS1 (Stress response suppressor 1) and STRS2 encoding DEAD box RNA helicases were shown to function as negative regulators of ABA-dependent and ABA-independent signaling networks [48]. Zegzouti [49] isolated an Ethylene-responsive 68 (ER68, corresponding to Arabidopsis thaliana RNA helicase 20), their results indicated the potential for ER68 RNA helicase activity to be involved with ethylene-regulated gene expression at either the transcriptional or post-transcriptional level. Similar to their results, S11_CY23_contig99 encoding ethylene-responsive RNA helicase, was only induced in incompatible interaction, which should attribute to resistance to Pst, however, the precise role played by S11_CY23_contig99 still needs to be further elucidated.

Caffeoyl-CoA O-methyltransferases (CCoAOMTs) is an important enzyme that participates in lignin biosynthesis especially in the formation of cell wall ferulic esters of plants. CCoAOMT was proposed to play a pivotal role in cell wall reinforcement during the induced disease resistance response. Lignin is often deposited at the sites of wounding or pathogen invasion, which may provide a physical barrier for protection of adjacent tissues from further damage. In the previous study [24], immunogold localization of lignin revealed a markedly higher labeling density in host cell walls of the infected wheat leaves of the resistant cultivar than in cell walls of the infected wheat leaves of the susceptible cultivar. In this study, S11_CY23_360-3 encoding CCoAOMT, induced at 24 hpi, and peaked at 48 hpi in the incompatible interaction, the transcripts accumulations occurred after the resistance gene was triggered, which indicated that lignification appears to be also an active resistance mechanism in the wheat-Pst panthosystem.

Suwon 11 showed HR to CY23 infection. Plant cells involved in the HR generate an oxidative burst by producing reactive oxygen species (ROS), superoxide anions, hydrogen peroxide, hydroxyl radicals, and nitrous oxide [50]. Peroxidases were thought to play an important role in ROS production [51,52]. In the present study, two genes (S11_CY23_contig32 and S11_CY23_ contig46) were predicted to encode peroxidase and peroxisomal membrane protein, respectively. Their transcripts peaked as early as 12 and 24 hpi, respectively. Our gene expression data were coincident with the previous report of a rapid increase of $O2^{--}$ and $H2O2$ at infection sites and a strong accumulation of $H2O2$ in mesophyll cells from 12-24 hpi using histochemical staining in the Suwon 11 leaves inoculated by CYR23 [19]. In contrast, $O2-$ and $H2O2$ could not be detected in most of the infection sites in the compatible interaction between Suwon 11 and CYR31. In the present study, we found that the expression of these two genes were much less in the compatible interaction than in the incompatible interaction.

Several studies have provided evidence supporting that the PR-5 protein plays an active role in resistance mechanisms in cereals [53-56]. Transcripts of four PR protein genes were analyzed during Fusarium graminearum infection, with PR-5 transcript accumulated as early as 6 to 12 hpi and peaked at 48 hpi [56]. We also found S11_CY23_112 homologous to a wheat PR-5-like protein gene. S11_CY23_112 transcript accumulated strongly at 24-48 hpi. A stronger induction of this gene could be observed in the incompatible interaction than in the compatible interaction, suggesting a general role of this protein in wheat resistance to stripe rust. Similarly, gene S11_CY23_274-5 from wheat was deduced to encode a receptor related to antifungal PR proteins. The predicted protein contained an extracellular domain related to the PR 5 protein, a central transmembrane spanning domain, and an intracellular protein serine/threonine kinase. Wang et al. isolated a PR5K gene from Arabidopsis thaliana and found that PR5K transcript accumulated at low levels in all tissues examined [57]. They suggested a possible interaction of PR5K with common or related microbial targets. Nevertheless, the interrelation of PR5K and PR5 protein during the interaction between wheat and Pst need further studies.

Protein kinase is known to play a central role in signaling during pathogen recognition and the subsequent activation of plant defense mechanisms [58]. We identified six TDFs encoding different protein kinases. S11_CY23_contig90 was highly homologous to the Arabidopsis MKP1 gene, which was showed to be induced at the transcriptional level during the interaction of wheat-Pst. The Arabidopsis genome contains 20 genes encoding mitogen-activated protein kinases (MAPKs), which interact with MKP1 [59]. Using expression profiling, a specific group of genes that probably represent targets of MKP1 regulation was also identified [59]. Surprisingly, the identity of these genes and interacting MAPKs suggested involvement of MKP1 in salt stress responses. Indeed, mkp1 plants have increased resistance to salinity [60]. Accordingly, the gene S11_CY23_contig90 may play a role in the integration and fine-tuning of plant responses to stripe rust pathogen challenging.

This study uncovered a number of new candidate genes possibly involved in the interactions of wheat and Pst. More than 42% of the sequenced TDFs had no homologous sequences in the EST databases. De Torres et al. [61] reported that the plant response to pathogen challenge evoked a large number of transcriptomic components not yet present in EST libraries. Over 60% of the differentially regulated transcripts in their cDNA-AFLP were absent from standard 8,200 feature Affymetrix Gene Chips [61]. Therefore, cDNA-AFLP analysis is a suitable tool for discovering new potential genes that are differentially expressed during the wheat-Pst interactions. Because most of the molecular mechanisms involved in the pathosystem interactions are yet to be determined, the large number of TDFs identified in this study will serve as candidates for further studies to determine their functions and dissect the molecular networks involved in the plant-pathogen interactions.

Conclusion

In this study, we have obtained a broad overview of the behaviour of the wheat transcriptomes to the stripe rust fungus and this has provided many interesting clues to the interaction of the wheat and Pst. We have also seen the differences between the incompatible (resistant) and compatible (susceptible) interaction. With regard to many genes, we observed patterns of transcript accumulation that reflect those observed by other groups in their studies. However, as a consequence of this study, we have also made observations that give additional insight in the complexities of the interactions that occur in both interactions, and uncovered a number of new candidate genes possibly involved in the interactions of wheat and Pst. Especially, 11 TDFs expressed specifically in the incompatible interaction were observed, and these genes should play important roles in the interaction of wheat and Pst. However, how and when they function in the infection process, need to be further studied.

Methods

Plant Materials and Inoculation

Wheat genotype Suwon 11 and Pst pathotype CYR23 were used for the cDNA-AFLP analysis of an incompatible interaction. For the qRT-PCR analysis of candidate genes, pathotype CYR31 was used with Suwon 11 to form a compatible interaction as in the previous study [34]. Suwon 11 contains YrSu that provides seedling resistance against stripe rust, [62,63] showing a typical HR when being challenged by CYR23. Plants were grown, inoculated and maintained following the procedures and conditions described by Kang & Li [64]. Control plants corresponding to each time point were brushed with sterile water, referred to as mock inoculation. Samples of the mock inoculated leaves taken just after water-inoculation (near 0 hpi) were treated as the initial control and those from mock-inoculated leaves at each time point served as the control for that time point. Leaf tissues were harvested at 6, 12, 18, 24, 36, 48, 72, 96, 120, 144 and 168 h post-inoculation (hpi) and quickly frozen in liquid nitrogen and stored at -80°C prior to total RNA extraction. The time points were selected based on the microscopic study of the incompatible interaction between Suwon 11 and pathotype CYR23 or CYR31 [19]. Plants were rated for symptom development 15 days after inoculation. Three biological replications were performed independently for each time point.

Histological Observation of Stripe Rust Fungus

To ensure the suitability of leaf samples for incompatible and compatible interactions, detection of stripe rust fungus infection process in different interactions was carried out using the Calcofluor White (Sigma Co., USA.) staining method as described by Kang et al. [65].

For microscopy, staining hyphae and autofluorescence of attacked mesophyll cells was observed using a fluorescence microscopy (excitation filter 485 nm, dichromic mirror 510 nm, barrier filter 520 nm). All microscopical examinations were done with an Olympus BX-51 microscope (Olympus Corporation, Japan). At least 50 penetration sites on each of four leaf specimens per treatment were scored.

cDNA-AFLP Analysis and TDFs Isolation

Total RNA was isolated from about 200 mg of the frozen wheat leaves using the Trizol™reagent (Invitrogen, Carlsbad, CA, USA) according to the manufacture's protocol. Twenty micrograms of total RNA was used initially for the first strand synthesis, followed by the second strand synthesis using the SMART™ PCR cDNA Synthesis Kit (Clontech, Mountain View, CA, USA) following the manufacturer's instruction. About 100 ng of double-stranded cDNA was subjected to standard AFLP template production. cDNA-AFLP analysis was performed with 64 primer combinations using the IRDye® Fluorescent 800 AFLP expression analysis kit (LI-COR) and TDFs were isolated with protocols as previously described [34].

Sequence Analysis

In order to efficiently analyze the large-scale EST data, a local stand-alone EST analysis platform was set up with the Linux operation system described by Wang et al. [34]. In addition, the wheat stem rust pathogen (P. graminis f. sp. tritici) whole genome database, http://www.broad.mit.edu/annotation/genome/puccinia_graminis, was used for the BLASTN searching to determine possible Pst genes.

In order to compare TDFs between the incompatible and compatible reactions, all 186 uniseqs from the previous compatible interaction [34] were formatted to form one local compatible database using the command of "formatdb" in the BLAST system. Then we did the TBLASTN seaching against local compatible database for all 255 uniseqs identified in the present study from the incompatible interaction as query sequences using an E-value of 1e-20 as the high stringent cut-off point.

Origin Confirmation for Genes Homologous to the Stem Rust Pathogen

Genomic DNA was extracted from the urediniospores of CYR23 and leaves of Suwon 11 according to the protocol described by Wang et al. [66]. Specific primers for possible Pst genes were designed using the software of Primer Premier 5.0.

Standard PCR amplification was performed separately with genomic DNA from the urediniospores and wheat leaves as templates. The PCR products were run on an agarose gel along with a molecular size marker (DL2000, TaKaRa Biotechnology Co., Ltd), followed by sequencing analyses of all amplified bands.

qRT-PCR and Data Analyses

Leaf tissues challenged by CYR23 or CYR31 were sampled at 12, 18, 24, 48, 72 and 120 hpi, as well as samplings of the control plants mock-inoculated with sterile water at each of the corresponding time points, and near 0 hpi for the mock-inoculated plants. Three independent biological replications were performed for both inoculated and control plants. Primer design, reverse transcription, and qRT-PCR reaction were conducted as described in our previous study [34]. To standardize the data, the amount of target gene transcript was normalized over the constitutive abundance of wheat 18S rRNA (GenBank accession no. AY049040). All reactions were performed in triplicate, including three non-template as the negative controls. Quantification of gene expression was performed using a 7500 Real-Time PCR System (Applied Biosystems, Forst City, CA). Dissociation curves were generated for each reaction to ensure specific amplification. Threshold values (CT) generated from the ABI PRISM 7500 Software Tool (Applied Biosystems, Foster City, CA, USA) were employed to quantify relative gene expression using the comparative 2-ΔΔCT method [67].

Authors' Contributions

XJW: designed experiments, analyzed data and wrote manuscript. WL: conducted qRT-PCR and collected and analyzed data. CLT: analyzed data and prepared manuscript. YLD: provided assistance in various experiments. JBM: conducted bioinformatical analysis. XLH and GRW: prepared samples and collected data. LLH: coordinated the experiments and data analyses. XMC: provided advices for experiments and revised manuscript. ZSK: conceived the project, designed the experiments and wrote manuscript. All authors read and approved the final manuscript.

Acknowledgements

This study was supported by grants from the National 863 Research Program (2006AA10A104), the earmarked fund for Modern Agro-industry Technology Research System, Nature Science Foundation of China (No. 30671350), the

Program for Changjiang Scholars and Innovative Research Team in Universities, Ministry of Education of China (No.200558) and the 111 Project from the Ministry of Education of China (B07049).

References

1. Yang YN, Shah J, Klessig DF: Signal perception and transduction in plant defense responses. Genes & Development 1997, 11:1621–1639.

2. Ebel J, Cosio EG: Elicitors of plant defense responses. International review of cytology 1994, 148:1–36.

3. Dixon RA, Harrison MJ, Lamb CJ: Early events in the activation of plant defense responses. Annual Review of Phytopathology 1994, 32:479–501.

4. Jeffery LD, Jones JD: Plant pathogens and integrated defense responses to infection. Nature 2001, 411:826–833.

5. Zhu Q, Dröge-Laser W, Dixon RA, Lamb C: Transcriptional activation of plant defense genes. Current Opinion in Genetics & Development 1996, 6:624–630.

6. Durner J, Shah J, Klessig DF: Salicylic acid and disease resistance in plants. Trends in Plant Science 1997, 2:266–274.

7. Hammond-Kosack KE, Jones JD: Resistance gene-dependent plant defense responses. Plant Cell 1996, 8:1773–1791.

8. Goodman RN, Novacky AJ: The hypersensitive reaction in plants to pathogens. A resistance phenomenon. American Phytopathological Society Press, St. Paul MN; 1994.

9. Dangl JL, Dietrich RA, Richberg MH: Death don't have no mercy: Cell death programs in plant-microbe interactions. Plant Cell 1996, 8:793–1807.

10. Greenberg JT: Programmed cell death: A way of life for plants. Proc Natl Acad Sci USA 1996, 93:12094–12097.

11. Ryals JA, Neuenschwander UH, Willits MG, Molina A, Steiner HY, Hunt MD: Systemic acquired resistance. Plant Cell 1996, 8:1809–1819.

12. Delaney TP: Genetic dissection of acquired resistance to disease. Plant Physiology 1997, 113:5–12.

13. Wan AM, Chen XM, He ZH: Wheat stripe rust in China. Australian Journal of Agricultural Research 2007, 58:605–619.

14. Zeng SM, Luo Y: Systems analysis of wheat stripe rust epidemics in China. European Journal of Plant Pathology 2008, 121:425–438.

15. Zheng WM, Liu F, Kang ZS, Chen SY, Li ZQ, Wu LR: AFLP analysis of pre-dominant races of Puccinia striiformis in china. Progress in Natural Science 2000, 10:532–537.

16. Xu SC, Zhang JY, Zhao WS, Wu LR, Zhang JX, Yuan ZD: Genetic analysis of major and minor gene(s) resistant to stripe rust in important resource wheat line Jinghe891-1. Agricultural Sciences in China 2002, 1:364–369.

17. Yin XG, Shang XW, Pang BS, Song JR, Cao SQ, Li JC, Zhang XY: Molecular mapping of two novel stripe rust resistant genes YrTp1 and YrTp2 in A-3 de-rived from Triticum aestivum × Thinopyrum ponticum. Agricultural Sciences in China 2006, 5:483–490.

18. Liu FH, Niu YC, Deng H, Tan GJ: Mapping of a major stripe rust resistance gene in Chinese native wheat variety chike using Microsatellite Markers. Jour-nal of Genetics and Genomics 2007, 34:1123–1130.

19. Wang CF, Huang LL, Buchenauer H, Han QM, Zhang HC, Kang ZS: His-tochemical studies on the accumulation of reactive oxygen species (O_2—and H_2O_2) in the incompatible and compatible interaction of wheat-Puccinia striiformis f. sp. tritici. Physiological and Molecular Plant Pathology 2007, 71:230–239.

20. Ling P, Wang MN, Chen XM, Kimberly GC: Construction and characteriza-tion of a full-length cDNA library for the wheat stripe rust pathogen (Puccinia striiformis f. sp. tritici). BMC Genomics 2007, 8:145.

21. Coram TE, Matthew L, Settles, Chen XM: Transcriptome analysis of high-temperature adult-plant resistance conditioned by Yr39 during the wheat-Puccinia striiformis f. sp. tritici interaction. Molecular Plant Pathology 2008, 9:479–493.

22. Coram TE, Wang MN, Chen XM: Transcriptome analysis of the wheat-Puccinia striiformis f. sp. tritici interaction. Molecular Plant Pathology 2008, 9:157–169.

23. Zhang YH, Qu ZP, Zheng WM, Liu B, Wang XJ, Xue XD, Xu LS, Huang LL, Han QM, Zhao J, Kang ZS: Stage-specific gene expression during uredinio-spore germination in Puccinia striiformis f. sp tritici. BMC Genomics 2008, 9:203.

24. Kang ZS, Huang L, Buchenauer H: Ultrastructural changes and localization of lignin and callose in compatible and incompatible interactions between wheat and Puccinia striiformis. Journal Plant Disease Protection 2002, 109:25–37.

25. Kang ZS, Wang Y, Huang LL, Wei GR, Zhao J: Histology and ultrastructure of incompatible combination between Puccinia striiformis and wheat with low reaction type resistance. Agricultural Sciences in China 2003, 2:1102–1113.

26. Fasters MK, Daniels U, Moerschbacher BM: A simple and reliable method for growing the wheat stem rust fungus, Puccinia graminis f. sp. tritici, in liquid culture. Physiological and Molecular Plant Pathology 1993, 42:259–265.

27. Blumwald E, Aharon GS, Lam BC: Early signal transduction pathways in plant-pathogen interactions. Trends in Plant Science 1998, 3:342–346.

28. Coram TE, Settles ML, Chen XM: Large-scale analysis of antisense transcription in wheat using the Affymetrix GeneChip wheat genome array. BMC Genomics 2009, 10:253.

29. Hulbert SH, Bai J, Fellers JP, Pacheco MP, Bowden RL: Gene expression patterns in near isogenic lines for wheat rust resistance gene Lr34/Yr18. Phytopathology 2007, 97:1083–1093.

30. Durrant WE, Rowland O, Piedras P, Hammond-Kosack KE, Jones JD: cDNA-AFLP reveals a striking overlap in race-specific resistance and wound response gene expression profiles. Plant Cell 2000, 12:963–977.

31. Polesani M, Desario F, Ferrarini A, Zamboni A, Pezzotti M, Kortekamp A, Polverari A: cDNA-AFLP analysis of plant and pathogen genes expressed in grapevine infected with Plasmopara viticola. BMC Genomics 2008, 9:142.

32. Zhang L, Meakin H, Dickinson M: Isolation of genes expressed during compatible interactions between leaf rust (Puccinia triticina) and wheat using cDNA-AFLP. Molecular Plant Pathology 2003, 4:469–477.

33. Bachem CW, Hoeven RS, de Bruijn SM, Vreugdenhil D, Zabeau M, Visser RG: Visualization of differential gene expression using a novel method of RNA fingerprinting based on AFLP: analysis of gene expression during potato tuber development. Plant Journal 1996, 9:745–753.

34. Wang XJ, Tang CL, Zhang G, Li YC, Wang CF, Liu B, Qu ZP, Zhao J, Han QM, Huang LL, Chen XM, Kang ZS: cDNA-AFLP analysis reveals differential gene expression in compatible interaction of wheat challenged with Puccinia striiformis f. sp. tritici. BMC Genomics 2009, 10:289.

35. Bevan M, Bancroft I, Bent E: Analysis of 1.9 Mb of contiguous sequence from chromosome 4 of Arabidopsis thaliana. Nature 1998, 391:485–493.

36. Grenville-Briggs LJ, van West P: The biotrophic stages of oomycete-plant interactions. Advances in Applied Microbiology 2005, 57:217–243.

37. Catanzariti AM, Dodds PN, Ellis JG: Avirulence proteins from haustoria-forming pathogens. FEMS microbiology letters 2007, 269:181–188.

38. Voegele RT, Mendgen K: Rust haustoria: nutrient uptake and beyond. New Phytologist 2003, 159:93–100.

39. Birch PR, Rehmany AP, Pritchard L, Kamoun S, Beynon JL: Trafficking arms: Oomycete effectors enter host plant cells. Trends in Microbiology 2006, 14:8–11.

40. Mortel M, Recknor J, Graham M, Nettleton D, Dittman J, Nelson R, Godoy C, Abdelnoor R, Almeida Á, Baum T, Whitham S: Distinct biphasic mRNA changes in response to asian soybean rust infection. Molecular Plant-Microbe Interaction 2007, 20:887–899.

41. Tao Y, Xie ZY, Chen WQ, Glazebrook J, Chang HS, Han B, Zhu T, Zou GZ, Katagiri F: Quantitative nature of Arabidopsis responses during compatible and incompatible interactions with the bacterial pathogen Pseudomonas syringae. The Plant Cell 2003, 15:317–330.

42. Caldo RA, Nettleton D, Wise RP: Interaction-dependent gene expression in Mla-specified response to barley powdery mildew. Plant Cell 2004, 16:2514–2528.

43. Gosti F, Beaudoin N, Serizet C, Webb AAR, Vartanian N, Giraudat J: ABI1 Protein Phosphatase 2C is a negative regulator of Abscisic Acid signaling. The Plant Cell 1999, 11:1897–1910.

44. Adie BAT, Pérez-Pérez J, Pérez-Pérez MM, Godoy M, Sánchez-Serrano JJ, Schmelz EA, Solano R: ABA is an essential signal for plant resistance to pathogens affecting JA biosynthesis and the activation of defenses in Arabidopsis. The Plant Cell 2007, 19:1665–1681.

45. Ton J, Mauch-Mani B: Beta-amino-butyric acid-induced resistance against necrotrophic pathogens is based on ABA-dependent priming for callose. Plant Journal 2004, 38:119–130.

46. Li D, Zhang H, Wang X, Song F: OsBIRH1, a DEAD-box RNA helicase with functions in modulating defence responses against pathogen infection and oxidative stress. Journal of Experimental Botany 2008, 59:2133–2146.

47. Gong ZZ, Dong CH, Lee HJ, Zhu JH, Xiong LM, Gong DM, Stevenson B, Zhu JK: A DEAD box RNA helicase is essential for mRNA export and important for development and stress responses in Arabidopsis. Plant Cell 2005, 17:256–267.

48. Kant P, Kant S, Gordon M, Shaked R, Barak S: STRESS RESPONSE SUPPRESSOR1 and STRESS RESPONSE SUPPRESSOR2, two DEAD box RNA helicases that attenuate Arabidopsis responses to multiple abiotic stresses. Plant Physiologist 2007, 145:814–830.

49. Zegzouti H, Jones B, Frasse P, Marty C, Maitre B, Latch A, Pech JC, Bouzayen M: Ethylene-regulated gene expression in tomato fruit: characterization

of novel ethylene-responsive and ripening-related genes isolated by differential display. Plant Journal 1999, 18:589–600.

50. Heath MC: Hypersensitive response-related death. Plant Molecular Biology 2000, 44:323–334.

51. Bestwick CS, Brown IR, Mansfield JW: Localized changes in peroxidase activity accompany hydrogen peroxide generation during the development of a nonhost hypersensitive reaction in lettuce. Plant Physioloy 1998, 118:1067–1078.

52. Do HM, Hong JK, Jung HW, Kim SH, Ham JH, Hwang BK: Expression of peroxidase-like genes, H2O2 production, and peroxidase activity during the hypersensitive response to Xanthomonas campestris pv. Vesicatoria in Capsicum annuum. Molecular Plant-Microbe Interactions 2003, 16:196–205.

53. Rebman G, Mauch F, Dudler R: Sequence of a wheat cDNA encoding a pathogen-induced thaumatin-like protein. Plant Molecular Biology 1991, 17:283–285.

54. Lin KC, Bushnell WR, Szabo LJ, Smith AG: I solation and expression of a host response gene family encoding thaumatin-like proteins in incompatible oat-stem rust fungus interactions. Molecular Plant-Microbe Interactions 1996, 9:511–522.

55. Schaffrath U, Freydl E, Dudler R: Evidence for different signaling pathways activated by inducers of acquired resistance in wheat. Molecular Plant-Microbe Interactions 1997, 10:779–783.

56. Pritsch C, Muehlbauer GJ, Bushnell WR, Somers DA, Vance CP: Fungal development and induction of defense response genes during early infection of wheat spikes by Fusarium graminearum. Molecular Plant-Microbe Interactions 2000, 13:159–169.

57. Wang X, Zafian P, Choudhary M, Lawton M: The PR5K receptor protein kinase from Arabidopsis thaliana is structurally related to a family of plant defense proteins. Proc Natl Acad Sci USA 1996, 93:2598–2602.

58. Romeis T: Protein kinases in the plant defence response. Current Opinion in Plant Biology 2001, 4:407–414.

59. Ulm R, Revenkova E, di Sansebastiano GP, Bechtold N, Paszkowski J: Mitogen-activated protein kinase phosphatase is required for genotoxic stress relief in Arabidopsis. Genes & Development 2001, 15:699–709.

60. Ulm R, Ichimura K, Mizoguchi T, Peck SC, Zhu T, Wang X, Shinozaki K, Paszkowski J: Distinct regulation of salinity and genotoxic stress responses by

Arabidopsis MAP kinase phosphatase 1. The EMBO Journal 2002, 21:6483–6493.

61. De Torres M, Sanchez P, Fernandez-Delmond I, Grant M: Expression profiling of the host response to bacterial infection: The transition from basal to induced defense responses in RPM1-mediated resistance. Plant Journal 2003, 33:665–676.

62. Cao ZJ, Jing JX, Wang MN: Relation analysis of stripe rust resistance gene in wheat important cultivar suwon11, suwon92 and hybrid 46. Acta Bot Boreal-Occident Sin 2003, 23:64–68.

63. Li ZF, Xia XC, Zhou XC, Niu YC, He ZH, Zhang Y, Li GQ, Wan AM, Wang DS, Chen XM, Lu QL, Singh RP: Seedling and slow rusting resistance to stripe rust in Chinese common wheat. Plant Disease 2006, 90:1302–1312.

64. Kang ZS, Li ZQ: Discovery of a normal T. type new pathogenic strain to Lovrin10. Acta Cllegii Septentrionali Occidentali Agriculturae 1984, 4:18–28.

65. Kang ZS, Shang HS, Li ZQ: The technique of tissues fluorescence staining of wheat and stripe rust fungus. Plant protection 1993, 19:27.

66. Wang XJ, Zheng WM, Buchenauer H, Zhao J, Han QH, Huang LL, Kang ZS: Development of a PCR-based detection of Puccinia striiformis in latent infected wheat leaves. European Journal of Plant Pathology 2008, 120:241–247.

67. Livak KJ, Schmittgen TD: Analysis of relative gene expression data using real-time quantitative PCR and the 2-ΔΔCT method. Methods 2001, 25:402–408.

CITATION

Generation and Analysis of Expression Sequence Tags from Haustoria of the Wheat Stripe Rust Fungus *Puccinia striiformis f. sp. Tritici*

Chuntao Yin, Xianming Chen, Xiaojie Wang, Qingmei Han, Zhensheng Kang and Scot H. Hulbert

ABSTRACT

Background

Stripe rust, caused by Puccinia striiformis f. sp. tritici (Pst), is one of the most destructive diseases of wheat (Triticum aestivum L.) worldwide. In spite of its agricultural importance, the genomics and genetics of the pathogen are poorly characterized. Pst transcripts from urediniospores and germinated urediniospores have been examined previously, but little is known about genes expressed during host infection. Some genes involved in virulence in other rust fungi have been found to be specifically expressed in haustoria. Therefore, the objective of this study was to generate a cDNA library to characterize genes expressed in haustoria of Pst.

Results

A total of 5,126 EST sequences of high quality were generated from haustoria of Pst, from which 287 contigs and 847 singletons were derived. Approximately 10% and 26% of the 1,134 unique sequences were homologous to proteins with known functions and hypothetical proteins, respectively. The remaining 64% of the unique sequences had no significant similarities in Gen-Bank. Fifteen genes were predicted to be proteins secreted from Pst haustoria. Analysis of ten genes, including six secreted protein genes, using quantitative RT-PCR revealed changes in transcript levels in different developmental and infection stages of the pathogen.

Conclusions

The haustorial cDNA library was useful in identifying genes of the stripe rust fungus expressed during the infection process. From the library, we identified 15 genes encoding putative secreted proteins and six genes induced during the infection process. These genes are candidates for further studies to determine their functions in wheat-Pst interactions.

Background

Rust fungi are a large group of obligately biotrophic basidiomycete fungi that completely depend on their living host tissue for growth and reproduction. Wheat (Triticum aestivum L.) is a host to three different rust fungi, causing stripe (yellow), leaf (brown) and stem (black) rust. Wheat stripe rust, caused by Puccinia striiformis Westend. f. sp. tritici Eriks. (Pst), is a serious problem in all major wheat growing countries [1,2]. In the United States, the disease is most destructive in the western United States and has become increasingly important in the south-central and south-eastern states [1,3,4]. Unlike the stem rust (P. graminis f. sp. tritici) and leaf rust (P. triticina) fungi, Pst does not have a known alternate host to complete the sexual cycle. During infection, urediniospores of Pst germinate on wheat leaf surfaces to produce germ tubes. Depending upon the isolate, Pst forms noticeable or unnoticeable appressoria [5,6], from which an infection peg is formed and penetrates a leaf stoma, followed by infection hyphae that form haustorial mother cells, and a specialized infection structure called the haustorium forms and an intimate feeding relationship is established. Haustoria are essential for rust fungi to take nutrients from their host [7-10] and have also been shown to be involved in vitamin synthesis [11].

Plant disease resistance relies on the recognition of pathogen avirulence (Avr) gene products by host resistance (R) genes through either direct (receptor-ligand model) or indirect (guard model) association, which induces defense responses. Haustoria play an essential role in the reactions of plants with rust fungi. For example, four avirulence genes from Melampsora lini, the flax rust pathogen, have been cloned and found to encode small secreted proteins expressed in the fungal haustoria [12,13]. A large number of plant-induced and haustorium-specific genes have been identified in the bean rust fungus Uromyces fabae [14,15]. To date, there are no reports of cloning and molecular characterization of either virulence or avirulence genes from any of the cereal rust pathogens.

The stripe rust fungus lacks several features to be an ideal model system for genetic analysis. It does not have a known alternate host for completing the sexual cycle. Like the other cereal rusts, Pst is very difficult to culture in vitro and stable transformation systems are yet to be developed. While molecular and genetic approaches are currently lacking, some advances are being made in genomics. Recently Ling et al. [16] constructed a full-length cDNA library of Pst from RNA extracted from urediniospores and identified some genes encoding protein products that maybe involved in virulence or infection. Some genes highly expressed in germinated urediniospores of the fungus were also reported [17]. Our understanding of the molecular mechanisms underlying infection and development within host tissue is still very limited. A haustorium is a hub of cellular communication between the host and the pathogen for the establishment of a biotrophic relationship. To gain some insights into haustorium-related functions and investigate Pst virulence mechanisms, we constructed a Pst haustorial cDNA library based on a protocol for the preparative isolation of haustoria from rust-infected leaves [18] and searched for expressed sequence tags (ESTs) encoding putative secreted proteins. More than 5,000 ESTs from the haustorial cDNA library were generated. Fifteen unique sequences were predicted to encode proteins secreted from haustoria. Quantitative real-time PCR (qRT-PCR) studies revealed that some cDNAs were specifically expressed in planta.

Results

Construction of a Haustorial cDNA Library

Stripe rust haustoria were isolated from heavily infected wheat leaves. Total RNA was extracted from haustoria of race PST-78 of Pst and a cDNA library was constructed with the pDNR-LIB vector. Most of the cloned cDNA inserts in this library were between 300-1,500 bp in size. A total of 6,000 random cDNA clones were sequenced from the 5' end, from which 5,126 high quality ESTs were obtained. While the sequencing reactions covered the full inserts of many of the smaller clones, 687 of the clones with larger inserts were also sequenced from the 3' end.

EST Sequence Analysis

The EST sequences were subjected to BLAST searches (described in methods). Of the 5,126 sequences, 1,420 sequences were found to be likely of plant origin as indicated by significant BLAST scores (E value ≤ 10-5) to plant sequences but little or no homology to other organisms. After removing contaminating plant sequences, 3,706 sequences were assembled into 1,134 unique sequences, of which 847 were singletons and 287 were contigs represented by multiple clones at frequencies ranging from 2 to 873. A majority of the contigs contained two, three or four sequences. The frequency of redundant ESTs was shown in Figure 1. The average G+C content of these unique sequences was 43.11%, which was similar to the G+C content of ESTs in P. graminis and Pst germinated urediniospores [17,19]. The sequences were deposited in the NCBI dbEST sequence database (Accession numbers GH737012–GH738498).

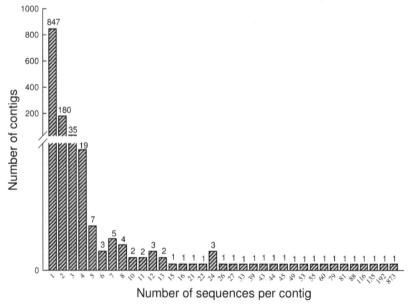

Figure 1. Frequency distribution of sequences from the haustorial cDNA library of Puccinia striiformis f. sp. tritici belonging to the same contig or unisequence. A total of 1,134 unisequences were used in this analysis.

All unique sequences were used in homology searches of the NCBI non-redundant protein sequences and the P. graminis genome database using the BLASTX and BLASTn algorithm. Unique sequences with significant homology (E ≤ 10-5) to known proteins were grouped according to their putative functions. Of the 1,134 unique sequences, only 109 (10%) showed significant similarities to proteins of known function, 296 (26%) showed significant similarities to predicted proteins of unknown function and 729 (64%) showed no significant similarity to a database entry. Based on the examination of the significant sequence similarity to a database entry, a putative functional category was assigned to the specific unisequence. The majority of the genes were predicted to code for proteins of unknown function (Figure 2). The largest group of genes with known functions showed similarities to ribosomal proteins, followed by the group of genes with similarities to proteins involved in primary metabolism and energy production. Some unisequences matched ESTs from Pst urediniospores [16] and germinated urediniospores [17] deposited in the Genbank. Some had high homology to ESTs from a haustorium-specific cDNA library of U. fabae [15].

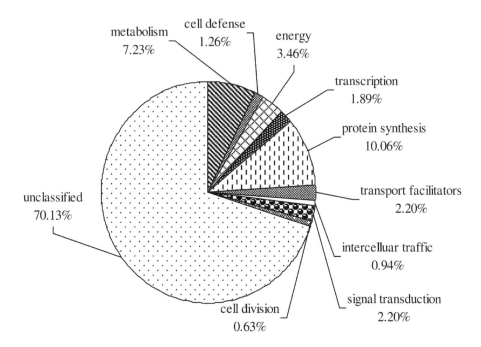

Figure 2. Functional classification of the Puccinia striiformis f. sp. tritici unisequence from the haustorial cDNA library showing significant similarities to proteins in public databases

Prediction of ESTs Encoding Secreted Proteins

To identify putative secreted proteins from the haustorial cDNA library, we selected unisequences that appeared to code for full-length open reading frames and predicted their translation products. Predicted proteins from open reading frames with in-frame stop codons before the start codon were analyzed with the signal P 3.0 algorithm [20] and iPSORT [21]. The analysis identified 15 unisequences encoding proteins with secretion signal peptides at the N-terminus (Table 1). These sequences were predicted to encode proteins ranging in size from 56-289 amino acids. Seven of them did not show significant homology to known protein sequences, seven matched predicted proteins of unknown function and one had significant homology to sulfate transporters. Eight encoded Cys-rich proteins with more than 5% cystein residues in the predicted protein.

Table 1. Genes from the haustorial cDNA library of Puccinia striiformis f. sp. tritici predicted to encode secreted proteins

Unisequence	GenBank accession	Size (aa)	No. of Cysteine residues	Homology in databases	E value
PSTha2a5	GH737102	117	6	predicted protein of *Puccinia graminis*	5.23E-23
PSTha9F18	GH737274	259	14	hypothetical protein of *Puccinia graminis*	8.96E-41
PSTha12a4	GH737444	289	10	predicted protein of *Puccinia graminis*	4.00E-23
PSTha12j12	GH737467	133	6	no homology	-
PSTha15N21	GH737567	98	7	no homology	-
PSTha21O8	GH737139	92	1	putative sulfate transporter	1.42E-38
PSTha5a23	GH737046	108	4	no homology	-
PSTha6i16	GH737950	73	5	predicted protein of *Puccinia graminis*	3.00E-10
PSTha8F13	GH738007	116	6	no homology	-
PSTha2c7	GH737231	204	13	predicted protein of *Puccinia graminis*	1.15E-15
PSTha16B3	GH737129	87	2	no homology	-
PSTha10F24	GH737323	56	3	no homology	-
PSTha16D6	GH737598	66	3	hypothetical protein of *Aspergillus niger*	2.00E-16
PSTha9C13	GH738022	65	1	predicted protein of *Puccinia graminis*	6.00E-10
PSTha12h2	GH737173	70	5	no homology	-

Expression Patterns of Genes from Haustorial cDNA Library

To examine developmental stage-specific gene expression, ten unisequences were selected from the haustorial cDNA library to assay their transcript levels during different developmental and infection stages through qRT-PCR. As determined by the GeNorm analysis, elongation factor-1, β-tubulin and actin showed M values of 0.699, 0.779 and 1.838. Elongation factor-1 was the most stable gene and was therefore used as a reference to normalize gene expression across different samples. Gene expression of the ten genes in uninfected wheat leaves, urediniospores, in vitro germinated urediniospores and infected wheat leaves was shown in Table 2. Among these genes, six were putative secreted proteins (PSTha2a5, PSTha12j12, PSTha5a23, PSTha12a4, PSTha12h2 and PSTha9F18). PSTha5a23 was specifically expressed in plants and was either not expressed in urediniospores and germinated urediniospores or expressed very weekly. Expression levels of PSTha2a5 and PSTha12j12 were weak in urediniospores, slightly increased in germinated urediniospores and high in Pst-infected leaves. PSTha12h2 was expressed in all stages of development, with the highest levels in infected leaves. PSTha12a4 transcripts only increased in germinated urediniospores but not in infected leaves. In contrast, PSTha9F18 was expressed strongly in urediniospores and germinated urediniospores, and its expression decreased dramatically in infected leaves.

Table 2. Expression of P. striiformis f. sp. tritici genes in different developmental stages

Unisequence	GenBank accession	Relative expression		
		GerUred/Ured[a]	InfW/GerUred[b]	InfW/Ured[c]
PSTha2a5	GH737102	4.70 ± 1.19	25.81 ± 9.88	119.89 ± 48.27
PSTha12j12	GH737467	9.40 ± 3.03	11.40 ± 3.06	104.31 ± 28.23
PSTha14i19	GH737527	0.55 ± 0.10	350.07 ± 56.80	192.38 ± 41.06
PSTha5a23	GH737046	3.15 ± 0.70	1165.64 ± 229.00	3617.32 ± 607.41
PSTha5a1	GH737242	3.09 ± 0.54	1948.90 ± 238.26	5980.69 ± 867.12
PSTha12a4	GH737444	5.70 ± 2.87	0.71 ± 0.31	3.68 ± 1.22
PSTha12h2	GH737173	0.47 ± 0.05	50.40 ± 5.14	23.48 ± 3.10
PSTha 9F18	GH737274	1.42 ± 0.34	8.11E-04 ± 1.42E-04	1.14E-03 ± 3.08E-04
PSTha5G19	GH737890	1.42 ± 0.82	0.2 ± 0.11	0.26 ± 0.16
PSTha11n16	GH737421	0.94 ± 0.22	0.22 ± 0.06	0.20 ± 0.05

[a] ratio of expression in germinated urediniospores (GerUred) vs. expression in urediniospores (Ured)
[b] ratio of expression in infected leaves (InfW) vs. expression in germinated urediniospores (GerUred)
[c] ratio of expression in infected leaves (InfW) vs. expression in urediniospores (Ured)

The other four genes for which expression patterns were examined were a THI2P homolog (PSTha14i19), a calcium/calmodulin-dependent protein kinase 2 homolog (PSTha11n16), a predicted chitinase (PSTha5a1) and a ubiquitin ligase E3C homolog (PSTha5G19). PSTha14i19 was expressed with the highest level in infected leaves. PSTha14i19 was homologous to U. fabae PIG4, which was thought to be involved in vitamin B1 biosynthesis [11,14,15]. A similar in planta induced rust gene was also reported in P. triticina, a wheat leaf rust pathogen [22-24]. Transcripts for PSTha5a1 were detected only in plants, while PSTha5G19 and PSTha11n16 were down regulated in infected leaves. None of these genes had transcripts in uninfected wheat leaves (data not shown), indicating that they were indeed Pst genes.

Discussion

Haustoria are specialized structures that are formed within the living cell of a host by biotrophic fungal pathogens during infection. Previous analyses of haustorial transcripts from other rust fungi [13-15,25] indicated that they are rich in sequences induced in planta and involved in virulence as well as other aspects of parasitism, like nutrient uptake. In the present study, we constructed a cDNA library from the wheat pathogen P. striiformis f. sp. tritici haustoria. A total of

5,126 randomly chosen ESTs were generated which represented 1,134 unique transcripts once plant sequences were removed and redundancies eliminated. Most of the highly redundant sequences, like several coding for ribosomal proteins, were very similar to other fungal sequences in databases and/or sequences found in Pst urediniospore and germinated urediniospore cDNA libraries and a U. fabae haustorium-specific cDNA library [15-17]. Proteins involved in protein synthesis, primary metabolism and energy production were most prevalent in predicted proteins with known functions. Similar findings were also reported in other pathogenic interactions during infection [19,22,24]. Overall, the less redundant sequences were surprisingly unique, especially considering the database searches included nucleotide and predicted protein searches of the P. graminis f. sp. tritici genome sequence and nucleotide searches of the Melampsora larici-populina genome sequence. Approximately 64% of the 1,134 unique sequences did not show significant similarities to known genes in databases. The frequency of sequences with no matches would likely have been lower if the full sequence of the cDNAs were available; some sequences appeared mostly non-coding. However, this frequency was not only high compared to other fungi [15,26] but also higher than that observed for Pst libraries made from urediniospores [16] or germinated urediniospores [17]. Avirulence proteins from eukaryotic plant pathogens described to date have indicated that they are diverse in function and have "novel" sequences with little sequence similarity to proteins of known function. If genes expressed in haustoria are more likely involved in virulence, which is probably more specific to particular fungal species, this might explain why a high percentage of genes expressed in haustoria share little or no sequence similarity to genes in other fungi.

Several of the haustorial transcripts were predicted to encode products homologous to potential virulence-related proteins based on sequence comparison with other plant or animal pathogens. For example, PSTha15i2 was homologous to glutamine synthetase. Several studies have indicated that glutamine metabolism is important for the virulence of various pathogens [27-29]. Glutamine synthetase enzyme activities were detected in pathogenic species of Mycobacterium, but were not detected in non-pathogenic species, indicating that this activity is potentially involved in the pathogenicity [30]. The glutamine synthetase of U. fabae was more strongly expressed in planta than in germinated urediniospores [15]. PSTha15B19 is a putative lpd gene encoding dihydrolipoyl dehydrogenase. The mutant of the lpd gene in Mycoplasma gallisepticum resulted in reduced virulence [31].

Several unisequences were candidates for genes encoding the key components of conserved signaling pathways. PSTha18c1, PSTha10L9 and PSTha12p13 were found to encode cAMP-dependent protein kinase type 2, protein ras-1 precursor and protein phosphatase PP2A regulatory subunit B, respectively. cAMP-dependent protein kinase pathway elements are remarkably conserved and effects on virulence have been the focus of many studies. In Cryptococcus neoformans, the cAMP signaling cascade is required for both melanin and capsule production, and mating filaments. All elements of the cAMP cascade are essential for the serum-induced switch of yeast to hyphal growth, which is important for the virulence of this fungus [32]. The cAMP cascade regulates pathogenicity of Ustilago maydis[33]. RAS proteins belong to the Rho family (a superfamily of GTPases). The ras1 mutants of C. neoformans were avirulent in animals [34]. Protein phosphatase PP2A is involved in several signal transduction pathways. Disruption of rgb1 gene, a subunit of PP2A, in Sclerotinia sclerotiorum reduced pathogenesis [35]. Further investigations of these proteins in Pst haustoria would be necessary to elucidate their functions in the infection process.

The fifteen unigenes predicted to encode secreted proteins are likely the best candidates for genes involved in specific virulence. Many plant pathogens manipulate their hosts through delivery of effector proteins [36-39]. In contrast to host resistant proteins, rust and mildew avirulence gene products described to date often share no significant sequence similarity to proteins of known function [36]. About 30 Avr genes in the flax rust pathogen have been identified by genetic analysis [40]. Recently Ellis and coworkers identified 21 HESPS (haustorially expressed secreted proteins) genes by examining a M. lini haustorium transcripts for secretion signals [12]. Among these HESPS, three co-segregated with the independent AvrM, AvrP4 and AvrP123 loci. Transient expression assays have shown that these genes function as avirulence determinants to induce R gene-dependent cell death in flax [12]. This indicates that avirulence proteins are very abundant among proteins secreted from haustoria and thus the haustorial libraries are very useful tools for identifying them. In the present study, we identified 15 putative secreted proteins from haustoria. The sequences of these proteins provided few clues to their functions except for PSTha21O8, which had significant sequence similarity to a sulfate transporter. Understanding the functions of these genes should shed light on the mechanisms of Pst virulence and biotrophism. Ongoing efforts to establish methods for stable transformation and transient expression assays in this biotrophic fungus are therefore a high priority.

Conclusions

A cDNA library was constructed from RNA of haustoria isolated from Pst-infected wheat leaves. A total of 5,126 EST sequences of high quality were generated and assembled into 1,134 unique sequences. Approximately 64% of them showed no significant similarities in public databases, indicating that many are likely specific to certain Puccinia taxa and valuable for future genomic studies of the stripe rust pathogen. Most of the transcripts with known functions were predicted to encode ribosomal proteins involved in protein synthesis, followed by proteins of primary metabolism and energy production. Some of the unisequences were predicted to encode products that exhibited high similarities to proteins potentially associated with virulence from other fungi. The 15 haustorium-specific genes predicted to encode secreted proteins are candidates for future studies to determine their potential functions in the wheat-Pst interactions.

Methods

Plant Genotypes and Pst Isolates

A wheat line carrying the Yr8 resistance gene in the 'Avocet Susceptible' background was inoculated with race PST-78 for production of spores and haustoria. Inoculation and culturing the host and pathogen was performed as described by Chen and Line [41]. To extract RNA from urediniospores, germinated urediniospores and infected wheat leaves, fresh urediniospores were harvested from infected leaves 15 days post inoculation (dpi). Urediniospores were germinated as previously described [17]. Briefly, fresh urediniospores were suspended in sterile distilled water in glass petri dishes and incubated in the dark for 12-15 h at 10°C. Infected wheat leaves were harvested at 8 dpi with PST-78. Fresh urediniospores, germinated urediniospores, uninfected leaves and infected leaves were frozen in liquid nitrogen, and stored at -80°C for further use.

Isolation of Pst Haustoria

Haustoria were isolated from heavily infected wheat leaves at 8 dpi (just prior to sporulation) using ConA affinity chromatography as described by Hahn and Mendgen [18]. Twenty-five g of infected wheat leaves were gently washed with chilled distilled water and homogenized in 160 ml of homogenization buffer [0.3 M sorbitol, 20 mM MOPS pH7.2, 0.1% BSA, 0.2% 2-mercaptoethanol, 0.2% PEG 6000] using a blender at maximum speed for 10 s. The homogenate was passed through a 20 μm nylon mesh and centrifuged at 5,000 g for 5 min. The resulting pellet was resuspended in 8 mL of suspension buffer. The suspension was centrifuged at 5,000 g for another 5 min. The pellet was resuspended in 4 mL of suspension buffer. Two 2 mL aliquots of the suspension were loaded onto two columns each filled with 4-5 mL of a sepharose 6MB coupled to ConA. The columns were incubated for 15 min after the aliquots entered the columns. After extensive washing of the columns with suspension buffer, haustoria were released from the columns by agitation using a wide-bore sterile pipette. The binding and washing steps were repeated 3-4 times in fresh columns, until most of the chloroplasts had been washed away. The haustoria in the suspension buffer were then transferred to a 15 mL Eppendorf tube and centrifuged for 1 min at 10,000 g and the pellet frozen in liquid nitrogen and stored at -80°C. The whole process was carried out with reagents at 4°C.

Isolation of RNA

For RNA isolation, fresh urediniospores, germinated urediniospores, infected leaves, uninfected leaves and haustorial cells were ground separately in a mortar in liquid nitrogen. Total RNA was isolated from frozen powder using the Qiagen Plant RNeasy kit (Qiagen, Chatsworth, GA) according to the manufacturer's instruction. For qRT-PCR analysis, RNA samples were treated with DNase I and purified with phenol/chloroform. The absence of genomic DNA contamination was subsequently confirmed by the null PCR amplification of RNA samples with primers designed for the Pst βtubulin gene and wheat GAPDH gene. The quantity and purity of isolated total RNA was analyzed by 2% agarose gel electrophoresis as well as by using a spectrophotometer.

Construction and Sequencing of a Haustorial cDNA Library

A cDNA library was prepared from 1 μg of total RNA from haustoria using the SMART™ cDNA library construction kit (Clontech, USA) according to the manufacturer's instruction. The colonies were subsequently picked and arrayed into 384-well micro-titer plates. Each well on the culture plate contained 75 μL of LB freezing storage medium [360 mM K2HPO4, 132 mM KH2PO4, 17 mM Na citrate, 4 mM MgSO4, 68 mM (NH4)2SO4, 44% (v/v) glycerol, 12.5 μg/ml of chloramphenicol, LB]. Colonies were incubated at 37°C overnight, and then stored at -80°C. Prior to sequencing, 52 clones were randomly picked to check for the presence of inserts by colony-PCR using the M13 forward and reverse primers. cDNA clones were sequenced with primer seq1 (5' CGACTCTAGACTC-GAGCAAG 3') from the 5'-end. Some clones with larger insert sizes were also sequenced with primer seq2 (5' AACAGCTATGACCATG 3') from the 3'-end using an ABI 3130-XL DNA sequencer.

Sequence analysis and database searches

Raw sequences were processed using cross-match [42], which resulted in the removal of poor quality sequences and vector sequences. Assembly of individual sequences into overlapping contigs was done as described [17]. Contaminating wheat sequences were removed from sequences by BLASTn of the NCBI non-redundant database and the 'dbEST_Others' (non-mouse, non-human) and BLASTX of the NCBI non-redundant protein sequences. The remaining ESTs were further compared with the NCBI non-redundant protein database and P. graminis f. sp. tritici genomic database using the BLASTX and BLASTn program http://www.broadinstitute.org/annotation/genome/puccinia_graminis/Blast.html, and Melampsora larici-populina genomic database using the BLASTn program http://genomeportal.jgi-psf.org/cgi-bin/runAlignment. E-values of less than 10-5 were considered significant matches to database sequences.

Secreted Protein Prediction

To search for potential secreted proteins, the sequences from the haustorial cDNA library were analyzed with the SignalP 3.0 algorithm [[20], http://www.cbs.dtu.dk/services/SignalP/]. To support the SignalP result, the protein sequences with predicted secretory signal peptides were also examined using iPSORT, http://hc.ims.u-tokyo.ac.jp/iPSORT/.

Quantitative Real-Time PCR Analysis

Using 1 μg of RNA isolated from fresh uninfected leaves, urediniospores, germinated urediniospores and infected leaves, reverse transcriptions were performed using superscript reverse transcriptase with an oligo (dT15) primer (Promega. Madison, WI. USA) according to the manufacturer's instructions. qRT-PCR was carried out using the Bio-Rad iQ5 Real-Time PCR system. Specific primers for each gene selected were designed using primer design software (Integrated DNA Technologies) and listed in Table 3. Real Time PCR was conducted in 20 μL volumes using SYBR Green PCR master mix (sigma). PCR conditions used were 95°C for 15 min, followed by 50 cycles of 95°C for 20 s, 55-60°C for 30 s and 72°C for 30 s, followed by a melting curve program. To identify a housekeeping gene with minimal variability in different rust developmental and infection stages, a qRT-PCR assay was established for three candidate reference genes (βtubulin, elongation factor-1 and actin) to analyze their transcription levels. Subsequently, the reference gene stability measures (M) were calculated by using the GeNorm tool as previously described [43]. A minimum of three biologically independent samples were used for each developmental stage, and two technical replicates were performed on every sample. Standard curves were generated for each gene. The quantification of gene expression was performed using the relative standard curve method by comparing the data with the reference gene.

Table 3. Primers used in qRT-PCR analysis to determine expression patterns of ESTs from the Puccinia striiformis f. sp. tritici haustorial library in various developmental and infection stages

Primer	Sequence(5' to 3')	Products size (bp)
RT-PSTha2a5-F	TGAATGGGTCGGTTGCCACAGATA	180
RT-PSTha2a5-R	GGCCCAAAGGGAATGGTCGAATTT	
RT-PSTha12j12-F	GCTTCGTTCGGGATTCAAAGCAAC	138
RT-PSTha12j12-R	ACATCTTGGGAACAGGCAGTTTCG	
RT-PSTha14i19-F	AAGTGCTCGAATGGGTCCTACCTT	135
RT-PSTha14i19-R	TGTGACGTTCACTTAGCCGATCCA	
RT-PSTha12h2-F	ACGTCAGTCAAAGATGTCGGCGAA	120
RT-PSTha12h2-R	TTCCTATCAATTAGCGCGGGAGCA	
RT-PSTha5a23-F	TTCCTACTCTGGCGACCAACATCA	194
RT-PSTha5a23-R	AAATCCGACTGACCGACATCCGTT	
RT-PSTha5a1-F	ACCGTATCGAAAGTGGTGTACGCT	82
RT-PSTha5a1-R	TGTCGTCCATTGGTCCCATAGTGT	
RT-PSTha12a4-F	GTTCACCAAAGCCACCTTCAACCA	128
RT-PSTha12a4-R	ATTAGACGGCGGCGTTCTTAGGAT	
RT-PSTha9F18-F	ATTCGAGATTAACGCGACCAACGG	169
RT-PSTha9F18-R	GAAAGGTCAATGACAACGGCGTCT	
RT-PSTha5G19-F	AGGTCTCGATTACCTTCCGCTTCT	127
RT-PSTha5G19-R	AAGAAAGATCGAAACCAGCACCAG	
RT-PSTha11n16-F	TGGGCATCTTCAGCTAGTTGGACT	181
RT-PSTha11n16-R	TCAACACATTCAGACCACCTCCGA	
RT-EF1-F	TTCGCCGTCCGTGATATGAGACAA	159
RT-EF1-R	ATGCGTATCATGGTGGTGGAGTGA	
RT-GAPDH-F	CAACGCTAGCTGCACCACTA	161
RT-GAPDH-R	TTCCACCTCTCCAGTCCTTG	
RT-TUB-F	CCGATCAATTCACGGCCATGTTCA	174
RT-TUB-R	AACCCTCTTCAACTTCCTCGTCGT	
RT-ACT-F	TGTCGGGTGGAACGACCATGTATT	146
RT-ACT-R	AGCCAAGATAGAACCACCGATCCA	

Authors' Contributions

CY isolated haustoria, constructed the haustorial cDNA library, participated in EST sequence analysis, conducted qRT-PCR and drafted the manuscript; XC contributed materials and resources, and wrote and revised the manuscript; XW, QH and ZK contributed to EST sequencing and BLAST searches. SH conceived and coordinated the study, interpreted the data, and wrote and revised the manuscript. All authors read and approved the final manuscript.

Acknowledgements

PPNS No.0524, Department of Plant Pathology, College of Agricultural, Human, and Natural Resource Sciences, Agricultural Research Center, Project no. WNPO0663, Washington State University, Pullman, WA 99164-6430. This research was supported in part by the 111 Project from the Ministry of Education of China (B07049). We thank Dr. Timothy Paulitz for critical review of the manuscript.

References

1. Chen XM: Epidemiology and control of stripe rust on wheat. Can J Plant Pathol 2005, 27:314–337.

2. Stubbs RW, Roelfs AP, Bushnell WR: Stripe rust. In The Cereal Rusts: Diseases, distribution, epidemiology and control. Volume II. Academic Press, Orlando, FL; 1985:61–101.

3. Chen XM, Moore MK, Milus EA, Long DL, Line RF, Marshall D, Jackson L: Wheat stripe rust epidemics and races of Puccinia striiformis f. sp. tritici in the United States in 2000. Plant Dis 2002, 86:39–46.

4. Chen XM: Challenges and solutions for stripe rust control in the United States. Austral J of Agri Res 2007, 58:648–655.

5. Allen RE: A cytological study of Puccinia glumarum on Bromus marginatus and Triticum vulgare. J Agri Res 1928, 36:487–513.

6. Kang ZS: Ultrastructure of plant pathogenic fungi. Beijing: China Science & Technology Press; 1995.

7. Hahn M, Mendgen K: Signal and nutrient exchange at biotrophic plant-fungus interfaces. Curr Opin Plant Biol 2001, 4:322–327.

8. Mendgen K: Nutrient uptake in rust fungi. Phytopathology 1981, 71:983–989.

9. Staples RC: Nutrients for a rust fungus: the role of haustoria. Trends in plant science 2001, 6:196–198.

10. Voegele RT, Mendgen K: Rust haustorium: Nutrient uptake and beyond. New Phytol 2003, 159:93–100.

11. Sohn J, Voegele RT, Mendgen K, Hahn M: High level activation of vitamin B1 biosynthesis genes in haustoria of the rust fungus Uromyces fabae. Mol Plant Microbe Interaction 2000, 13:629–636.

12. Catanzariti AM, Dodds PN, Lawrence GJ, Ayliffe MA, Ellis JG: Haustorially-expressed secreted proteins from flax rust are highly enriched for avirulence elicitors. Plant Cell 2006, 18:243–256.

13. Dodds PN, Lawrence GJ, Catanzariti AM, Ayliffe MA, Ellis JG: The Melampsora lini Avr567 avirulence genes are expressed in haustoria and their products are recognized inside plant cell. Plant Cell 2004, 16:755–768.

14. Hahn M, Mendgen K: Characterization of in plant-induced rust genes isolated from a haustorium-specific cDNA library. Mol Plant Microbe Interact 1997, 10:427–437.

15. Jakupovic M, Heintz M, Reichmann P, Mendgen K, Hahn M: Microarray analysis of expressed sequence tags from haustoria of the rust fungus Uromyces fabae. Fungal Genet Biol 2006, 43:8–19.

16. Ling P, Wang MN, Chen XM, Campbell KG: Construction and characterization of a full-length cDNA library for the wheat stripe rust pathogen (Puccinia striiformis f. sp. tritici). BMC Genomics 2007, 8:145–158.

17. Zhang Y, Qu Z, Zheng W, Liu B, Wang X, Xue X, Xu L, Huang L, Han Q, Zhao J, Kang Z: Stage-specific gene expression during urediniospore germination in Puccinia striiformis f. sp tritici. BMC Genomics 2008, 9:203–212.

18. Hahn M, Mendgen K: Isolation by ConA binding of haustoria from different rust fungi and comparison of their surface qualities. Protoplasma 1992, 170:95–103.

19. Broeker K, Bernard F, Moerschbacher BM: An EST library from Puccinia graminis f. sp. tritici reveals genes potentially involved in fungal differentiation. FEMS Microbiol Lett 2006, 256:273–281.

20. Bendtsen JD, Nielsen H, von Heijne G, Brunak S: Improved prediction of signal peptides: SignalP 3.0. J Mol Biol 2004, 340:783–795.

21. Bannai H, Tamada Y, Maruyama O, Nakai K, Miyano S: Extensive feature detection of N-terminal protein sorting signals. Bioinformatics 2002, 18:298–305.

22. Hu GG, Linning R, Mccallum B, Banks T, Cloutier S, Butterfield Y, Liu J, Kirkpatrick R, Stott J, Yang G, Smailus D, Jones S, Marra M, Schein J, Bakkeren G: Generation of a wheat leaf rust, Puccinia triticina, EST database from stage-specific cDNA libraries. Mol Plant Pathol 2007, 8:451–467.

23. Rampitsch C, Bykova NV, McCallum B, Beimcik E, Ens W: Analysis of the wheat and Puccinia triticina (leaf rust) proteomes during a susceptible host-pathogen interaction. Proteomics 2006, 6:1897–1907.

24. Thara KV, Fellers JP, Zhou JM: In planta induced genes of Puccinia triticina. Mol Plant Pathology 2003, 4:51–56.

25. Voegele RT, Struck C, Hahn M, Mendgen K: The role of haustoria in sugar supply during infection of broad bean by the rust fungus Uromyces fabae. Proc Natl Acad Sci USA 2001, 98:8133–8138.

26. Ebbole DJ, Jin Y, Thon M, Pan H, Bhattarai E, Thomas T, Dean R: Gene discovery and gene expression in the rice blast fungus, Magnaporthe grisea: analysis of expressed sequence tags. Mol Plant Microbe Interact 2004, 17:1337–1347.

27. Hava DL, Camilli A: Large-scale identification of serotype 4 Streptococcus pneumoniae virulence factors. Mol Microbiol 2002, 45:1389–1406.

28. Kloosterman TG, Hendriksen WT, Bijlsma JJ, Bootsma HJ, van Hijum SA, Kok J, Hermans PW, Kuipers OP: Regulation of glutamine and glutamate metabolism by GlnR and GlnA in Streptococcus pneumoniae. J Biol Chem 2006, 281:25097–25109.

29. Lau GW, Haataja S, Lonetto M, Kensit SE, Marra A, Bryant AP, McDevitt D, Morrison DA, Holden DW: A functional genomic analysis of type 3 Streptococcus pneumoniae virulence. Mol Microbiol 2001, 40:555–571.

30. Raynaud C, Etienne G, Peyron P, Lanéelle MA, Daffé M: Extracellular enzyme activities potentially involved in the pathogenicity of Mycobacterium tuberculosis. Microbiology 1998, 144:577–587.

31. Hudson P, Gorton TS, Papazisi L, Cecchini K, Frasca S Jr,, Geary SJ: Identification of a virulence-associated determinant, dihydrolipoamide dehydrogenase (lpd), in Mycoplasma gallisepticum through in vivo screening of transposon mutants. Infect Immun 2006, 74:931–939.

32. Fernandes L, Araújo MA, Amaral A, Reis VC, Martins NF, Felipe MS: Cell signaling pathways in Paracoccidioides brasiliensis-inferred from comparisons with other fungi. Genet Mol Res 2005, 4:216–231.

33. D'Souza CA, Heitman J: Conserved cAMP signaling cascades regulate fungal development and virulence. FEMS Microbiol Rev 2001, 25:349–364.

34. Waugh MS, Nichols CB, DeCesare CM, Cox GM, Heitman J, Alspaugh JA: Ras1 and Ras2 contribute shared and unique roles in physiology and virulence of Cryptococcus neoformans. Microbiology 2002, 148:191–201.

35. Erental A, Harel A, Yarden O: Type 2A phosphoprotein phosphatase is required for asexual development and pathogenesis of Sclerotinia sclerotiorum. Mol Plant Microbe Interact 2007, 20:944–954.

36. Catanzariti AM, Dodds PN, Ellis JG: Avirulence proteins from haustoria-forming pathogens. FEMS Microbiol Lett 2007, 269:181–188.

37. Collmer A, Badel JL, Charkowski AO, Deng WL, Fouts DE, Ramos AR, Rehm AH, Anderson DM, Schneewind O, van Dijk K, Alfano JR: Pseudomonas syringae Hrp type II I secretion system and effector proteins. Proc Natl Acad Sci USA 2000, 97:8770–8777.

38. Kamoun S: A catalogue of the effector secretome of plant pathogenic oomycetes. Annu Rev Phytopathol 2006, 44:41–60.

39. Petnicki-Ocwieja T, Schneider DJ, Tam VC, Chancey ST, Shan L, Jamir Y, chechter LM, Janes MD, Buell CR, Tang X, Collmer A, Alfano JR: Genome-wide identification of proteins secreted by the Hrp type III protein secretion

system of Pseudomonas syringae pv. tomato DC3000. Proc Natl Acad Sci USA 2002, 99:7652–7657.

40. Ellis JG, Dodds PN, Lawrence GJ: Flax rust resistance gene specificity is based on direct resistance-avirulence protein interactions. Annu Rev Phytopathol 2007, 45:289–306.

41. Chen XM, Line RF: Inheritance of stripe rust resistance in wheat cultivars used to differentiate races of Puccinia striiformis in North America. Phytopathology 1992, 82:633–637.

42. Green Group [http://www.phrap.org/].

43. Vandesompele J, De Preter K, Pattyn F, Poppe B, Van Roy N, De Paepe A, Speleman F: Accurate normalization of real-time quantitative RT-PCR data by geometric averaging of multiple internal control genes. Genome Biol 2002, 3:RESEARCH0034.

CITATION

Originally published under the Creative Commons Attribution License. Yin C, Chen X, Wang X, Han Q, Kang Z, Hulbert SH. Generation and analysis of expression sequence tags from haustoria of the wheat stripe rust fungus Puccinia striiformis f. sp. Tritici. BMC Genomics 2009, 10:626. doi:10.1186/1471-2164-10-626.

Living the Sweet Life: How does a Plant Pathogenic Fungus Acquire Sugar from Plants?

Nicholas J. Talbot

Plant diseases are an important constraint on worldwide crop production, accounting for losses of 10–30% of the global harvest each year [1]. As a consequence, crop diseases represent a significant threat to ensuring global food security. To feed the growing human population it will be necessary to double food production by 2050, which will require the sustainable intensification of world agriculture in an era of unpredictable climate change [2],[3]. Controlling the most important plant diseases represents one of the best means of delivering as much of the current productivity of crops as possible. To accomplish this task, a fundamental understanding of the biology of plant infection by disease-causing agents, such as viruses, bacteria, and fungi will be necessary [1],[2].

Fungal pathogens can broadly be divided into two groups—the biotrophs and necrotrophs [4],[5]. Biotrophic pathogens are parasites that have evolved the means to grow within living plant cells without stimulating plant defence mechanisms [6]. This means that they are able to spread rapidly throughout plant tissue while, at the same time, diverting nutrients from the living plant to fuel their own growth at the expense of plant productivity. In contrast, the necrotrophic pathogens use toxins and depolymerising enzymes to kill and degrade plant cells, consuming the resulting products [7]. These modes of nutrition are highly distinctive and plants have evolved independent defence mechanisms to contend with such different pathogens [5],[7]. To compound this challenge to plants, some pathogens exhibit both types of nutrition, switching from biotrophic growth to a rapid killing of plant cells as disease symptoms occur [7]–[9]. Because of their rather sophisticated nature, biotrophic pathogens cause some of the most pervasive plant diseases, which are difficult to control. Powdery mildew of barley caused by Blumeria graminis, for instance, continues to be one of the most important temperate cereal diseases [10], while yellow and brown rust diseases cause significant losses to worldwide wheat production [11]. The spread of the UG99 strain of wheat stem rust, which is highly virulent against most elite cultivars of wheat grown around the world, throughout Africa and the Middle East, shows how vulnerable existing cereal production is to attack by these sophisticated plant parasites [12].

In order to grow, a plant pathogenic fungus must secure an organic carbon source from the plant. In most plant diseases, however, we have little idea of what constitutes the major carbon source for an invading fungus during growth in plant tissue. Fungi are osmotrophic organisms, which means that they proliferate in a substrate by secreting a large diversity of extracellular enzymes that depolymerise polymers, such as cellulose, lignin, proteins, and lipids and then deliver the resulting simple sugars, amino acids, and fatty acids into fungal hyphae by means of plasma membrane-localised transporters [8],[13]. It is clear from analysing the genome sequences of both plant pathogenic and free-living fungi that they possess large numbers of extracellular enzymes and transporter-encoding genes, although, as might be expected, extracellular enzymes appear more restricted in number in biotrophic species [14]. The transporters, which are of the major facilitator transporter family, allow fungi to grow on an extremely diverse set of materials and are one of the reasons why fungi can occupy such a large number of ecological niches.

How do biotrophic plant pathogens acquire nutrients efficiently from a living plant cell? To understand this process it is essential to understand how plant pathogenic fungi enter living plant tissue. A large number of plant pathogenic fungal species develop specialised cells called appressoria that are able to breach the outer cuticle of plants and thereby gain entry to epidermal cells [8],[9]. The plant cells are not ruptured in this process, but instead the fungus is able to invaginate the plant plasma membrane and grow within the apoplast—the space between the plant plasma membrane and the plant cell wall [13]. This allows the fungus to occupy intact, living plant cells and set up a specialised interface to allow sequestration of nutrients directly from host cells [13]. A study published in this issue of PLoS Biology [15] provides a significant advance in understanding the mechanism by which a plant pathogenic fungus is able to acquire nutrients in planta. Ustilago maydis is a biotrophic pathogen which causes corn smut—a disease that is characterised by production of tumours on the stems and leaves of maize plants and, ultimately, by the liberation of large numbers of black teliospores that allow the fungus to be disseminated to new maize plants [16]. Corn smut can be a serious disease in maize-growing regions of Mexico and the United States [16]. In the study, the authors identified a novel plasma membrane-localised sucrose transporter encoded by the SRT1 gene, and have shown its contribution to fungal virulence. SRT1 encodes a plasma membrane protein with 12 membrane-spanning domains and is unusual because, in contrast to the relatively broad spectrum hexose transporters previously identified and characterised in fungi, Srt1 appears to be specific for the transport of sucrose. The conclusions of this study are that sucrose, which constitutes the most abundant storage sugar within plants and the product of photosynthesis, is directly utilised by invading pathogens without the need for its extra-cellular degradation by fungal secreted invertases.

The authors present a number of independent lines of evidence to support these conclusions. Srt1 was expressed in the yeast Saccharomyces cerevisiae, where they were able to study both its substrate specificity and affinity for sucrose. Srt1 is a sucrose-transporter with an extremely low KM of 26 ± 4.3 µM, which suggests that the fungus is able to out-compete plant sucrose transporters, such as the SUC family of energy-dependent H+ symporters that have an affinity for sucrose that is 20–200-fold lower than Srt1 [17],[18], and thereby gain an advantage in its proliferation within plant cells (Figure 1). The unusually high affinity of Srt1 for sucrose was measured by uptake experiments with 14C-labelled sucrose that also revealed its specificity for sucrose compared to other disaccharides, such as maltose, raffinose, and trehalose, as well as monosaccharides. Srt1 was also shown to be a plasma membrane–localised protein when expressed in S. cerevisiae, consistent with its transporter function [15]. Importantly, the authors also demonstrated that SRT1 is expressed specifically during invasion of plant tissue and fails to be expressed by U. maydis when cultured away from a plant, either in the presence of sucrose or in the absence of glucose as sole carbon source. This suggests that SRT1 expression is not under glucose catabolite repression, or substrate induction, as might be expected, but is instead induced by signals from the plant, allowing it to be specifically expressed when required during the biotrophic growth of U. maydis within maize tissue. To test their hypothesis that Srt1 is necessary for sucrose degradation in plants, the authors first carried out a targeted deletion of SRT1 and showed that it was necessary for virulence of the fungus. They then expressed an Arabidopsis sucrose transporter gene, SUC9 [17], and showed that this was able to restore the ability of a srt1 mutant to cause disease. When considered together, these results indicate that transport of sucrose by Srt1 is an essential requirement for the ability of this fungus to cause disease. As Arabidopsis SUC9 is unlikely to have additional functions within the fungus [17], it seems most likely that the role of Srt1 in planta is therefore specifically associated with its ability to transport sucrose.

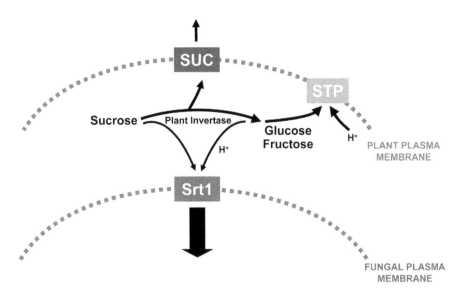

Figure 1. Efficient acquisition of sucrose by a plant pathogenic fungus. Model for the uptake of sucrose by the biotrophic plant pathogenic fungus Ustilago maydis based on a new study by Wahl and colleagues [15]. U. maydis possesses a high-affinity sucrose H+-symporter, Srt1, which is present in the plasma membrane of its invasive hyphae. The plant–fungus interface is established by invagination of the plant plasma membrane during intracellular invasive growth by the fungus. Srt1 competes for apoplastic sucrose with the plant SUC (or SUT) sucrose transporters [17],[18] and with plant invertases, which result in glucose and fructose generation. This reduces direct uptake of sucrose by plant cells or hexose uptake by the STP transporters, allowing the fungus to derive its primary carbon source from living plant cells without eliciting plant defence mechanisms. Adapted from Figure 8 of Wahl et al. [15].

Previous studies in pathogenic fungi have suggested a role for infection-specific hexose transporters in plant infection. The broad bean rust fungus Uromyces fabae, for instance, expresses a H+ symporter with specificity for glucose and fructose in its haustoria—the specialised invasive hyphae made by rusts and powdery mildews [19]. Electrophysiological studies showed its substrate specificity, whereas cytological analysis showed that it is very specifically localised at the haustorial plasma membrane. Similarly, the mycorrhizal fungus Geosiphon pyriformis, which forms a symbiotic relationship with plant roots allowing them to acquire nutrients more efficiently from soil, possesses a novel hexose transporter with highest affinity for glucose (followed by mannose, galactose, and fructose), which may be involved in the fungus–root cell mutualistic symbiosis that is of such importance to plant growth in diverse terrestrial ecosystems [20]. However, the role of these transporters in each respective plant–fungal interaction could not be determined by gene functional analysis because both U. fabae and G. pyriformis are obligate symbionts, which cannot be cultured away from their plant hosts, so the significance of these transporters to growth of each fungus in living plant tissue could not be directly assessed. For the first time, Wahl et al. [15], have therefore provided direct evidence of the utilisation of sucrose by a plant pathogen, challenging a widely held view that sucrose utilisation by fungi is the result of extracellular invertase activity from the fungus followed by specific transport of glucose [21].

An additional strategic advantage of directly utilising sucrose from the plant is also apparent from this work. Recently, a number of studies have indicated that glucose signalling may be intimately connected with the elicitation of plant defence responses [22],[23]. By direct utilisation of sucrose from the apoplast, the fungus may be able to evade recognition much more effectively while at the same time taking advantage of the major carbon source available within plant tissue. How widespread this form of growth will prove to be is, at present, unclear. However, close homologues of Srt1 are found in a range of both biotrophicnd free-living fungi, including closely related biotrophic Ustilago species, which are also amenable to gene function analysis [15]. This enables a direct test of this hypothesis within a range of both plant pathogenic and saprotrophic species to determine the relevant contribution of Srt1 homologues to the ability of fungi to grow either as pathogens, or saprotrophically. In this way it may be possible to investigate the evolution of sucrose transporters in fungi and shed light on whether biotrophic pathogens, such as U. maydis, have evolved specific mechanisms to allow utilisation of carbon substrates from living tissue that limit their ability to be detected by the host and whether this contrastswith necrotrophic and free-living fungal species.

The current study allows the formulation of a model by which U. Maydis is able to acquire apoplastic sucrose via the activity of a very high-affinity sucrose transporter that is present specifically in the fungal plasma membrane and, by means of its high-substrate specificity and extremely high affinity for sucrose, is able to efficiently out-compete plant apoplastic sucrose transporters and the presence of plant extracellular invertases and plant membrane–localised hexose transporters, thereby allowing efficient growth and development of the plant pathogen while preventing the induction of plant defences.

References

1. Strange R. N, Scott P. R (2005) Plant disease: a threat to global food security. Annu Rev Phytopathol 43: 83–116.

2. FAO (2006) World agriculture: towards 2030/2050. Interim Report. Global Perspective Studies Unit, Food and Agriculture Organization of the United Nations: Rome, Italy. Available: http://www.fao.org/es/ESD/AT2050web. Accessed 14 January 2010.

3. Evans N, Baierl A, Semenov M. A, Gladders P, Fitt B. D. L (2008) Range and severity of a plant disease increased by global warming. J R Soc Interface 5: 525–531.

4. Jones J. D. G, Dangl J. L (2006) The plant immune system. Nature 444: 323–329.

5. Glazebrook J (2005) Contrasting mechanisms of defense against biotrophic and necrotrophic pathogens. Annu Rev Phytopathol 43: 205–227.

6. Mendgen K, Hahn M (2002) Plant infection and the establishment of fungal biotrophy. Trends Plant Sci 7: 352–356.

7. Oliver R. P, Ipcho S. V. S (2004) Arabidopsis pathology breathes new life into the necrotrophs-vs.-biotrophs classification of fungal pathogens. Molec Plant Pathol 5: 347–352.

8. Tunlid A, Talbot N. J (2002) Genomics of parasitic and symbiotic fungi. Curr Opin Microbiol 5: 513–519.

9. Wilson R. A, Talbot N. J (2009) Under pressure: investigating the biology of plant infection by the rice blast fungus Magnaporthe oryzae. Nat Rev Microbiol 7: 185–195.

10. Jørgensen I. H (1992) Discovery, characterization and exploitation of Mlo powdery mildew resistance in barley. Euphytica 63: 141–152.

11. McIntosh R. A, Brown G. N (1997) Anticipatory breeding for resistance to rust diseases in wheat. Annu Rev Phytopathol 35: 311–326.

12. Singh R. P, Hodson D. P, Huerta-Espino J, Jin Y, Njau P, et al. (2008) Will stem rust destroy the world's wheat crop? Advances in Agronomy 98: 271–309.

13. Mendgen K, Hahn M (2002) Plant infection and the establishment of fungal biotrophy. Trends Plant Sci 7: 352–356.

14. Cornell M, Alam I, Soanes D. M, Wong H. M, Hedeler C, et al. (2007) Comparative genome analysis across a kingdom of eukaryotic organisms: specialization and diversification of the fungi. Genome Res 17: 1809–1822.

15. Wahl R, Wippel K, Goos S, Kämper J, Sauer N (2010) A novel high-affinity sucrose transporter is required for virulence of the plant pathogen Ustilago maydis. PLoS Biol 8(2): e1000303.

16. Kahmann R, Kämper J (2004) Ustilago maydis: How its biology relates to pathogenic development. New Phytol 164: 31–42.

17. Sauer N, Ludwig A, Knoblauch A, Rothe P, Gahrtz M, et al. (2004) AtSUC8 and AtSUC9 encode functional sucrose transporters, but the closely related AtSUC6 and AtSUC7 genes encode aberrant proteins in different Arabidopsis ecotypes. Plant J 40: 120–130.

18. Carpaneto A, Geiger D, Bamberg E, Sauer N, Fromm J, et al. (2005) Phloem-localized, proton-coupled sucrose carrier ZmSUT1 mediates sucrose efflux under the control of the sucrose gradient and the proton motive force. J Biol Chem 280: 21437–21443.

19. Voegele R. T, Struck C, Hahn M, Mendgen K (2001) The role of haustoria in sugar supply during infection of broad bean by the rust fungus Uromyces fabae. Proc Natl Acad Sci USA 98: 8133–8138.

20. Schüßler A, Martin H, Cohen D, Fritz M, Wipf D (2006) Characterization of a carbohydrate transporter from symbiotic glomeromycotan fungi. Nature 444: 933–936.

21. Bisson L. F, Coons D. M, Kruckeberg A. L, Lewis D. A (1993) Yeast sugar transporters. Crit Rev Biochem Molec Biol 28: 259–308.

22. Roitsch T, Balibrea M. E, Hofmann M, Proels R, Sinha A. K (2003) Extracellular invertase: key metabolic enzyme and PR protein. J Exp Bot 54: 513–524.

23. Rolland F, Baena-Gonzalez E, Sheen J (2006) Sugar sensing and signaling in plants: conserved and novel mechanisms. Annu Rev Plant Biol 57: 675–709.

CITATION

Originally published under the Creative Commons Attribution License. Talbot NJ (2010). Living the Sweet Life: How Does a Plant Pathogenic Fungus Acquire Sugar from Plants? PLoS Biol 8(2): e1000308. doi:10.1371/journal.pbio.1000308.

FRAP Analysis on Red Alga Reveals the Fluorescence Recovery is Ascribed to Intrinsic Photoprocesses of Phycobilisomes Rather Than Large-Scale Diffusion

Lu-Ning Liu, Thijs J. Aartsma, Jean-Claude Thomas,
Bai-Cheng Zhou and Yu-Zhong Zhang

ABSTRACT

Background

Phycobilisomes (PBsomes) are the extrinsic antenna complexes upon the photosynthetic membranes in red algae and most cyanobacteria. The PBsomes in the cyanobacteria has been proposed to present high lateral mobility on the thylakoid membrane surface. In contrast, direct measurement of PBsome motility in red algae has been lacking so far.

Methodology/Principal Findings

In this work, we investigated the dynamics of PBsomes in the unicellular red alga Porphyridium cruentum in vivo and in vitro, using fluorescence recovery after photobleaching (FRAP). We found that part of the fluorescence recovery could be detected in both partially- and wholly-bleached wild-type and mutant F11 (UTEX 637) cells. Such partial fluorescence recovery was also observed in glutaraldehyde-treated and betaine-treated cells in which PBsome diffusion should be restricted by cross-linking effect, as well as in isolated PBsomes immobilized on the glass slide.

Conclusions/Significance

On the basis of our previous structural results showing the PBsome crowding on the native photosynthetic membrane as well as the present FRAP data, we concluded that the fluorescence recovery observed during FRAP experiment in red algae is mainly ascribed to the intrinsic photoprocesses of the bleached PBsomes in situ, rather than the rapid diffusion of PBsomes on thylakoid membranes in vivo. Furthermore, direct observations of the fluorescence dynamics of phycoerythrins using FRAP demonstrated the energetic decoupling of phycoerythrins in PBsomes against strong excitation light in vivo, which is proposed as a photoprotective mechanism in red algae attributed by the PBsomes in response to excess light energy.

Introduction

Photosynthetic organisms are able to capture solar energy using light-harvesting antennae in the photosynthesis process [1], [2]. In the eukaryotic red algae and most of the prokaryotic cyanobacteria, the light-harvesting antennae are extrinsic supramolecular complexes, designated phycobilisomes (PBsomes). PBsomes are arranged on the surface of photosynthetic membranes [3]–[5], absorbing light and transferring it to the reaction center with high efficiency [6]–[8].

The major components of PBsomes are the phycobiliproteins (PBPs). According to the different properties, PBPs are commonly divided into four main classes: allophycocyanin (APC), phycocyanin (PC), phycoerythrin (PE) and phycoerythrocyanin (PEC) [1], [3], [9]. In the presence of linker polypeptides, PBPs are assembled into two subcomplexes: the core that combines with the membrane-bound photosynthetic reaction centers, and the peripheral rods that attach to the core [7], [10]–[12]. The former is mainly composed of APCs, and the latter contain PCs, PCs+PEs or PCs+PECs, depending on the species [1], [3], [4]. In the unicellular red alga Porphyridium (P.) cruentum, the main components of PBPs are PEs, which consist of B-PE and b-PE [13], [14]. These two types of PEs are both heterooligomers, but they differ by the presence/absence of the γ subunit [13], [15], [16].

Fluorescence recovery after photobleaching (FRAP) has been exploited to study the protein diffusion since 1970s [17]–[19]. It has been applied in exploring the mobility of PBsomes in cyanobacteria, suggesting that PBsomes could diffuse laterally on the surface of thylakoid membrane [20]–[22]. Such rapid mobility of PBsomes was further proposed to be a prerequisite for the light-state transition [23], a physiological mechanism that adapts to distribute the excitation light between photosystem I and photosystem II [24]–[26]. To date, the measurement of PBsome mobility in red algae has been lacking. Structural explorations have revealed larger size of the hemiellipsoidal PBsomes in P. cruentum compared to the hemidiscoidal PBsomes in cyanobacteria, and significant macromolecular crowding of PBsomes on the thylakoid surface in red algae [14], [27], [28]. These findings raise the question: the rapid and long-range diffusion of PBsomes in red algae may be highly restricted by a combination of macromolecular crowding of membrane surface and specific protein-protein interactions upon the thylakoid surface [29], [30].

In this paper, using FRAP we directly investigated the dynamics of PBsomes in P. cruentum wild-type (WT) and mutant F11 cells, as well as the isolated PBsomes in vitro. Our findings revealed that the fluorescence recovery observed is due to the intrinsic photoprocesses of PBsomes in situ, rather than large-scale PBsome diffusion in vivo. Furthermore, in situ light-induced energetic decoupling of PBsomes investigated by FRAP was proposed as an energy dissipation mechanism in vivo in response to excess light energy.

Results

Fluorescence Recovery of Partially Photobleached WT Cell

FRAP has been performed to study the mobility of PBsomes on cyanobacterial thylakoid membrane [20], [31]–[33]. Here we applied the same strategy to individual cells of the red alga P. cruentum for analyzing the diffusion dynamics of PBsomes upon the thylakoid membranes. In the experiments, confocal fluorescence images were acquired of a particular plane in the cell. We excited the PBsomes in P. cruentum cells with a 568 nm laser which was mostly absorbed by PEs in the peripheral PBsome rods. The fluorescence emission of PBsome terminal emitters was detected in the range of 650–750 nm. For pre- and post-scanning 5% of the laser power was applied, and photobleaching was performed by zooming-in scan at 100% laser power (10 mW). Figure 1A shows a typical FRAP image sequence of PBsome emission in a single P. cruentum cell. Part of the cellular area was bleached to a depth of 65% and the fluorescence recovery of the bleached area was detected by a post-scan. The recovery of the fluorescence in bleached cellular region is depicted in Figure 1B (dashed line) and adequately fitted to an exponential function (Equation 1, see Materials and Methods). The fluorescence recovery levels off at 66% of the initial fluorescence intensity only (from 35% after bleaching). The recovery rate (R) was calculated to be 39.5 s-1 with Equation 1. Furthermore, we found that more intense illumination could induce less fluorescence recovery. When the photobleaching reaches 80–90% depth, the fluorescence recovery of PBsomes was much less, though still visible. This was corroborated with previous finding in cyanobacterium Thermosynechococcus elongates [33].

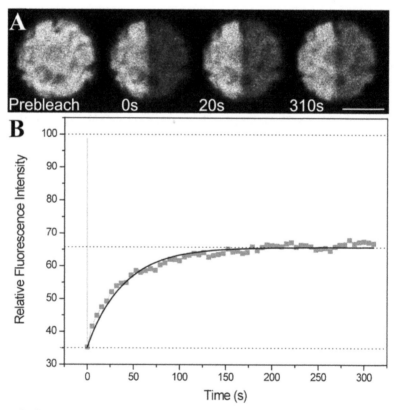

Figure 1. Qualitative FRAP experiments of red alga P. cruentum WT cell. A, selected fluorescence images from typical sequences recorded before bleaching, immediately after bleaching of PBsomes, and at various time lapses. Excitation is at 568 nm and detection range is from 650 to 750 nm. Scale bar: 5 μm; B, total fluorescence intensity of bleached cell region as a function of time. The recovery of the fluorescence is presented as square spots and fitted to an exponential function (solid line).

Fluorescence Recovery of Wholly Photobleached WT Cell

To further survey the photodynamics of PBsome complexes, we explored a series of comparative experiments with the aforementioned FRAP procedure to whole cells. Figure 2 shows the time-course photobleaching of the whole P. cruentum cell. Three continuous cycles of photobleaching were carried out on the same cell. All the PBsomes were generally bleached by whole-cell scanning with intense laser power (100%). Generally, no fluorescence recovery by the lateral mobility of PBsomes was expected in this case. However, the partial recovery of fluorescent intensity from the whole cell was still observed. It was also found that stronger bleaching leads to less recovery, which has been observed in partially photobleached cell. There is one probability that such fluorescence recovery may be ascribed to the diffusion of PBsomes from neighboring lamellae surfaces. However, we found the fact that scanning with full power caused bleaching of PBsomes not only in the focusing plane, but also perpendicularly, in a few-micrometer flanking areas through the cell (data not shown).

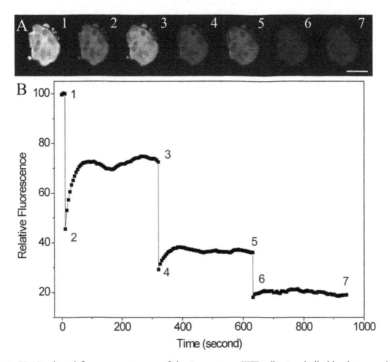

Figure 2. FRAP selected fluorescence images of the P. cruentum WT cells. A, wholly bleaching on the cells. Excitation is at 568 nm and detection range is from 650 to 750 nm. Selected fluorescence images from sequences recorded before bleaching, immediately after bleaching of PBsomes, and at various time lapses. Cycle 1: 1-2-3; cycle 2: 3-4-5; cycle 3: 5-6-7; Scale bar: 5 μm; B, one-dimensional bleaching profiles derived from the sequences of fluorescence images of panel A.

Fluorescence Recovery of Glutaraldehyde/Betaine-Treated WT Cell

Furthermore, FRAP experiments were performed on whole cells pre-treated with 1% glutaraldehyde or 0.5 M betaine, which have been documented to be able to chemically promote cross-linking between PBsomes and thylakoid membrane [27], [34]–[39]. Figure 3 shows that the fluorescence recovery was unexpectedly detected in all cases. The recovery rates in all conditions were calculated to be constantly 40.0±1.0 s–1, similar to the results of the native cells mentioned above.

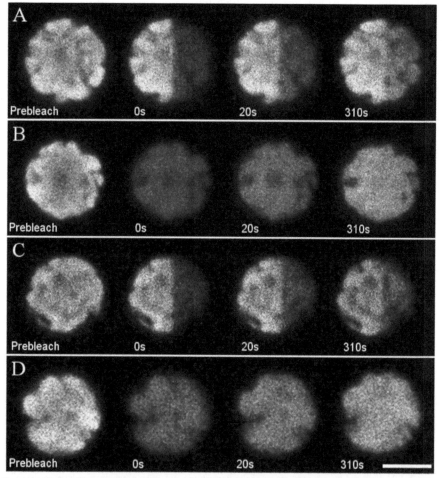

Figure 3. FRAP selected fluorescence images of P. cruentum WT cells pretreated with glutaraldehyde/betaine. Excitation is at 568 nm and detection range is from 650 to 750 nm. A, partially-bleached cell pretreated with betaine; B, wholly-bleached cell pretreated with betaine; C, partially-bleached cell pretreated with glutaraldehyde; D, wholly-bleached cell pretreated with glutaraldehyde. Scale bar: 5 μm.

Analysis of The Fluorescence Profiles

Investigations on one-dimensional fluorescence profiles obtained by summing pixel values across the edge of the bleached area (along the X direction in Figure 4) may allow us to analyze the PBsome mobility in real time. As the PBsome complexes diffuse upon the membrane surface, one should observe: (1) a fluorescence loss in the unbleached area if the PBsomes diffuse across the cells (Figure 4A–C); (2) successive flattening of the slope of the fluorescence intensity gradient (Figure 4A–C); (3) changes of the recovery rate as a function of distance from the edge of the bleached area (Figure 4D–F). However, as depicted in Figure 4, the above-mentioned effects were all not obvious, indicating that there was no remarkable difference of the fluorescence behaviors between all native cells, and glutaralde-hyde- or betaine-treated cells.

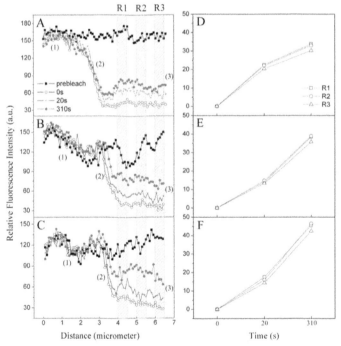

Figure 4. Selected analysis of the fluorescence profiles across the edge of the bleached area in partially-bleached cells, allowing the real-time exploration of the fluorescence dynamics of PBsomes correlated to their lateral diffusion. A, untreated cell; B, betaine-treated cell; C, glutaraldehyde-treated cell. Area (1), the unbleached cellular area; Area (2), the slope of the fluorescence intensity gradient; Area (3), the fluorescence recovery as a function of distance from the edge of the bleached area. The mean fluorescence intensities of three different regions (R1, R2 and R3 in Figure 4A–C) at distinct recovery time were analyzed: D, from untreated cell; E, from betaine-treated cell; F, from glutaraldehyde-treated cell. Fluorescence intensities are normalized at zero second. As shown, no remarkable difference of the fluorescence intensities in these three cellular areas could be observed, indicating the fluorescence recovery of PBsomes does not differ as a function of distance from the edge of the bleached area of theses cells.

Fluorescence Recovery of F11 Cells

In vitro single-molecule result has elaborated that two types of PEs in the PBsomes of P. cruentum play different roles in the energy transfer [40]. Unlike B-PEs, b-PEs are not involved in the energy decoupling of PBsomes. We measured the cells of P. cruentum mutant F11 strain (UTEX 637) which is depleted of B-PEs using FRAP. In Figure 5, the fluorescence recovery of PBsomes in all partially-bleached, wholly-bleached, as well as chemical-fixed cells by glutaraldehyde or betaine was obviously observed, similar to the results of WT cells.

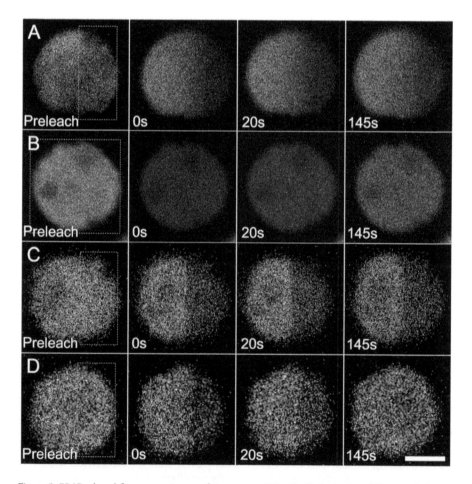

Figure 5. FRAP selected fluorescence images of P. cruentum F11 cells. Excitation is at 568 nm and detection range is from 650 to 750 nm. Open squares indicate bleached areas by zooming-in illumination. Scale bar: 5 μm. A, partially-bleached cell; B, wholly-bleached cell; C, partially-bleached cell pretreated with glutaraldehyde; D, partially-bleached cell pretreated with betaine.

Moreover, PBsomes from the mutant P. cruentum F11 cell have a declined dimension due to lower level of PE content [40]–[42]. PBsomes with a smaller size were expected previously to enhance their mobility [31]. However, by examining the FRAP profiles of mutant F11 cell, we found that the fluorescence recovery has similar recovery rates as that of the native cell, suggesting that the fluorescence recovery is not affected by the dimension of PBsomes in red algae.

Fluorescence Recovery of Ensemble PBsomes in Vitro

FRAP experiment was also exploited on isolated PBsomes from P. cruentum. Intact PBsomes functionally isolated from the thylakoid membrane were spread on the clean surface of glass cover slide. Photobleaching was performed on the pool of immobilized PBsome complexes by zooming-in on a small area of the sample (Figure 6, square). The fluorescence recovery of bleached PBsome aggregations were unexpectedly observed in vitro in the first 150 s timespan of our measurements. The fluorescence feature, more specifically the shape of the aggregate, remained fairly stable in the course of the experiment. It revealed that the fluorescence recovery could be detected even in the immobile PBsomes. Taken together, our experimental data strongly demonstrated that the fluorescence recovery observed is most likely ascribed to the intrinsic photoprocesses of the bleached PBsomes in situ, rather than the diffusion of PBsomes.

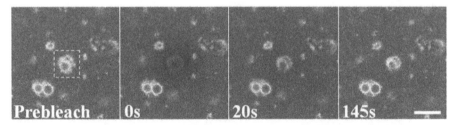

Figure 6. Fluorescence images of isolated PBsomes from P. cruentum WT cells. Excitation is at 568 nm and detection range is from 650 to 750 nm. Selected fluorescence images from sequences recorded before bleaching, immediately after bleaching of PBsomes, and at various time lapses. The positions of PBsome aggregations imaged out of the bleaching area did not change, indicating the immobilization of PBsomes above the substrate. Scale bar: 10 μm.

Fluorescence Properties of PE Emission in Vivo

In the meanwhile, we studied the fluorescence properties of PE during photobleaching by altering the detecting region from 650–750 nm into 550–600 nm. A lower power of scanning laser (10%) was applied in the prebleaching and postbleaching in order to record continuous fluorescence images, and maximum laser intensity (100%) was introduced to bleach the samples of interest by zooming-in scanning.

Figure 7 illustrates selected images of PE fluorescence in the cells in response to intense illumination. Interestingly, wholly (Figure 7A) and partially bleaching (Figure 7B) on the WT cells of P. cruentum both resulted in the increase of PE fluorescence intensity. Such increase was also observed in the adjacent non-bleached area near the edges of bleaching region in the partially bleached cells, probably due to the scattered bleaching light. We further studied the PE fluorescent behavior by using glutaraldehyde-treated cell as a control. As a result, the fluorescence increase was not detected (Figure 7C), demonstrating that the increase of PE fluorescence imaged is due to the energetic decoupling of PE molecules within the PBsome rods. Furthermore, experimental results of P. cruentum mutant F11 strain, which only contain b-PE assemble with neighboring R-PC [40], did not show the fluorescence increase either (Figure 7D), implying the different roles of B-PE and b-PE in the energetic decoupling. These are corroborated with the previous conclusion obtained from single-molecule experiment [40]. Furthermore, it provides the direct evidence of the energetic decoupling of PBsomes against intense exciting energy in vivo. Similar fluorescence increase of PE has also been observed in the FRAP bleach in cryptophyte [43].

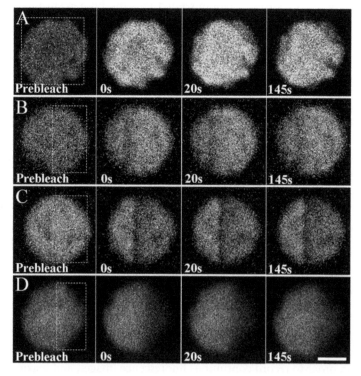

Figure 7. Selected PE fluorescence images in cells imaged with confocal microscopy by detecting fluorescence at 550–600 nm. A, B, native WT cells; C, WT cell pretreated with glutaraldehyde; D, F11 cell. Scale bar: 5 μm.

Figure 8 shows quantitatively the fluorescence intensities of PE in cells as a function of time. It is seen that the bleaching on WT natural cell induced the increase of PE emission (curves 1 and 2), and longer irradiation time of photo-bleaching was capable to trigger more remarkable fluorescence reduction (curves 2, 3, 4 and 5). The PE fluorescence intensity of the bleached areas in WT cells pretreated with glutaraldehyde and F11 cells is relatively constant, lower than the initial fluorescence intensities before photobleaching (curves 6 and 7). The plots revealed that there is no significant recovery of PE, either from increased or declined fluorescence levels. This suggested, on the other hand, that there is no significant mobility of PEs that are the major components of PBsomes.

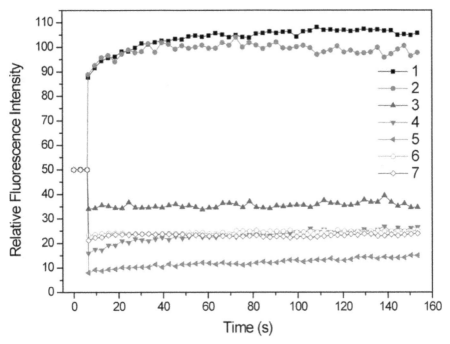

Figure 8. Time-lapse PE fluorescence intensity in cells in the process of photobleaching. Powerful irradiation can cause initial increase of PE and prolonged illumination time may induce bleaching of PE. (1), whole bleaching of native WT cell; (2), 5 times scan; (3), 10 times scan; (4), 20 times scan; (5), 30 times scan; (6), WT cell pretreated with glutaraldehyde; (7), F11 mutant cell.

Fluorescence Recovery of PEs in Ensemble PBsomes in vitro

Figure 9 shows the fluorescence dynamics of PE in isolated PBsomes upon intense illumination at 514 nm. Immediately after photobleaching, continuous scan recorded a dramatic fluorescence increase in the wholly-bleached area of WT PB-somes (Figure 9A), whereas PE fluorescence was bleached in both WT PBsomes treated with glutaraldehyde (Figure 9B) and B-PE-lacking PBsomes from the F11 mutant (Figure 9C). This further confirmed the energetic decoupling of PBsomes in response to intense exciting light.

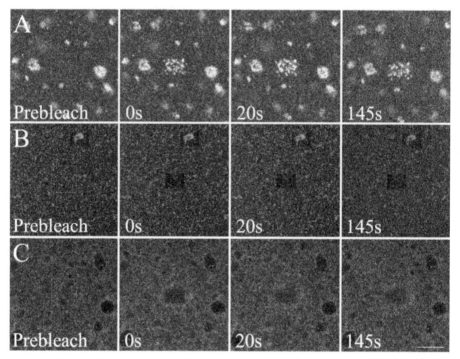

Figure 9. Selected PE fluorescence images of ensemble PBsome isolated from P. cruentum WT cells (A), WT cell pretreated with glutaraldehyde (B) and F11 mutant (C) obtained with confocal microscopy by detecting 550–600 nm. Scale bar: 5 μm.

Discussion

Thin-section electron microscopic images of P. cruentum cell present a globular shape with a typical diameter of 5–8 μm [44], [45]. The large chloroplasts consist of parallel-arranged and unstacked lamellae that occupy most of the intra-cellular space in vivo. Such a regular thylakoid organization is supposedly suitable for FRAP measurements [21], [22].

In cyanobacteria, PBsomes have been suggested to diffuse rapidly on the surface of thylakoid membrane [20]. However, recent investigations have shown that the thylakoid membrane in the unicellular red alga P. cruentum exhibits dense crowding of PBsome complexes [27], [28]. In addition to determining the arrangements of PBsomes on the thylakoid membrane, the supramolecular crowding has also profound implications for the dynamic behavior of PBsomes in P. cruentum: the rapid and long-range movement of PBsomes may be inhibited by the dense packing of membrane surface, and the limited free vertical spacing between opposite thylakoid layers, as well as remarkable steric hindrance, taking into account the large size of individual PBsomes. Furthermore, PBsomes attach quite strongly to the membrane surfaces [46], whereas the rapid diffusion of PBsomes requires looser association of PBsomes to the photosynthetic reactions centers.

This contradiction was explored in the present work by studying the fluorescence dynamics of PBsomes on thylakoid membranes from P. cruentum using FRAP. First, FRAP results of WT cells present partial fluorescence recovery. Assuming the fluorescence recovery is attributed to the diffusion of PBsome complexes [20], it then indicated that some PBsomes are presumably immobile on the thylakoid membranes of red algae. However, this probability was not supported by the finding that the fluorescence recovery of PBsomes declines as the illumination strength increases. The partial fluorescence recovery was also observed in partially-bleached F11 cells, wholly-bleached WT and F11 cells, and even in immobilized PBsome complexes in vitro. We also detected, similar fluorescence recovery behavior of P. cruentum cells pretreated with protein cross-linking agents glutaraldehyde or betaine with the fixation effect. By comparative analysis of the fluorescence feature and recovery rate, no significant difference of the fluorescence behavior between those cells was observed. All these data strongly indicate the observed recovery is not likely ascribed to the lateral diffusion of PBsome complexes on the thylakoid membranes.

Another evidence of the PBsome diffusion in red algae was provided by recording the PE fluorescent signal instead of the fluorescence emission of PBsome terminal emitters. After photobleaching, rapid diffusion of PBsomes was expected to cause the recovery of PE fluorescence to the initial fluorescence intensity. Alternatively, we could not observe such tendency in either the area of WT cells having increased PE fluorescence intensity or the area of glutaraldehyde/betaine-pretreated WT cells and F11 cells showing decreased PE fluorescence intensity (Figures 7 and 8).

Based on all the observations we conclude that, lateral diffusion of red algal PBsomes is not the predominant mechanism which could potentially explain the fluorescence recovery that we observed by means of FRAP. Instead, the fluorescence recovery is probably ascribed to the intrinsic photoprocesses of the bleached PBsomes in situ, rather than the rapid diffusion of PBsomes on thylakoid in vivo, although the mechanism of the intrinsic photoprocesses still remains to be determined.

The mechanism for light-state transition in red algae remains controversial. Unlike the state transition in higher plant, light-dependent phosphorylation was absent in cyanobacteria and rhodophytes [47]. Rapid mobility of PBsomes in cyanobacteria has been proposed to be required for the PBsome-dependent state transition [31]. In red alga P. cruentum, the long-distance diffusion of PBsome complexes may be highly restricted, however potential diffusion of PBsomes confined to small domains at sub-optical scales would not be excluded. For instance the local conformational changes and movement of PBsomes between adjacent photosystem II and photosystem I complexes [48] can probably be involved in the energy redistribution in red algae.

By monitoring the fluorescence of PEs in the PBsomes, we studied the fluorescence dynamics of ensemble PBsomes in situ during photobleaching. We demonstrated directly the energetic decoupling of PBsomes in vivo in response to the bleaching laser. The PEs (specifically B-PEs), rather than PC/APC, are mainly involved in the energetic decoupling. This is in agreement with single-molecule observations [40] and ensemble results of isolated PBsomes (data not shown). Such an energetic decoupling of PBsomes is proposed to have important implication on the photoprotective role of PBsomes: It may allow excess photon energy from PE to photosynthetic reaction centers to be modulated to minimize the risk of chlorophyll photooxidation. Therefore, like the light-state transition, this mechanism might contribute to prevent PE-containing algae from excessive photoinhibition.

Photosynthetic membrane has been considered as an ideal model to study the protein mobility in view of their naturally fluorescent properties [21]. However, the complexity of photosynthetic membrane should be taken into account in FRAP, as numbers of fluorescent protein are located densely in the membrane with a close functional association benefiting for efficient energy transfer. Not only does the spectral overlap make it hard to distinguish the individual contribution of each complex [20], but also the photoreactions among distinct cooperated protein complexes can be triggered by exposure laser, for instance the energetic decoupling in the PBsomes as we observed and the photoblinking behaviors revealed in single-molecule studies on phycobiliproteins [49]–[52]. In addition, it was proposed that all the fluorescent molecules can undergo reversible photobleaching [53]. Therefore, investigations on each functional protein complex involved may be prerequisite for understanding the fluorescence dynamics of photosynthetic membrane.

Materials and Methods

Growth of Cells and Separation of Intact PBsomes

P. cruentum WT and mutant F11 strain UTEX 637 were grown in an artificial sea water medium [54]. Flasks were supplied with 3% CO_2 in air through a plug of sterile cotton at a constant temperature of 20°C. Cultures were illuminated continuously with light provided by daylight fluorescent lamps at 6 W·m-2. Intact PBsomes were separated following the previous protocol [27].

FRAP Measurements

Growing cells or cells with corresponding pre-treatments by 1% (v/v) glutaraldehyde or 0.5 M betaine buffer were immobilized on a slide covered with a glass slip and located under objective lens of laser scanning confocal fluorescence microscopy (LSCM).

FRAP experiments were carried out with a Leica TCS NT confocal microscope. Samples were excited with a 568 nm Yellow Krypton laser (10 mW) or a 514 nm Argon laser (10 mW). PBsome emission was detected at 650–750 nm or 550–600 nm for detecting PBsome emission and PE emission, respectively. After recording a prebleaching image, the sample was bleached by zooming in and scanning with maximum laser power for three seconds. Subsequently, images were captured with the same power as that of pre-bleaching image. No detectable photobleaching during the recording of successive image scans was observed with optimized laser power.

The fluorescence recovery curve is fitted by single exponential function, given by

$$F(t) = A(1-e^{-t/\tau}) + B \quad (1)$$

where $F(t)$ is the intensity as time t; A and B are the amplitudes of the time-dependent and time-independent terms, respectively; τ is the lifetime of the exponential term (time constant), and the recovery rate is given by $R = 1/\tau$.

Competing Interests

The authors have declared that no competing interests exist.

Authors' Contributions

Conceived and designed the experiments: LNL TA BCZ YZZ. Performed the experiments: LNL. Analyzed the data: LNL TA BCZ YZZ. Contributed reagents/materials/analysis tools: JCT. Wrote the paper: LNL TA.

References

1. Glazer AN (1989) Light guides directional energy transfer in a photosynthetic antenna. J Biol Chem 264: 1–4.

2. Mullineaux CW (2005) Function and evolution of grana. Trends Plant Sci 10: 521–525.

3. MacColl R (1998) Cyanobacterial phycobilisomes. J Struct Biol 124: 311–334.

4. Bald D, Kruip J, Rögner M (1996) Supramolecular architecture of cyanobacterial thylakoid membranes: How is the phycobilisome connected with the photosystems? Photosynth Res 49: 103–118.

5. Zilinskas BA, Greenwald LS (1986) Phycobilisome structure and function. Photosynth Res 10: 7–35.

6. Adir N (2005) Elucidation of the molecular structures of components of the phycobilisome: reconstructing a giant. Photosynth Res 85: 15–32.

7. Grossman AR, Schaefer MR, Chiang GG, Collier JL (1993) The phycobilisome, a light-harvesting complex responsive to environmental conditions. Microbiol Rev 57: 725–749.

8. Biggins J, Bruce D (1989) Regulation of excitation energy transfer in organisms containing phycobilins. Photosynth Res 20: 1–34.

9. Colyer CL, Kinkade CS, Viskari PJ, Landers JP (2005) Analysis of cyanobacterial pigments and proteins by electrophoretic and chromatographic methods. Anal Bioanal Chem 382: 559–569.

10. Lundell DJ, Williams RC, Glazer AN (1981) Molecular architecture of a light-harvesting antenna: in vitro assembly of the rod substructures of Synechococcus 6301 phycobilisomes. J Biol Chem 256: 3580–3592.

11. Redlinger T, Gantt E (1981) Phycobilisome structure of Porphyridium cruentum: polypeptide composition. Plant Physiol 68: 1375–1379.

12. Liu L-N, Chen X-L, Zhang Y-Z, Zhou B-C (2005) Characterization, structure and function of linker polypeptides in phycobilisomes of cyanobacteria and red algae: An overview. Biochim Biophys Acta 1708: 133–142.

13. Gantt E, Lipschultz CA (1974) Phycobilisomes of Porphyridium cruentum: pigment analysis. Biochemistry 13: 2960–2966.

14. Gantt E, Lipschultz CA, Zilinskas B (1976) Further evidence for a phycobilisome model from selective dissociation, fluorescence emission, immunoprecipitation, and electron microscopy. Biochim Biophys Acta 430: 375–388.

15. Ficner R, Huber R (1993) Refined crystal structure of phycoerythrin from Porphyridium cruentum at 0.23-nm resolution and localization of the gamma subunit. Eur J Biochem 218: 103–106.

16. Lundell DJ, Glazer AN, DeLange RJ, Brown DM (1984) Bilin attachment sites in the alpha and beta subunits of B-phycoerythrin. Amino acid sequence studies. J Biol Chem 259: 5472–5480.

17. Axelrod D, Koppel DE, Schlessinger J, Elson E, Webb WW (1976) Mobility measurement by analysis of fluorescence photobleaching recovery kinetics. Biophys J 16: 1055–1069.

18. Zhang F, Lee GM, Jacobson K (1993) Protein lateral mobility as a reflection of membrane microstructure. Bioessays 15: 579–588.

19. Lippincott-Schwartz J, Altan-Bonnet N, Patterson GH (2003) Photobleaching and photoactivation: following protein dynamics in living cells. Nat Cell Biol: Suppl S7–14.

20. Mullineaux CW, Tobin MJ, Jones GR (1997) Mobility of photosynthetic complexes in thylakoid membranes. Nature 390: 421–424.

21. Mullineaux CW (2004) FRAP analysis of photosynthetic membranes. J Exp Bot 55: 1207–1211.

22. Mullineaux CW, Sarcina M (2002) Probing the dynamics of photosynthetic membranes with fluorescence recovery after photobleaching. Trends Plant Sci 7: 237–240.

23. Joshua S, Mullineaux CW (2004) Phycobilisome diffusion is required for light-state transitions in cyanobacteria. Plant Physiology 135: 2112–2119.

24. Mullineaux CW, Emlyn-Jones D (2005) State transitions: an example of acclimation to low-light stress. J Exp Bot 56: 389–393.

25. Schluchter WM, Shen G, Zhao J, Bryant DA (1996) Characterization of psaI and psaL mutants of Synechococcus sp. strain PCC 7002: a new model for state transitions in cyanobacteria. Photochem Photobiol 64: 53–66.

26. Bruce D, Brimble S, Bryant DA (1989) State transitions in a phycobilisomeless mutant of the cyanobacterium Synechococcus sp. PCC 7002. Biochim Biophys Acta 974: 66–73.

27. Arteni AA, Liu L-N, Aartsma TJ, Zhang Y-Z, Zhou B-C, et al. (2008) Structure and organization of phycobilisomes on membranes of the red alga Porphyridium cruentum. Photosynth Res 95: 169–174.

28. Liu L-N, Aartsma TJ, Thomas J-C, Lamers GEM, Zhou B-C, et al. (2008) Watching the native supramolecular architecture of photosynthetic membrane

in red algae: ography of phycobilisomes, and their crowding, diverse distribution patterns. J Biol Chem 283: 34946–34953.

29. Kirchhoff H (2008) Molecular crowding and order in photosynthetic membranes. Trends Plant Sci 13: 201–207.

30. Kirchhoff H, Haferkamp S, Allen JF, Epstein D, Mullineaux CW (2008) Protein diffusion and macromolecular crowding in thylakoid membranes. Plant Physiol 146: 1571–1578.

31. Sarcina M, Tobin MJ, Mullineaux CW (2001) Diffusion of phycobilisomes on the thylakoid membranes of the cyanobacterium Synechococcus 7942. Effects of phycobilisome size, temperature, and membrane lipid composition. J Biol Chem 276: 46830–46834.

32. Kondo K, Mullineaux CW, Ikeuchi M (2009) Distinct roles of CpcG1-phycobilisome and CpcG2-phycobilisome in state transitions in a cyanobacterium Synechocystis sp. PCC 6803. Photosynth Res 99: 217–225.

33. Yang S, Su Z, Li H, Feng J, Xie J, et al. (2007) Demonstration of phycobilisome mobility by the time- and space-correlated fluorescence imaging of a cyanobacterial cell. Biochim Biophys Acta 1767: 15–21.

34. Brimble S, Bruce D (1989) Pigment orientation and excitation energy transfer in Porphyridium cruentum and Synechococcus sp. PCC 6301 cross-linked in light state 1 and light state 2 with glutaraldehyde. Biochim Biophys Acta 973: 315–323.

35. Li D, Xie J, Zhao J, Xia A, Li D, et al. (2004) Light-induced excitation energy redistribution in Spirulina platensis cells: "spillover" or "mobile PBSs"? Biochim Biophys Acta 1608: 114–121.

36. Biggins J (1983) Mechanism of the light state transition in photosynthesis. I. Analysis of the kinetics of cytochrome f oxidation in state 1 and state 2 in the red alga, Porphyridium cruentum. Biochim Biophys Acta 724: 111–117.

37. Bruce D, Biggins J, Steiner T, Thewalt M (1985) Mechanism of the light state transition in photosynthesis. IV. Picosecond fluorescence spectroscopy of Anacystis nidulans and Porphyridium cruentum in state 1 and state 2 at 77 K. Biochim Biophys Acta 806: 237–246.

38. Dilworth MF, Gantt E (1981) Phycobilisome-thylakoid topography on photosynthetically active vesicles of Porphyridium cruentum. Plant Physiol 67: 608–612.

39. Mustardy L, Cunningham FX Jr, Gantt E (1992) Photosynthetic membrane topography: quantitative in situ localization of photosystems I and II. Proc Natl Acad Sci USA 89: 10021–10025.

40. Liu L-N, Elmalk AT, Aartsma TJ, Thomas J-C, Lamers GEM, et al. (2008) Light-induced energetic decoupling as a mechanism for phycobilisome-related energy dissipation in red algae: a single molecule study. PLoS ONE 3: e3134.

41. Sivan A, Arad SM (1993) Induction and characterization of pigment mutants in the red microalga Porphyridium sp (Rhodophyceae). Phycologia 32: 68–72.

42. Sivan A, Thomas JC, Dubacq JP, Moppes D, Arad S (1995) Protoplast fusion and genetic complementation of pigment mutations in the red microalga Porphyridium sp. J Phycol 31: 167–172.

43. Kaňa R, Prášil O, Mullineaux CW (2009) Immobility of phycobilins in the thylakoid lumen of a cryptophyte suggests that protein diffusion in the lumen is very restricted. FEBS Lett 583: 670–674.

44. Gantt E, Conti SF (1965) The ultrastructure of Porphyridium cruentum. J Cell Biol 26: 365–381.

45. Gantt E, Conti SF (1966) Granules associated with the chloroplast lamellae of Porphyridium cruentum. J Cell Biol 29: 423–434.

46. Mullineaux CW (2008) Phycobilisome-reaction centre interaction in cyanobacteria. Photosynth Res 95: 175–182.

47. Gantt E, Grabowski B, Cunningham FX, Jr, Green BR, Parson WW (2003) Antenna systems of red algae: phycobilisomes with photosystem II and chlorophyll complexes with photosystem I. In: Green BR, Parson WW, editors. Light-Harvesting Antennas in Photosynthesis. Dordrecht: Kluwer Academic Publishers. pp. 307–322.

48. McConnell MD, Koop R, Vasil'ev S, Bruce D (2002) Regulation of the distribution of chlorophyll and phycobilin-absorbed excitation energy in cyanobacteria. A structure-based model for the light state transition. Plant Physiol 130: 1201–1212.

49. Ray K, Chowdhury MH, Lakowicz JR (2008) Single-molecule spectroscopic study of enhanced intrinsic phycoerythrin fluorescence on silver nanostructured surfaces. Anal Chem 80: 6942–6948.

50. Zehetmayer P, Hellerer T, Parbel A, Scheer H, Zumbusch A (2002) Spectroscopy of single phycoerythrocyanin monomers: dark state identification and observation of energy transfer heterogeneities. Biophys J 83: 407–415.

51. Loos D, Cotlet M, De SF, Habuchi S, Hofkens J (2004) Single-molecule spectroscopy selectively probes donor and acceptor chromophores in the phycobiliprotein allophycocyanin. Biophys J 87: 2598–2608.

52. Zehetmayer P, Kupka M, Scheer H, Zumbusch A (2004) Energy transfer in monomeric phycoerythrocyanin. Biochim Biophys Acta 1608: 35–44.

53. Verkman AS (2002) Solute and macromolecule diffusion in cellular aqueous compartments. Trends Biochem Sci 27: 27–33.

54. Jones RF, Speer HL, Kury W (1963) Studies on the growth of the red alga Porphyridium cruentum. Physiol Plant 16: 636–643.

CITATION

Originally published under the Creative Commons Attribution License. Liu L-N, Aartsma TJ, Thomas J-C, Zhou B-C, Zhang Y-Z (2009). FRAP Analysis on Red Alga Reveals the Fluorescence Recovery Is Ascribed to Intrinsic Photoprocesses of Phycobilisomes than Large-Scale Diffusion. PLoS ONE 4(4): e5295. doi:10.1371/journal.pone.0005295.

Distinct, Ecotype-Specific Genome and Proteome Signatures in the Marine Cyanobacteria *Prochlorococcus*

Sandip Paul, Anirban Dutta, Sumit K. Bag,
Sabyasachi Das and Chitra Dutta

ABSTRACT

Background

The marine cyanobacterium Prochlorococcus marinus, having multiple eco-types of distinct genotypic/phenotypic traits and being the first documented ex-ample of genome shrinkage in free-living organisms, offers an ideal system for studying niche-driven molecular micro-diversity in closely related microbes. The present study, through an extensive comparative analysis of various ge-nomic/proteomic features of 6 high light (HL) and 6 low light (LL) adapted strains, makes an attempt to identify molecular determinants associated with their vertical niche partitioning.

Results

Pronounced strand-specific asymmetry in synonymous codon usage is observed exclusively in LL strains. Distinct dinucleotide abundance profiles are exhibited by 2 LL strains with larger genomes and G+C-content ≈ 50% (group LLa), 4 LL strains having reduced genomes and G+C-content ≈ 35-37% (group LLb), and 6 HL strains. Taking into account the emergence of LLa, LLb and HL strains (based on 16S rRNA phylogeny), a gradual increase in average aromaticity, pI values and beta- & coil-forming propensities and a decrease in mean hydrophobicity, instability indices and helix-forming propensities of core proteins are observed. Greater variations in orthologous gene repertoire are found between LLa and LLb strains, while higher number of positively selected genes exist between LL and HL strains.

Conclusion

Strains of different Prochlorococcus groups are characterized by distinct compositional, physicochemical and structural traits that are not mere remnants of a continuous genetic drift, but are potential outcomes of a grand scheme of niche-oriented stepwise diversification, that might have driven them chronologically towards greater stability/fidelity and invoked upon them a special ability to inhabit diverse oceanic environments.

Background

Evolution of a microbe is often driven by its environment or life-style. Microorganisms adapted to some specialized environmental conditions have been reported to display conspicuous genome and/or proteome features [1-8]. Species of widely varying taxonomic origins, but thriving in same/similar environmental conditions such as high temperature or high salinity, may converge to similar genome and/or proteome composition. In contrast, closely related bacterial species inhabiting distinct ecological niches may display substantial genomic diversity [1-3,6,8-11]. Unveiling the plausible causes/consequences, at the genome and proteome levels, of such niche-dependent evolution of the microbial world poses a major challenge to the present-day life-scientists. The marine cyanobacterium Prochlorococcus marinus [12], having multiple ecotypes exhibiting distinct niche-specific phenotypic as well as genotypic characteristics, offers a useful system to address this issue.

Prochlorococcus are one of the most abundant life forms on this planet and more importantly, are major contributors to global photosynthesis [13-15]. A variety of Prochlorococcus strains, each specialized to dwell in different conditions of light, temperature and nutrient abundances [14,16-18] dominate the euphotic zones of the ocean–mostly between latitudes 40°S and 40°N, and sometimes beyond. To date, the complete genomes of 12 different strains of P. marinus have been sequenced (listed in Table 1) and wide variations have been observed in genetic architectures, genome sizes and genomic G+C-content of these strains [19]. On the basis of vertical niche partitioning, these 12 strains are classified into two major Prochlorococcus ecotypes: high light adapted (HL) ecotype being most abundant in surface waters and low light adapted (LL) ecotype dominating deeper waters [19,20]. 6 of the sequenced strains have been identified to belong to the LL group and the other 6 have been found to survive at HL conditions [19]. The phenomenon of oceanic niche differentiation for P. marinus has been previously investigated into—and several inferences regarding the effects of light adaptation, nutrient availability and predator influence on their genome evolution and diversification has been arrived at [21-24]. However, the full expanse of ecologically relevant differences in genomic, physicochemical and physiological characteristics among these strains are yet to be explored. In the present study, we have attempted to identify novel niche-specific molecular signatures in the genome and proteome compositions of 12 different Prochlorococcus strains, and also investigated the adaptive strategies of different Prochlorococcus strains for their survival in diverse oceanic environments.

Table 1. General features of 12 Prochlorococcus strains under study

Organism	Accession no. (Ref_Seq)	Genome Size (Mb)	G+C- content (%)	Abbreviation
Low light adapted strains				
P. marinus str. MIT 9313	NC_005071.1	2.41	50.74	LL1
P. marinus str. MIT 9303	NC_008820.1	2.70	50.01	LL2
P. marinus subsp. marinus str.CCMP1375 (SS120)	NC_005042.1	1.75	36.44	LL3
P. marinus str. MIT 9211	NC_009976.1	1.70	37.01	LL4
P. marinus str. NATL1A	NC_008819.1	1.90	34.98	LL5
P. marinus str. NATL2A	NC_007335.1	1.84	35.12	LL6
High light adapted strains				
P. marinus str. AS9601	NC_008816.1	1.70	31.32	HL1
P. marinus str. MIT 9312	NC_007577.1	1.71	31.21	HL2
P. marinus subsp. pastoris str.CCMP1986 (MED 4)	NC_005072.1	1.70	30.73	HL3
P. marinus str. MIT 9515	NC_008817.1	1.70	30.79	HL4
P. marinus str. MIT 9215	NC_009840.1	1.70	31.15	HL5
P. marinus str. MIT 9301	NC_009091.1	1.60	31.34	HL6

It is worth mentioning in this context that Prochlorococcus is the first documented example of genome shrinkage along with A+T enrichment in a free-living organism [25]. Earlier examples of genome reduction had been restricted to endosymbionts or pathogens with a host-dependent lifestyle, which evolve under the constraint of frequent population bottlenecks with a subsequent increase in genetic drift [2,3,26-29]. Considering the abundance of P. marinus in the marine ecosystem, their reductive genome evolution might not be influenced by similar population bottlenecks and resulting genetic drifts, and thus seems to be a more complex phenomenon to explain. Although P. marinus genome evolution has been investigated previously [25,30] and the event of genome shrinkage have been ascribed to various factors related to their growth in oligotrophic waters [20,23,31], selection for metabolic economy [25,31,32], loss of low fitness genes [33], and smaller cell sizes [25], it is still unclear to what extent it has been driven by any random genetic drift and/or other specific selection force(s). Our analyses indicate that the ecotype-specific molecular signatures exhibited by P. marinus strains under study are not mere remnants of a continuous genetic drift, but a potential outcome of niche-oriented stepwise diversification of Prochlorococcus, orchestrated by an array of interplaying adaptive forces.

Results

In an attempt to understand the trends in molecular evolution in Prochloro-coccus, we have analyzed various genome and proteome characteristics of 6 LL and 6 HL strains of P. marinus. The analyses of genome/proteome in P. marinus include the study of trends in codon, dinucleotide and amino acid usages, gene synteny of orthologous sequences, intergenic sequence composi-tion, physicochemical properties of the encoded proteins and the extent of positive selection among different strains. These analyses were primarily di-rected towards the identification of niche-specific variations within different Prochlorococcus strains.

Strand-Specific Asymmetry in Synonymous Codon Usage in Low Light Adapted P. Marinus Genomes

In order to find out the trends in synonymous codon usage, we have carried out correspondence analysis (COA) on relative synonymous codon usage (RSCU) of 12 different strains of P. marinus. Figure 1 shows the positions of the indi-vidual genes on the planes defined by the first and second major axes generated by COA on RSCU values of coding sequences of respective Prochlorococcus genomes under study. Interestingly, in cases of all LL strains, the genes tran-scribed from the leading and the lagging strands of replication are segregat-ed in two distinguishable clusters (with little overlap between them), either along Axis1 or Axis2 or both. Similar scatter plots with two distinct clusters of genes were observed earlier in cases of microbial genomes with pronounced strand-specific mutational bias [2,3,34,35]. When the positions of all synony-mous codons are plotted on the plane defined by the first and second major axis of COA on RSCU for genes of all LL strains (plot not shown), a clear separation between G-/U- ending codons and A-/C- ending codons is observed, indicating the presence of asymmetric mutational bias at synonymous codon usage level in low light (LL) adapted genomes of P. marinus. In contrast, for each of the HL strains, a single cluster of genes (i.e., no segregation between the leading and lagging strand genes) is found on the Axis1-Axis2 plane of COA on their RSCU values (Figure 1), suggesting the absence of any pronounced strand-specific asymmetry in their synonymous codon usage. For all LL strains, GT3-content of genes exhibit significant correlations with Axis1 and/or Axis2 values (Table 2). 3 of the 6 LL strains, viz. LL1, LL3 and LL4 show highly significant cor-relations between Axis1 and GT3-content of genes, whereas LL2, LL5 and LL6 strains show significantly high correlations between Axis2 and GT3-content (Table 2). In LL strains the positions of the leading and lagging strand genes on the planes defined by Axis1 and Axis2 of COA (Figure 1), therefore, clearly indicate the overall GT3-richness of the leading strand genes. However, for the HL strains of Prochlorococcus, the percentage of variance explained by the first two principal axes are relatively low and the correlation values of these axes with GC3- or GT3-contents of genes are either insignificant or much lower than those observed for LL strains (Table 2). This suggests that no single axis and/or parameter can explain strand-specific variations in synonymous codon usage of HL strains.

Table 2. General features and correlations of GC3 and GT3 content with first two axes of COA on RSCU values of genes in 12 Prochlorococcus genomes

Organism		ORFs under study	G+C- content (%)	% of total variation explained by COA on RSCU		Correlation coefficient (r)			
						Axis 1 vs.		Axis 2 vs.	
				Axis 1	Axis 2	GC3	GT3	GC3	GT3
P. marinus str. MIT 9313	(LL1)	1947	52.1	14.33	10.89	-0.59	**0.85**	-0.69	-0.45
P. marinus str. MIT 9303	(LL2)	2080	51.6	15.93	12.29	**-0.89**	0.43	-0.26	**-0.86**
P. marinus subsp. marinus str. CCMP1375	(LL3)	1492	37.1	7.47	6.23	-0.15	**-0.81**	-0.58	0.25
P. marinus str. MIT 9211	(LL4)	1486	38.6	7.31	6.10	-0.24	**-0.78**	0.52	-0.36
P. marinus str. NATL1A	(LL5)	1541	35.8	6.73	5.67	-0.50	-0.24	-0.20	**0.71**
P. marinus str. NATL2A	(LL6)	1517	35.9	6.81	5.80	0.54	0.07	-0.10	**0.71**
P. marinus str. AS9601	(HL1)	1463	31.9	5.51	5.07	-0.07	0.06	0.11	-0.13
P. marinus str. MIT 9312	(HL2)	1503	31.8	5.85	5.20	0.23	-0.09	-0.09	0.10
P. marinus subsp. pastoris str. CCMP1986	(HL3)	1451	31.5	5.96	5.68	-0.05	-0.06	0.17	0.03
P. marinus str. MIT 9515	(HL4)	1465	31.6	5.78	5.49	0.11	-0.12	-0.26	0.00
P. marinus str. MIT 9215	(HL5)	1506	31.8	5.67	5.25	-0.21	-0.07	0.09	-0.14
P. marinus str. MIT 9301	(HL6)	1469	32.0	5.54	5.12	0.09	0.00	-0.13	-0.11

Correlations significant at $p < 10^{-6}$ are indicated in boldface.

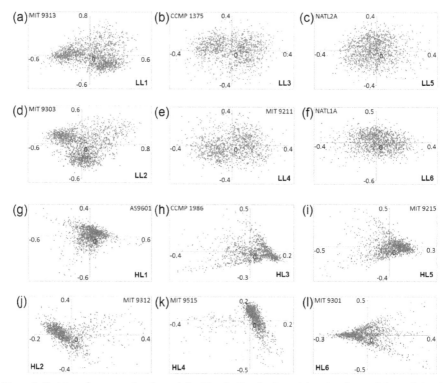

Figure 1. Position of genes on the planes defined by the first (horizontal Axis1) and second (vertical Axis2) major axes generated by COA on RSCU values of coding sequences for each of the 12 Prochlorococcus strains (a to l). Genes transcribed from the leading and lagging strands are represented by red and blue coloured dots respectively.

Chi-square tests on occurrences of different codons on two replicating strands of representative LL strains (LL1 and LL6) further reveal significant overrepresentation of 28 and 22 G-/U- ending codons on the leading strands of LL1 and LL6 respectively (p < 0.001); while 28 and 25 A-/C- ending codons are overrepresented in the genes encoded on the lagging strands of LL1 and LL6 respectively (p < 0.001). The codon 'CUG' is the only exception, which, in spite of being G-ending, is significantly overrepresented in the lagging strand genes of LL1.

In microbial genomes characterized by pronounced strand asymmetry [2,3,34,35], replicational-transcriptional selection usually play a major role in shaping genome organization. The leading strands of replication of such organisms, in general, contain higher number of genes due to replicational selection, and are also enriched with highly expressed genes as an effect of transcriptional selection. However, in LL strains of P. marinus, the predicted protein coding sequences are found to be distributed almost equally in two strands. In fact, in three LL strains (LL1, LL2 and LL3), the number of predicted protein coding sequences are lower in the leading strands than in the lagging strands (approximately 48% in the leading strands and 52% in the lagging strands), indicating the absence of replicational selection. The lagging strands of the strains MIT9313 (LL1) and MIT9303 (LL2) are also found to be enriched in ribosomal proteins, which are typically highly expressed. Only two such genes are encoded by each of their leading strands, leaving 36 and 37 ribosomal proteins to be encoded by their lagging strands (LL1 and LL2 respectively). For other LL strains, the distribution of ribosomal genes is quite conventional ($\approx$ 25 on leading strands and $\approx$ 12 on lagging strands). However, most of the other potentially highly expressed genes (e.g., RNA polymerases, transcription and translation processing factors, etc.), are present in higher numbers in the leading strands of all LL strains. Hence it is difficult to arrive at any definite conclusion regarding the effects of transcriptional selection on the LL strains of Prochlorococcus.

Larger Extent of Genomic Rearrangements Between Small and Large P. Marinus Genomes

In order to understand the nature and extent of genomic rearrangements during the events of niche specific diversification, we have analyzed the conservation of the relative order or synteny of orthologs across the chromosomes of different P. marinus strains. Figure 2 represents graphically, the gene synteny of 519 orthologs (see methods) in three low light (LL1, LL3 and LL6) and two high light adapted (HL3 and HL4) strains of P. marinus. The low light strains of Prochlorococcus have considerable variations in their genomic size and G+C-content (Table 1). The LL1 strain here represents the high G+C and large genome containing LL strains, while the LL3 & LL6 strains represent the other members of the LL group which have lower G+C-content and reduced genome size. HL3 and HL4 strains are representatives of the HL group, which also share smaller genome and lower G+C content. Interestingly, the extent of variation in order and orientation of orthologous genes between two strains of similar light optima but of distinct genome sizes and G+C-content (viz., between LL1 and LL3, Figure 2), is significantly greater than that between two strains with distinct light optima, but having similar genome size and G+C-content (e.g., between LL6 and HL3). It is worth mentioning at this point that this conclusion holds good for all 12 P. marinus strains under study, i.e., the results will be similar even if the representative strains selected here are replaced with any other strains form the groups they represent. The analysis, therefore, suggests that genome reduction in P. marinus has been accompanied by numerous seemingly random genome rearrangements such as translocations and inversions. Amongst the reduced genomes, the order and orientation of orthologous gene clusters remain more or less conserved except some local reorientation of genome fragments, the number and extent of reorientations being higher between the strains with distinct light adaptation (Figure 2).

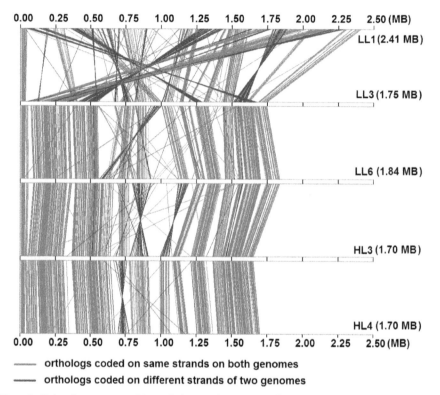

Figure 2. Order of arrangement of 519 orthologs on chromosomes of 5 representative Prochlorococcus strains (LL1, LL3, LL6, HL3 and HL4). The red and blue lines join the locations of pair of orthologs on the linearly depicted chromosomes. Red lines indicate presence of the orthologous pairs on the same strand (either +/+ or -/-) of the two chromosomes they join, while the blue colour indicates the presence of the orthologs on different strands (+/- or -/+). Chromosomal scales are shown in megabase pairs (MB) and 0 MB represents the predicted origins of replication for the 5 strains.

Niche-Specific Dinucleotide Abundance Values of P. Marinus Genomes

It has been reported previously that dinucleotide abundance values are usually similar in related species and can be regarded as a genome signature [36]. To understand the patterns in dinucleotide signature, we have calculated dinucleotide abundance values for all P. marinus genomes and also for E. coli (a representative outgroup). The genomic G+C-bias of E. coli is similar to that of LL1 and LL2 Prochlorococcus strains. The results are shown diagrammatically in Figure 3.

Some Prochlorococcus-specific trends are exhibited by all P. marinus organisms under study, irrespective of their G+C-bias or ecological adaptation. For instance, the dinucleotide CG is appreciably overrepresented in E. coli, but significantly underrepresented in all P. marinus, including the strains which have similar G+C-content to that of E. coli. Contrasting trends in E. coli and P. marinus strains are also observed for the dinucleotides AG/CT and GA/TC. The dinucleotide abundance values of AC/GT are also significantly underrepresented in all P. marinus strains, but not in E. coli.

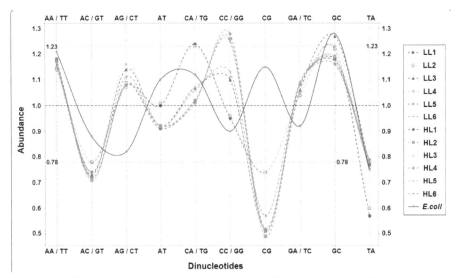

Figure 3. Plot of dinucleotide abundance profiles of E. coli and 12 Prochlorococcus strains. Differently coloured lines join the abundance values of dinucleotide pairs for each of the organisms. Abundance values ≥ 1.23 or ≤ 0.78 are significantly over or underrepresented (as described by Karlin et al., Theoretical population biology, 61, 367-390, 2002).

However, significant intra-Prochlorococcus differences are also present in dinucleotide abundance profiles, on the basis of which all P. marinus strains under study may be divided into three distinct groups:

(a) Group LLa, comprised of the two LL strains P. marinus MIT 9313 (LL1) and MIT 9303 (LL2)—both having larger genomes (≈ 2.5 MB) and average G+C-content ≈ 50%: Genomes of these two strains are characterized by significantly high values of CA/TG and low values of TA. The values for AT, AC/GT and CG are also relatively higher and that of CC/GG are lower, as compared to other P. marinus strains.

(b) Group LLb, consisting of other four LL strains (LL3, LL4, LL5 and LL6), characterized by relatively lower G+C-content (between 35%–37%) and small genome size (< 2 MB): These four LL strains exhibit highly similar patterns, which are visibly distinct mainly at CA/TG and CC/GG from the almost overlapping profile of HL strains (Figure 3).

(c) Group HL, including all 6 HL strains having reduced genome and G+C-content ≈ 31%: The dinucleotide CC/GG is significantly overrepresented only in the HL Prochlorococcus strains.

Clustering by Amino Acid Composition Reveals a Balance Between Genomic G+C-Bias and Prochlorococcus-Specific Selection Forces

In an attempt to investigate whether the strand-specific mutational bias has any impact on amino acid usage in gene products of LL strains of P. marinus in comparison to their HL counterparts, we performed correspondence analysis (COA) on relative amino acid usage (RAAU) of the encoded proteins of each organism. No clear segregation can be observed for proteins encoded by the leading and lagging strands in any of the P. marinus genomes under study (data not shown), implying that the strand-specific mutational bias has hardly any influence on the amino acid compositions of the gene products of LL strains of Prochlorococcus. In all the strains of P. marinus, the first three axes generated by COA on amino acid usage cumulatively explain about 39% of the total variability. Both mean hydrophobicity and aromaticity of the encoded proteins exhibit strong correlations with either of the first two principal axes and seem to be the major contributors to amino acid usage variation in P. marinus proteins (data not shown).

In order to check whether the amino acid usage patterns in LL and HL groups of Prochlorococcus follow any specific trends, we carried out a clustering analysis on relative abundances (with respect to E. coli) of different amino acid residues of each organism (Figure 4) in a dataset comprising of all P. marinus strains under study along with a cyanobacterial representative Cyanothece sp. (having G+C-content similar to those of E. coli and two LLa strains), and two non-cyanobacterial species—the bacteroidetes/chlorobi G. foresetii (average genomic G+C-content = 36.6%, similar to those of LLb strains) and the epsilon-proteobacteria C. jejuni (average genomic G+C-content = 30.3%, similar to those of HL strains). Branching patterns suggest an optimization between the G+C-bias and P. marinus group-specific selection for amino acid usage. As can be seen in Figure 4, a major branching between the organisms occurs according to the average genomic G+C-content. Organisms having average G+C-content around 50%—viz. E. coli, Cyanothece sp. and the two LLa strains—cluster together under the node 'a', while the organisms with lower G+C-content (30-37%) form a distinct cluster at the node 'b.' However, within the A+T-rich or relatively G+C-rich clusters, finer segregation occur according to taxonomy, i.e., between the cyanobacterial and non-cyanobacterial species. For instance, node 'a' acts as a bifurcation point between the gamma-proteobacteria E. coli and the three cyanobacterial species, followed by another bifurcation at node 'c' between the two LLa strains and non-Prochlorococcus species Cyanothece. Similarly, within the A+T-rich cluster, the HL strains of P. marinus with G+C-content ≈ 30-31% club together with their LLb counterparts (average G+C-content ≈ 35-37%) under the node 'f,' distinctly separated from C. jejuni and G. foresetii. Node 'f' also acts as the point of divergence of LLb and HL strains, which cluster separately under the nodes 'g' and 'j' (containing 4 LLb strains and 6 HL strains respectively). These observations indicate that though the amino acid usage patterns in Prochlorococcus are primarily guided by their directional mutational bias, other selection pressures must also have exerted some significant influence. Three cyanobacterial species, namely Cyanothece sp., P. marinus str. MIT9313 (LL1) and P. marinus str. MIT9303 (LL2)–all have nearly the same G+C-content as E. coli, yet their amino acid usage abundance patterns are quite distinct from that of E. coli. Similarly, the LLb and HL strains of P. marinus display significant differences in amino acid usage patterns from that of non- P. marinus species (G. foresetii and C. jejuni, respectively) having similar G+C-content. A careful examination of Figure 4 delineates the Prochlorococcus-specific features in amino acid usage patterns. For instance, as compared to E. coli, Thr is underrepresented in all strains of P. marinus, including those having ≈ 50% G+C-content (and also in C. jejuni, but not in Cyanothece

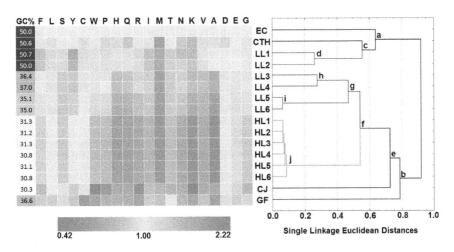

Figure 4. Single linkage (Euclidean distances) clustering based on standardized amino acid usages (with respect to E. coli) of 12 Prochlorococcus strains and 4 other microbes, accompanied by a heatmap representation of the standardized amino acid usage values. The overrepresentation or underrepresentation of amino acid residues in the organisms are shown in green and red colored blocks of varying colour intensities, respectively. [Abbreviations, EC → E. coli; CJ → C. jejuni; GF → G. foresetii and CTH → Cyanothece sp.]

and G. forsetti). Ser is overrepresented in all A+T-rich P. marinus species, especially in comparison to C. jejuni and G. foresetii. Tyr is typically underrepresented in all Prochlorococcus strains, compared to the non-Prochlorococcus species of similar G+C-content. Frequency of Phe is also relatively lower in Prochlorococcus in comparison to any other species of similar G+C-bias. All these observations suggest that the amino acid usage in P. marinus is a result orchestrated by the forces of species-specific selection acting on mutational bias and genetic drift.

Niche-Specific Variations in Physicochemical and Structural Features of Prochlorococcus Orthologs

In an attempt to have a better insight into the niche-specific physicochemical and structural properties of P. marinus proteins, if any, we performed a comparative analysis of the core proteins, i.e., proteins found to be present in all P. marinus strains under study. Different proteomic properties of 519 orthologous proteins between all 12 P. marinus strains are summarized in Table 3, which shows a gradual increase in average aromaticity and pI values and decrease in mean hydrophobicity and instability indices of orthologs, as one moves from the members of group LLa to group LLb to group HL. In other words, the core proteins of LL strains are, in general, more hydrophobic, more acidic, less aromatic and less stable than their HL orthologs. The comparison of structural properties reveals that among the three groups, members of the group LLa exhibit the highest propensity for helix formation and lowest propensity for beta-sheet and coil formation. LLb group orthologs are characterized by intermediate values for all three propensities, and group HL orthologs display trends opposite to that of group LLa–i.e. lowest propensity for helix formation and highest propensity for beta-sheet/coil formation (Table 3).

Table 3. Different amino acid indices and secondary structural traits of 519 orthologous proteins present in 12 Prochlorococcus strains

Organism	Amino acid indices (Mean)				Secondary structural traits (%)		
	Hydrophobicity	Aromaticity	Isoelectric Point (pI)	Instability Index (II)	Alpha helix	Beta sheet	Coil
LL1	-0.11	0.067	6.62	40.61	37.51	13.76	48.73
LL2	-0.11	0.067	6.65	40.71	37.45	13.74	48.81
LL3	-0.16	0.072	7.07	38.88	35.85	14.58	49.57
LL4	-0.16	0.072	7.08	38.81	36.13	14.26	49.61
LL5	-0.18	0.073	7.08	38.09	35.15	14.68	50.17
LL6	-0.18	0.073	7.07	38.10	35.15	14.69	50.16
HL1	-0.20	0.077	7.15	37.03	34.32	15.33	50.35
HL2	-0.20	0.077	7.18	37.19	34.31	15.36	50.33
HL3	-0.20	0.077	7.23	37.33	34.42	15.36	50.22
HL4	-0.20	0.077	7.32	37.36	34.54	15.27	50.19
HL5	-0.20	0.077	7.21	37.02	34.38	15.27	50.35
HL6	-0.20	0.077	7.14	37.15	34.35	15.20	50.45

However, one may argue that these inter-group variations in physicochemical properties and structural propensities of P. marinus strains can only be a reflection of their varying genomic G+C-bias rather than being niche specific. In order to address this issue, we carried out a comparative analysis of various proteomic features of orthologous sequences from three representative P. marinus strains LL1, LL3 and HL3 (from groups LLa, LLb and HL respectively) with those of three other caynobacterial species, Synechococcus elongatus (55.5% G+C-content), Synechocystis sp.(47.4% G+C-content) and Nostoc sp.(41.3% G+C-content) as well as three non-cyanobacterial species namely E. coli, Bacillus cereus and Francisella tularensis. The non-cyanobacterial species were chosen as reference organisms, due to their close average genomic G+C-content to those of LLa, LLb and HL strains (50.8%, 35.5% and 32.3% for E. coli, B. cereus &F. tularensis, respectively), and the significant number of orthologs they share with these P. marinus strains. Values of different physicochemical parameters and structural propensities of the orthologs from these reference species and the representative P. marinus strains are summarized in the Table 4. It reveals that the values of any specific parameter, say of hydrophobicity index or instability index, is in most cases, not comparable between orthologs from P. marinus strains and outgroup organisms of similar G+C-bias. For instance, the values observed for B. cereus are quite different from those of LL3—the genomic G+C-content of both being quite similar. The average pI value of B. cereus proteins is not only significantly less than that of their LL3 orthologs, it is even lesser than LL1 proteins. The average instability index of B. cereus proteins is also much less than that of the Prochlorococcus orthologs. The average helix forming propensity of B. cereus proteins is closer to that of LL1 proteins, while their beta sheet forming propensity is almost same to that of HL3 proteins. Similarly, the aromaticity and instability indices or helix forming propensities of E. coli proteins are significantly different from those of Prochlorococcus strain LL1, while most of the F. tularensis protein characteristics differ widely from those of the P. marinus HL3 strain with similar G+C-bias. Comparing between the cyanobacterial species, amino acid indices and secondary structural traits of Synechococcus (the closest taxonomic relative of Prochlorococcus) seem to be guided by its G+C-content (Table 4: Set IV). The hydrophobicity values and the helix-forming propensities of the other two cyanobacteria also gradually decrease with decrease in genomic G+C-content within the set. However, the other indices and structural traits of Synechocystis and Nostoc do not reflect any systematic co-variation with their G+C-bias. These observations suggest that the variations in proteomic features of Prochlorococcus might not be a mere outcome of their G+C-bias, there could be significant influence of other selection forces as well.

Table 4. Comparison between various amino acid indices and secondary structural traits of six sets of proteins of Prochlorococcus and non-Prochlorococcus orthologs

Organisms		Mean of amino acid indices				Secondary structural traits (%)		
		Hydrophobicity	Aromaticity	pI	Instability Index	Alpha helix	Beta sheet	Coil
Set I (303 pairs)	BC	-0.14	0.08	6.51	34.34	38.62	16.64	44.74
	LL1	-0.08	0.06	6.61	38.72	38.07	14.33	47.60
	LL3	-0.13	0.07	7.09	37.33	36.37	15.22	48.41
	HL3	-0.17	0.07	7.23	36.17	34.65	16.01	49.35
Set II (136 pairs)	EC	-0.05	0.08	6.91	36.58	40.26	14.43	45.31
	LL1	-0.07	0.07	6.57	38.94	37.44	14.76	47.80
	LL3	-0.12	0.07	7.13	37.65	35.59	16.17	48.24
	HL3	-0.15	0.08	7.30	35.63	33.87	16.63	49.49
Set III (265 pairs)	FT	-0.15	0.08	7.30	33.68	38.22	16.60	45.19
	LL1	-0.12	0.06	6.64	39.46	37.71	14.35	47.94
	LL3	-0.17	0.07	7.18	37.93	36.03	15.12	48.85
	HL3	-0.21	0.07	7.33	36.57	34.56	15.79	49.64
Set IV (175 pairs)	SN	-0.03	0.07	6.88	41.23	40.46	13.39	46.14
	LL1	-0.06	0.06	6.61	39.39	37.54	14.27	48.19
	LL3	-0.10	0.07	7.15	37.82	35.47	15.13	49.40
	HL3	-0.15	0.08	7.25	35.96	34.25	15.73	50.01
Set V (963 pairs)	SSP	-0.09	0.08	6.41	38.77	37.72	14.60	47.68
	LL1	-0.06	0.07	6.76	42.06	38.80	13.15	48.05
	LL3	-0.11	0.08	7.45	38.95	36.00	14.81	49.18
	HL3	-0.16	0.09	7.81	36.21	34.34	16.12	49.55
Set VI (961 pairs)	NOS	-0.08	0.08	6.58	38.59	38.35	15.05	46.61
	LL1	-0.06	0.07	6.81	41.96	39.00	13.11	47.89
	LL3	-0.11	0.08	7.49	38.87	36.25	14.74	49.01
	HL3	-0.17	0.09	7.82	36.30	34.27	16.09	49.62

BC - *Bacillus cereus* (35.5% G+C), EC - *Escherichia coli* (50.8% G+C), FT - *Francisella tularensis* (32.3% G+C), SN - *Synechococcus elongatus* (55.5% G+C), SSP - *Synechocystis sp.* (47.4% G+C), NOS - *Nostoc sp.* (41.3% G+C)

Higher Positive Selection Between Orthologs from Strains with Opposite Light Optima

In order to better understand the evolutionary trends in different P. marinus species having distinct genome composition and/or light adaptation, the rates of synonymous and non-synonymous substitutions (dS and dN) were calculated between 519 orthologous sequences of LL1, LL6 and HL3 strains (representatives of groups LLa, LLb and HL respectively), and the number of genes showing positive selection (dN > dS) between each possible pair of organisms were determined. Figure 5 depicts a Venn diagram for the number of positively selected genes among the strains under study. Out of 519 orthologs, maximum number of positively selected genes (90) is found between LL1 and HL3–the strains that differ in genome size, G+C-composition and light adaptation. The strains LL6 and HL3 having nearly similar genome size and G+C-bias, but distinct light optima come next with 78 positively selected genes among them and the two strains LL1 and LL6 of the same light group, but distinct genome size and G+C-bias, exhibit the minimum number (68) of positively selected genes. Among these three sets, there are several genes, which are positively selected between any two out of the three possible pairs of P. marinus strains under study. There are 25 genes selected positively between HL3 and either of the LL strains i.e. between the strains of two opposite light optima, irrespective of their G+C-bias and genome size. 17 genes are positively selected between the strains of distinct G+C-bias and genome size— between the relatively G+C-rich and large genome strain LL1 and either of the A+T-rich and reduced genome strains LL6 or HL3. Only 7 genes display common positive selection between LL6 and LL1 (the strains with similar light adaptation but of different genome size and G+C-bias) and between LL6 and HL3 (the strains having relatively lower G+C-bias and genome size). Thus, our study indicates the presence of a considerable positive selection pressure in diversification of the Prochlorococcus core genome, which in turn, suggests an appreciable role of random genetic drift in vertical niche partitioning of the strains.

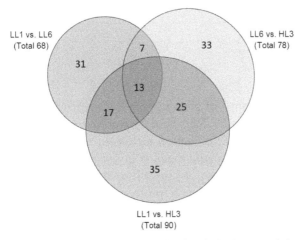

Figure 5. Venn diagram depicting the number of positively selected (dN/dS > 1) orthologs between pairs of different Prochlorococcus strains. The 519 core proteins of 3 representative organisms (LL1, LL6 & HL3) were considered for the analysis. dN/dS could be calculated for 414 × 3 pairs out of the 519 × 3 pairs considered.

Pronounced Effects of Directional Mutational Bias in the Intergenic Regions of HL P. Marinus Strains

In an attempt to examine whether the G+C-bias of intergenic regions of the different strains (with varying genomic G+C-content), follow trends similar to the respective coding regions, G+C-content of intergenic regions were calculated. The intergenic regions are, in general, more A+T rich than the overall genomic G+C-content of respective organisms. Also, the A+T bias of intergenic regions are more pronounced in HL strains than their LL counterparts.

Unconstrained intergenic regions are more prone to mutational change. Accumulation of unfavourable mutations may render a coding region nonfunctional, facilitating its removal from the genome in course of time. We have identified probable remnants of coding sequences within intergenic regions of two representative Prochlorococcus strains having reduced genomes (LL6 and HL3). The G+C-content of these remnants are, in most cases, higher than that of average intergenic DNA, but lower than the average G+C-content of the bona fide coding regions (Table 5). These particular non-coding sequences, therefore, may be remnants of coding sequences that are in the process of being eliminated from the genome.

Table 5. Putative remnants of coding regions and their G+C-content (%)

Organism	Putative remnants of coding regions			Overall G+C%		
	No. of hits	Average G+C%	Standard Deviation	Intergenic	Coding	Genomic
LL6	48*	33.78	4.87	28.76	35.98	35.12
HL3	93*	26.35	4.04	23.33	31.64	30.80

* for details see Additional file 5

Discussion

Exhibition of a wide range of genomic G+C-content (30.8% to 50.7%) and genome sizes (1.6 Mb to 2.7 Mb) by different strains of P. marinus, and also their adaptation to different ecological niches—a situation encountered rarely in the microbial world—demand detailed investigation. We have performed a large scale comprehensive study to critically analyze the direction and strength of mutational pressure and genomic/proteomic determinants associated with the adaptation of these strains to oceanic environments subject to different light intensities. From this study it appears that (a) low light adapted (LL) free living Prochlorococcus strains exclusively show strand asymmetry in synonymous codon usage, (b) general trends in amino acid usage in LLa, LLb and HL strains differ appreciably, (c) distinct dinucleotide abundance profiles are exhibited by LLa, LLb and HL strains, (d) higher number of genes have undergone positive selection between the strains with distinct light optima, i.e., between LL and HL strains and (e) there are definite trends in variations of different physicochemical and structural features in core proteomes of different groups of Prochlorococcus strains, which are not solely governed by their genomic G+C-bias. These observations, along with the findings on large-scale genome reduction associated with gradual increase in genomic A+T-content and extensive chromosomal rearrangements between different strains, strongly suggest a stepwise diversification of Prochlorococcus strains, in course of their adaptive evolution (Figure 6).

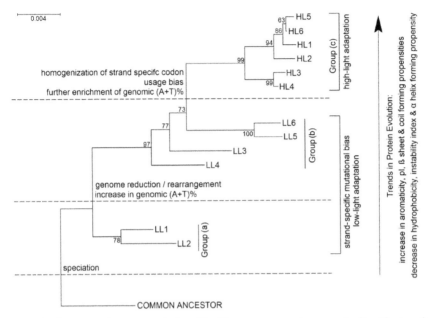

Figure 6. Trends in genome/proteome evolution of Prochlorococcus, suggesting a stepwise diversification of the ecotypes. The model is based on the 16S rRNA phylogeny of the 12 strains, inferred from a bootstrap consensus tree (500 replicates) generated using the Minimum Evolution method (CNI algorithm), with the software MEGA (version 4).

Among several genome/proteome signatures of P. marinus strains reported for the first time in this work, the most notable is the impact of pronounced replication-strand-specific asymmetry on synonymous codon usage, observed exclusively in the low light adapted strains of P. marinus (Figure 1). This is noteworthy for two reasons: (i) Presence of pronounced strand-specific mutational bias with detectable influence on codon usage was observed so far mostly for obligatory intracellular microorganisms having reduced genomes [2,3,5]. Interestingly, all 6 LL strains of P. marinus exhibiting strand-specific synonymous codon usage are free-living and two of them (LL1 and LL2) are characterized by relatively larger genome size. On the other hand, for the reduced genomes of 6 HL strains, no perceivable sign of strand asymmetry could be seen in their usage of synonymous codons. (ii) In most of the other microbial genomes with asymmetric mutational bias, the genes, especially the highly expressed ones, are present in the leading strands of replication in significantly higher numbers, the phenomenon referred to as replicational-transcriptional selection [2,3,34,35]. No such definite significant bias in gene distribution is observed in either of the strands of replication in the LL strains of P. marinus. Strand asymmetry in codon usage of Prochlorococcus, therefore, may not bear an explicit causality to the event of genome reduction or with replicational-transcriptional selection.

The homogenization of the strand asymmetric bias in the HL strains may be attributed, at least partially, to the absence of a specific type of DNA repair enzyme MutY. In previous studies of Rocap et al. [20] and Dufresne et al. [25] it have been shown that the enzyme MutY is absent in the strain P. marinus str. CCMP1986 (HL3), while it is present in P. marinus str. CCMP1375 (LL3) and P. marinus str. MIT9313 (LL1). MutY, an A/G-specific DNA glycosylase, acts with MutT (NTP pyrophosphohydrolase) and MutM (formamido-pyrimidine-DNA glycosylase) to avoid misincorporation of oxidized guanine (8-oxoG) in DNA and to repair the base mismatches A:8-oxoG [37]. Knocking out both mutM and mutY in E. coli results in a 1,000-fold increase of G:C to A:T transversions in comparison to the wild-type strain [38]. Our analysis reveals (through BLASTP search) that mutY is present only in the LL strains, but not in any of the 6 HL strains. The excess number of 'G's present in the leading strands of LL strains might have transversed to 'A's in the HL strains due to the absence of mutY in the later, and this in turn, caused a simultaneous increase of 'T's in the lagging strands, eventually leading to homogenization of the G+T and A+C frequencies in two strands of replication in the HL strains. Existing mutational drift towards A+T-enrichment in the HL strains might also have facilitated achieving the uniformity in those strains. Further insights may be accumulated in this regard with the availability of more completely sequenced Prochlorococcus genomes in future.

In the process of gradual genome reduction, mutations often accumulate in expendable genes, thereby transforming them, by degrees, to pseudogenes, to small fragments, to extinction [39]. In the reduced genomes of P. marinus, we have found some putative remnants of coding regions, the A+T-content of which are, in general, higher than that of coding regions, but lower than other non-coding regions. This is in agreement with the fact that the reduced genomes of P. marinus (especially those of HL strains) are subject to a strong mutational A+T-drift, and will therefore result in gradual A+T-enrichment of the genic remnants already released from amino-acid-coding constraints in recent past. The base composition of such remnants is expected to gradually approach the A+T-content of bona fide non-coding regions.

Comparison of orthologous gene synteny from five representative strains having different genome size and G+C-content clearly points at a high level of chromosomal rearrangement during genome shrinkage in Prochlorococcus. This finding is in agreement with earlier findings on association of chromosomal rearrangement events with higher rates of chromosomal evolution and/or the phenomenon of genome reduction, as in Arabidopsis thaliana [40] and different endoparasites/endosymbionts [41,42]. Intra-chromosomal recombination at duplicated sequences often results in deletion of intervening sequences, and rearrangement of flanking regions, thereby leading to genome shrinkage [39].

Previous analyses with endosymbiotic or endoparasitic organisms like Bartonella, Tropheryma, Buchnera, Wigglesworthia etc. [2,3,28,29] revealed that the phenomenon of genome reduction is normally associated with population bottlenecks or other mechanisms such as selective sweeps. In case of the hyperthermophile Nanoarchaeum equitans, extreme genome reduction is a feature of its thermoparasitic adaptation [1]. Although our knowledge of bacterial populations in open oceans is not exhaustive, it may certainly be assumed that P. marinus ecotypes, the most abundant free-living marine cyanobacteria and an important contributor to global photosynthesis, are not subject to small population sizes [13-15,30]. More importantly, the HL strains with reduced genomes are apparently biologically superior than their LL counterparts [21]. It is possible that the bias towards reduced A+T rich genomes in HL strains is consistent with cellular economy at regions with limited nitrogen and phosphorous near the ocean surface. Scarcity of these elements that are essential in DNA synthesis favors the incorporation of an AT base-pair containing seven atoms of nitrogen, one less than a GC base-pair. It is worth mentioning at this point that the trends in amino acid usage in different P. marinus strains, as observed in this study are quite compatible with the earlier report by Lv et al. [43] on influence of resource availability on proteome composition of these species. For instance, increase in overall aromaticity from LLa to LLb and HL strains is in full agreement with the observations by Lv et al. [43] on increased carbon-content in the encoded proteins of different HL strains as compared to that of LL strains. The average instability indices of the HL proteins are significantly lower than those of their LL orthologs, suggesting that the HL proteins, in general, may be more stable. Proteins characterized by higher percentages of helix structures, experience increased overall packing that imparts more rigidity [44] and, hence, a decrease in regions with helix-forming propensities with a subsequent increase in coiled structures in HL proteins probably makes them more flexible. It is also tempting to presume that higher values of aromaticity and pI in HL proteins, as compared to LL orthologs, might facilitate cation-pi interactions in the former, imparting more stability. The central issue in the adaptation of HL proteins to their environmental niches may, therefore, be the conservation of their functional state, characterized by a well-balanced optimization of stability and flexibility.

Conclusion

The current study advocates for the presence of adaptive selection forces that might have played significant role in governing Prochlorococcus evolution and fitness at the genome and proteome levels. An optimization between these adaptive forces and directional mutational bias has set definite trends in molecular evolution of P. marinus. This characterizes different P. marinus ecotypes with distinct niche-specific compositional, physicochemical and structural traits, thereby driving them chronologically towards increasing stability and/or fidelity.

Methods

Sequence Retrieval

All predicted protein coding sequences and the complete genome sequences of the 12 different strains of P. marinus were retrieved from the NCBI GenBank (listed in Table 1). For comparison, the predicted protein coding sequences of E. coli (NC_000913.2), Bacillus cereus (NC_003909.8), Francisella tularensis (NC_006570.1), Synechococcus elongatus (NC_006576.1), Synechocystis sp. (NC_000911.1), Nostoc sp. (NC_003272.1), Campylobacter jejuni (NC_003912.7), Cyanothece (NC_011884.1) and Gramella forsetii (NC_008571.1) were also retrieved from GenBank. Annotated ORFs, which encode proteins less than 100 amino acids long, were not considered for further analysis.

Determination of Leading and Lagging Strand Genes

In order to identify the replication origin (oriC) or termination (ter) sites we performed GC-skew (G-C/G+C) analysis using a sliding window of 10 Kb along the genome sequence. The sites were validated by checking the neighbouring gene organization (e.g. identified origins in Prochlorococcus genomes were flanked by DNA polymerase beta subunit III gene on the 3' side and the Threonine synthatase gene on the 5' side) and the presence of DnaA boxes in their vicinity [45]. Based on the predicted oriC and ter sites, the leading strands and lagging strands of replication for each genomes were identified along with the genes encoded on the two strands.

Multivariate Analyses on Synonymous Codon and Amino Acid Usage and Cluster Analysis on Amino Acid Usage

Correspondence analysis (COA) on relative synonymous codon usage (RSCU) and amino acid usage of genes/proteins were performed on individual genomes in order to identify any significant variation in the usage of codons or amino acids, if present, and help ascertain the underlying cause(s), using the program CODONW 1.4.2 [46].

To find out the variation in amino acid usage between LL and HL Prochlorococcus strains, a cluster analysis on standardized amino acid usage was carried out using STATISTICA (version 6.0, published by Statsoft Inc., USA) for all 12 Prochlorococcus organisms (Table 1) along with E. coli, Cyanothece, C. jejuni and G. forsetii having G+C-content nearly equal to the different LL and HL strains. The amino acid usage of E. coli was chosen as a well-defined reference for standardizing the amino acid composition for the analysis and to produce an accompanying heat map. With the help of a program developed in-house in Visual Basic, a 16×20 matrix (heatmap) was generated, where the rows and the columns correspond to data sources (i.e., organisms in the cluster) and standardized amino acid usage values, respectively. The overrepresentation or underrepresentation of standardized amino acid usage values of the organisms in the matrix are shown in green or red colored blocks (Figure 4) respectively, and their intensities varying in accordance with their deviation from the standard (yellow). The extreme left column represents the genomic G+C-content of the respective organisms.

Dinucleotide Analysis of DNA Sequences

For all Prochlorococcus genomes and E. coli, the dinucleotide abundance for each possible dinucleotide was calculated as the ratio between the observed and expected frequencies of the concerned dinucleotide in its genomic context [47]. Dinucleotide abundance values generally represent the genomic signature of any species [48] and here we were interested to see whether all Prochlorococcus genomes follow a similar trend or not.

Determination of Orthologs

Stand alone BLAST package (ver. 2.2.18) was downloaded from the NCBI FTP site and using the package, all-to-all BLASTN and BLASTP searches were performed with the genes from all the 12 strains of P. marinus. Orthologs across these organisms were defined for this study as protein coding genes having a BLASTP sequence Identity ≥ 60%, not more than 20% difference in length and E-value ≤ 1e-20. The resultant list of 'orthologs' were checked for consistency with the data obtained from Genplot, http://www.ncbi.nlm.nih.gov/sutils/geneplot.cgi, which houses a pair-wise list of genes giving mutually best BLASTP hits when all genes from the genomes of any two organisms are 'blasted' against each other. We have identified 519 orthologs present in all 12 P. marinus genomes as their core proteome. The stringent measures employed for the similarity search ensure that these orthologs have been sufficiently conserved throughout the adaptive evolution of P. marinus, and any niche-specific features deciphered from this dataset would certainly not be a trivial outcome. For comparative analysis with suitable outgroup organisms, we retrieved orthologs of nine organisms including three representative P. marinus strains (LL1, LL3 and HL3), E. coli, B. cereus, F. tularensis, S. elongatus, Synechocystis sp. and Nostoc sp. from NCBI GenePlot by filtering the symmetrical best hits of protein homologs.

Estimation of Synonymous and Non-Synonymous Substitution Patterns in Irthologous Sequences

Positive selection can be inferred from a higher proportion of non-synonymous over synonymous substitutions per site (dN/dS > 1). The dN and dS values were calculated for 519 orthologs of LL1, LL6 and HL3 using the software MEGA (version 4) [49]. The calculation was based on the modified Nei-Gojobori Jukes-Cantor method that considers deviations from an equal frequency of transitions and transversions [50,51].

Gene Synteny Visualization

Comparison of the gene repertoire or gene synteny between 5 representative Prochlorococcus strains (LL1, LL3, LL6, HL3 and HL4) were carried out using a Java program developed in-house. It can represent the arrangement of orthologous genes between two chromosomes by joining the locations of the orthologs by differently coloured lines. The red lines represent the genes present on the same strand (+/-) and blue lines represent orthologs coded on different strands of chromosomes being compared.

Calculation of Codon/Amino Acid Usage Indices and Estimation of Secondary Structure of Proteins

Indices like relative synonymous codon usage (RSCU) [52], G+C and G+T-content at third codon positions (GC3 & GT3 respectively), aromaticity and average hydrophobicity (Gravy score) [53] of protein coding sequences were calculated to find out the factors influencing codon and amino acid usages. The isoelectric point (pI) and instability index [54] of each protein were calculated using the Expasy proteomics server [55]. Secondary structures of the identified orthologs were computed using the software PREDATOR [56] and the varying percentages of the structural components (viz. helices, sheets, and coils) in proteins from different strains were also noted.

Identification of Intergenic Regions

The sequences coding for mRNAs and structural RNAs were noted from the protein table and structural RNA table respectively (available from NCBI) for each of the organisms. Intergenic regions were identified by subtracting the regions of these gene sequences from the whole genome. The overall G+C-content of the intergenic regions were calculated after concatenating all the intergenic sequences together, for each of the 12 Prochlorococcus strains. For identification of probable pseudogenes/remnants of coding DNA in LL6 and HL3 (two representative strains of groups LLb and HL having reduced genomes), their intergenic regions were subjected to a similarity search (tBlastX) against a pool of Prochlorococcal genes (consisting of sequences from three representative strains LL1, LL6, HL3). 48 hits for LL6 and 93 hits for HL3 were identified, having sequence identities ≥ 30%, aligned lengths ≥ 15 amino acids, and E-values < 1e-3.

Abbreviations

COA: correspondence analysis; LL: low light; HL: high light; pI: isoelectric point; RSCU: relative synonymous codon usage; GT3: G+T-content at third codon positions; GC3: G+C-content at third codon positions.

Authors' Contributions

SP and AD made substantial contributions to the design of the study, devised and carried out the overall strategy and drafted the manuscript. SKB developed relevant programs for data mining and analysis of genome sequences and also participated in sequence analysis. SD participated in the initial phase of the work, made thoughtful discussion during execution of the project and preparation of the manuscript. CD conceived and coordinated the study and revised the manuscript critically for important intellectual content. All authors read and approved the final manuscript.

Acknowledgements

We thank Sanjib Chatterjee and Avik Datta, IICB for giving technical support in calculating dinucleotide abundance and synonymous and non-synonymous divergence. This work was supported by the Department of Biotechnology, Government of India (Grant Number BT/BI/04/055-2001) and Council of Scientific and Industrial Research (Project no. CMM 0017). SP and AD are supported by

Senior Research Fellowships from Council of Scientific and Industrial Research, India.

References

1. Das S, Paul S, Bag SK, Dutta C: Analysis of Nanoarchaeum equitans genome and proteome composition: indications for hyperthermophilic and parasitic adaptation. BMC Genomics 2006, 7:186.

2. Das S, Paul S, Chatterjee S, Dutta C: Codon and amino acid usage in two major human pathogens of genus Bartonella—optimization between replicational-transcriptional selection, translational control and cost minimization. DNA Res 2005, 12(2):91–102.

3. Das S, Paul S, Dutta C: Evolutionary constraints on codon and amino acid usage in two strains of human pathogenic actinobacteria Tropheryma whipplei. J Mol Evol 2006, 62(5):645–658.

4. Eisenberg H: Life in unusual environments: progress in understanding the structure and function of enzymes from extreme halophilic bacteria. Arch Biochem Biophys 1995, 318(1):1–5.

5. Moran NA, Wernegreen JJ: Lifestyle evolution in symbiotic bacteria: insights from genomics. Trends Ecol Evol 2000, 15(8):321–326.

6. Paul S, Bag SK, Das S, Harvill ET, Dutta C: Molecular signature of hypersaline adaptation: insights from genome and proteome composition of halophilic prokaryotes. Genome Biol 2008, 9(4):R70.

7. Pikuta EV, Hoover RB, Tang J: Microbial extremophiles at the limits of life. Crit Rev Microbiol 2007, 33(3):183–209.

8. Singer GA, Hickey DA: Thermophilic prokaryotes have characteristic patterns of codon usage, amino acid composition and nucleotide content. Gene 2003, 317(1-2):39–47.

9. Bliska JB, Casadevall A: Intracellular pathogenic bacteria and fungi--a case of convergent evolution? Nat Rev Microbiol 2009, 7(2):165–171.

10. Merhej V, Royer-Carenzi M, Pontarotti P, Raoult D: Massive comparative genomic analysis reveals convergent evolution of specialized bacteria. Biol Direct 2009, 4:13.

11. Mongodin EF, Nelson KE, Daugherty S, Deboy RT, Wister J, Khouri H, Weidman J, Walsh DA, Papke RT, Sanchez Perez G, et al.: The genome of Salinibacter ruber: convergence and gene exchange among hyperhalophilic bacteria and archaea. Proc Natl Acad Sci USA 2005, 102(50):18147–18152.

12. Chisholm S, Olson R, Zettler E, Goericke R, Waterbury J, Welschmeyer N: A novel free-living prochlorophyte abundant in the oceanic euphotic zone. Nature 1988, 334:340–343.

13. Goericke R, Welschmeyer N: The marine prochlorophyte Prochlorococcus contributes significantly to phytoplankton biomass and primary production in the Sargasso Sea. Deep-sea research Part 1 Oceanographic research papers 1993, 40(11-12):2283–2294.

14. Partensky F, Blanchot J, Vaulot D: Differential distribution and ecology of Prochlorococcus and Synechococcus in oceanic waters: a review. Bulletin de l'Institut océanographique(Monaco) 1999, 457–475.

15. Partensky F, Hess WR, Vaulot D: Prochlorococcus, a marine photosynthetic prokaryote of global significance. Microbiol Mol Biol Rev 1999, 63(1):106–127.

16. Moore LR, Rocap G, Chisholm SW: Physiology and molecular phylogeny of coexisting Prochlorococcus ecotypes. Nature 1998, 393(6684):464–467.

17. Urbach E, Scanlan DJ, Distel DL, Waterbury JB, Chisholm SW: Rapid diversification of marine picophytoplankton with dissimilar light-harvesting structures inferred from sequences of Prochlorococcus and Synechococcus (Cyanobacteria). J Mol Evol 1998, 46(2):188–201.

18. West NJ, Scanlan DJ: Niche-partitioning of Prochlorococcus populations in a stratified water column in the eastern North Atlantic Ocean. Appl Environ Microbiol 1999, 65(6):2585–2591.

19. Kettler GC, Martiny AC, Huang K, Zucker J, Coleman ML, Rodrigue S, Chen F, Lapidus A, Ferriera S, Johnson J, et al.: Patterns and implications of gene gain and loss in the evolution of Prochlorococcus. PLoS Genet 2007, 3(12):e231.

20. Rocap G, Larimer FW, Lamerdin J, Malfatti S, Chain P, Ahlgren NA, Arellano A, Coleman M, Hauser L, Hess WR, et al.: Genome divergence in two Prochlorococcus ecotypes reflects oceanic niche differentiation. Nature 2003, 424(6952):1042–1047.

21. Coleman ML, Chisholm SW: Code and context: Prochlorococcus as a model for cross-scale biology. Trends Microbiol 2007, 15(9):398–407.

22. Garcia-Fernandez JM, Diez J: Adaptive mechanisms of nitrogen and carbon assimilatory pathways in the marine cyanobacteria Prochlorococcus. Res Microbiol 2004, 155(10):795–802.

23. Martiny AC, Coleman ML, Chisholm SW: Phosphate acquisition genes in Prochlorococcus ecotypes: evidence for genome-wide adaptation. Proc Natl Acad Sci USA 2006, 103(33):12552–12557.

24. Sullivan MB, Waterbury JB, Chisholm SW: Cyanophages infecting the oceanic cyanobacterium Prochlorococcus. Nature 2003, 424(6952):1047–1051.

25. Dufresne A, Garczarek L, Partensky F: Accelerated evolution associated with genome reduction in a free-living prokaryote. Genome Biol 2005, 6(2):R14.

26. Berg OG, Kurland CG: Evolution of microbial genomes: sequence acquisition and loss. Mol Biol Evol 2002, 19(12):2265–2276.

27. Lynch M, Blanchard JL: Deleterious mutation accumulation in organelle genomes. Genetica 1998, 102-103(1-6):29–39.

28. Wernegreen JJ: Genome evolution in bacterial endosymbionts of insects. Nat Rev Genet 2002, 3(11):850–861.

29. Wernegreen JJ, Moran NA: Evidence for genetic drift in endosymbionts (Buchnera): analyses of protein-coding genes. Mol Biol Evol 1999, 16(1):83–97.

30. Hu J, Blanchard JL: Environmental sequence data from the Sargasso Sea reveal that the characteristics of genome reduction in Prochlorococcus are not a harbinger for an escalation in genetic drift. Mol Biol Evol 2009, 26(1):5–13.

31. Garcia-Fernandez JM, de Marsac NT, Diez J: Streamlined regulation and gene loss as adaptive mechanisms in Prochlorococcus for optimized nitrogen utilization in oligotrophic environments. Microbiol Mol Biol Rev 2004, 68(4):630–638.

32. Dufresne A, Salanoubat M, Partensky F, Artiguenave F, Axmann IM, Barbe V, Duprat S, Galperin MY, Koonin EV, Le Gall F, et al.: Genome sequence of the cyanobacterium Prochlorococcus marinus SS120, a nearly minimal oxyphototrophic genome. Proc Natl Acad Sci USA 2003, 100(17):10020–10025.

33. Marais GA, Calteau A, Tenaillon O: Mutation rate and genome reduction in endosymbiotic and free-living bacteria. Genetica 2008, 134(2):205–210.

34. Lafay B, Lloyd AT, McLean MJ, Devine KM, Sharp PM, Wolfe KH: Proteome composition and codon usage in spirochaetes: species-specific and DNA strand-specific mutational biases. Nucleic Acids Res 1999, 27(7):1642–1649.

35. McInerney JO: Replicational and transcriptional selection on codon usage in Borrelia burgdorferi. Proc Natl Acad Sci USA 1998, 95(18):10698–10703.

36. Gentles AJ, Karlin S: Genome-scale compositional comparisons in eukaryotes. Genome Res 2001, 11(4):540–546.

37. Michaels ML, Cruz C, Grollman AP, Miller JH: Evidence that MutY and MutM combine to prevent mutations by an oxidatively damaged form of guanine in DNA. Proc Natl Acad Sci USA 1992, 89(15):7022–7025.

38. Horst JP, Wu TH, Marinus MG: Escherichia coli mutator genes. Trends Microbiol 1999, 7(1):29–36.

39. Andersson SG, Zomorodipour A, Andersson JO, Sicheritz-Ponten T, Alsmark UC, Podowski RM, Naslund AK, Eriksson AS, Winkler HH, Kurland CG: The genome sequence of Rickettsia prowazekii and the origin of mitochondria. Nature 1998, 396(6707):133–140.

40. Yogeeswaran K, Frary A, York TL, Amenta A, Lesser AH, Nasrallah JB, Tanksley SD, Nasrallah ME: Comparative genome analyses of Arabidopsis spp.: inferring chromosomal rearrangement events in the evolutionary history of A. thaliana. Genome Res 2005, 15(4):505–515.

41. Belda E, Moya A, Silva FJ: Genome rearrangement distances and gene order phylogeny in gamma-Proteobacteria. Mol Biol Evol 2005, 22(6):1456-1467.

42. Mira A, Ochman H, Moran NA: Deletional bias and the evolution of bacterial genomes. Trends Genet 2001, 17(10):589–596.

43. Lv J, Li N, Niu DK: Association between the availability of environmental resources and the atomic composition of organismal proteomes: evidence from Prochlorococcus strains living at different depths. Biochem Biophys Res Commun 2008, 375(2):241–246.

44. Fleming PJ, Richards FM: Protein packing: dependence on protein size, secondary structure and amino acid composition. J Mol Biol 2000, 299(2):487–498.

45. Mackiewicz P, Zakrzewska-Czerwinska J, Zawilak A, Dudek MR, Cebrat S: Where does bacterial replication start? Rules for predicting the oriC region. Nucleic Acids Res 2004, 32(13):3781–3791.

46. Penden J: Analysis of codon usage. PhD thesis. University of Nottingham, Department of Genetics; 1997.

47. Karlin S, Burge C: Dinucleotide relative abundance extremes: a genomic signature. Trends Genet 1995, 11(7):283–290.

48. Karlin S, Mrazek J, Campbell AM: Compositional biases of bacterial genomes and evolutionary implications. J Bacteriol 1997, 179(12):3899–3913.

49. Tamura K, Dudley J, Nei M, Kumar S: MEGA4: Molecular Evolutionary Genetics Analysis (MEGA) software version 4.0. Mol Biol Evol 2007, 24(8):1596–1599.

50. Nei M, Gojobori T: Simple methods for estimating the numbers of synonymous and nonsynonymous nucleotide substitutions. Mol Biol Evol 1986, 3(5):418–426.

51. Nei M, Kumar S: Molecular evolution and phylogenetics. Oxford University Press, USA; 2000.

52. Sharp PM, Li WH: The codon Adaptation Index--a measure of directional synonymous codon usage bias, and its potential applications. Nucleic Acids Res 1987, 15(3):1281–1295.

53. Kyte J, Doolittle RF: A simple method for displaying the hydropathic character of a protein. J Mol Biol 1982, 157(1):105–132.

54. Guruprasad K, Reddy BV, Pandit MW: Correlation between stability of a protein and its dipeptide composition: a novel approach for predicting in vivo stability of a protein from its primary sequence. Protein Eng 1990, 4(2):155–161.

55. Expasy Proteomics Server [http://expasy.org].

56. Frishman D, Argos P: Seventy-five percent accuracy in protein secondary structure prediction. Proteins 1997, 27(3):329–335.

CITATION

Originally published under the Creative Commons Attribution License. Paul S, Dutta A, Bag SK, Das S, Dutta C. Distinct, Ecotype-Specific Genome and Proteome Signatures in the Marine Cyanobacteria Prochlorococcus. BMC Genomics 2010, 11:103 doi:10.1186/1471-2164-11-103.

Global Expression Analysis of the Brown Alga *Ectocarpus siliculosus* (Phaeophyceae) Reveals Large-Scale Reprogramming of the Transcriptome in Response to Abiotic Stress

Simon M. Dittami, Delphine Scornet, Jean-Louis Petit,
Béatrice Ségurens, Corinne Da Silva, Erwan Corre,
Michael Dondrup, Karl-Heinz Glatting, Rainer König,
Lieven Sterck, Pierre Rouzé, Yves Van de Peer, J. Mark Cock,
Catherine Boyen and Thierry Tonon

ABSTRACT

Background

Brown algae (Phaeophyceae) are phylogenetically distant from red and green algae and an important component of the coastal ecosystem. They have developed unique mechanisms that allow them to inhabit the intertidal zone, an environment with high levels of abiotic stress. Ectocarpus siliculosus is being established as a genetic and genomic model for the brown algal lineage, but little is known about its response to abiotic stress.

Results

Here we examine the transcriptomic changes that occur during the short-term acclimation of E. siliculosus to three different abiotic stress conditions (hyposaline, hypersaline and oxidative stress). Our results show that almost 70% of the expressed genes are regulated in response to at least one of these stressors. Although there are several common elements with terrestrial plants, such as repression of growth-related genes, switching from primary production to protein and nutrient recycling processes, and induction of genes involved in vesicular trafficking, many of the stress-regulated genes are either not known to respond to stress in other organisms or are have been found exclusively in E. siliculosus.

Conclusions

This first large-scale transcriptomic study of a brown alga demonstrates that, unlike terrestrial plants, E. siliculosus undergoes extensive reprogramming of its transcriptome during the acclimation to mild abiotic stress. We identify several new genes and pathways with a putative function in the stress response and thus pave the way for more detailed investigations of the mechanisms underlying the stress tolerance of brown algae.

Background

The brown algae (Phaeophyceae) are photosynthetic organisms, derived from a secondary endosymbiosis [1], that have evolved complex multicellularity independently of other major groups such as animals, green plants, fungi, and red algae. They belong to the heterokont lineage, together with diatoms and oomycetes, and are hence very distant phylogenetically, not only from land plants, animals, and fungi, but also from red and green algae [2]. Many brown algae inhabit the intertidal zone, an environment of rapidly changing physical conditions due to the turning tides. Others form kelp forests in cold and temperate waters as well as in deep-waters of tropical regions [3,4]. Brown algae, in terms of biomass, are the primary organisms in such ecosystems and, as such, represent important habitats for a wide variety of other organisms. As sessile organisms, brown algae require high levels of tolerance to various abiotic stressors such as osmotic pressure, temperature, and light. They differ from most terrestrial plants in many aspects of their biology, such as their ability to accumulate iodine [5], the fact that they are capable of synthesizing both C18 and C20 oxylipins [6], their use of laminarin as a storage polysaccharide [7], the original composition of their cell walls, and the associated cell wall synthesis pathways [8-10]. Many aspects of brown algal biology, however, remain poorly explored, presenting a high potential for new discoveries.

In order to fill this knowledge gap, Ectocarpus siliculosus, a small, cosmopolitan, filamentous brown alga (see [11] for a recent review) has been chosen as a model [12], mainly because it can complete its life cycle rapidly under laboratory conditions, is sexual and highly fertile, and possesses a relatively small genome (200 Mbp). Several genomic resources have been developed for this organism, such as the complete sequence of its genome and a large collection of expressed sequence tags (ESTs). Although Ectocarpus is used as a model for developmental studies [13,14], no molecular studies have been undertaken so far to study how this alga deals with the high levels of abiotic stress that are a part of its natural environment. This is also true for intertidal seaweeds in general, where very few studies have addressed this question.

In the 1960s and 1970s several studies (reviewed in [15]) examined the effects of abiotic stressors such as light, temperature, pH, osmolarity and mechanical stress on algal growth and photosynthesis. However, only a few of the mechanisms underlying the response to these stressors—for example, the role of mannitol as an osmolyte in brown algae [16,17]—have been investigated so far. Developing and applying molecular and biochemical tools will help us to further our knowledge about these mechanisms—an approach that was suggested 12 years ago by Davison and Pearson [18]. Nevertheless, it was only recently that the first transcriptomic approaches were undertaken to investigate stress tolerance in intertidal seaweeds. Using a cDNA microarray representing 1,295 genes, Collén et al. [19,20] obtained data demonstrating the up-regulation of stress-response genes in the red alga Chondrus crispus after treatment with methyl jasmonate [19] and suggesting that hypersaline and hyposaline stress are similar to important stressors in natural environments [20]. Furthermore, in the brown alga Laminaria digitata, Roeder et al. [21] performed a comparison of two EST libraries (sporophyte and protoplasts) and identified several genes that are potentially involved in the stress response, including the brown alga-specific vanadium-dependent bromoperoxidases and mannuronan-C5-epimerases, which are thought to play a role in cell wall modification and assembly. These studies have provided valuable information about the mechanisms and pathways involved in algal stress responses, but they were nevertheless limited by the availability of sequence information for the studied organisms at the time.

With the tools and sequences available for the emerging brown algal model E. siliculosus, we are now in a position to study the stress response of this alga on the level of the whole transcriptome. For this, we have developed an EST-based microarray along with several tools and annotations (available on our Ectocarpus transcriptomics homepage [22]), and used this array to study the transcriptomic response of E. siliculosus to three forms of abiotic stress: hyposaline, hypersaline, and oxidative stress. Hypersaline stress is a stress experienced by intertidal seaweeds—for example, in rockpools at low tide (due to evaporation) or due to anthropogenic influences—and is comparable to desiccation stress. Hyposaline stress is also common in the intertidal zone, and can arise, for example, due to rain. Furthermore, organisms with a high tolerance to saline stress can inhabit a wide range of habitats. E. siliculosus strains have been isolated from locations covering a wide range of salinity. A specimen was found in a highly salt-polluted area of the Werra river in Germany, where chloride concentrations at times reached 52.5 grams per liter [23]. At the same time, E. siliculosus can be found in estuaries, in the Baltic sea, and one strain of E. siliculosus was isolated from freshwater [24]. Oxidative stress is commonly experienced by living organisms. Reactive oxygen species (ROSs) are produced intracellularly in response to various stressors due to malfunctioning of cellular components, and have been implicated in many different signaling cascades in plants [25]. In algae, several studies have demonstrated the production of ROSs in response to biotic stress (reviewed in [26]). Therefore, protection against these molecules is at the basis of every stress response and has been well studied in many organisms. We simulated this stress by the addition of hydrogen peroxide to the culture medium.

Results

Determination of Sub-Lethal Stress Conditions

The aim of this study was to determine the mechanisms that allow short-term acclimation to abiotic stress. To be sure to monitor the short-term response to stress rather than just cell death, the intensity of the different stresses needed to be chosen carefully. Using a pulse amplitude modulation fluorometer (see Materials and methods), we measured the effects of different stress intensities on photosynthesis. Figure 1 shows the change in quantum yield of photosynthesis in response to different intensities of the different stresses, where values of over 0.5 indicate low stress. The quantum yield can vary during the course of the day even under controlled conditions, as changes in light have a strong impact on this parameter. Stress conditions were chosen to have a clear effect on the photosynthesis rate, but to be sub-lethal, allowing the alga to acclimate and recover. The conditions that corresponded best to these criteria were 1.47 M NaCl (hypersaline condition, approximately three times the salinity of normal seawater), 12.5% seawater, and

1 mM H2O2 (oxidative stress condition), although, for this last stressor, we can assume that the H2O2 concentration in the medium decreases over the course of the experiment. Each stress was applied for 6 hours because this corresponds to the time span between high and low tide. In addition, experiments carried out on land plants [27] and red algae [19] have indicated that the application of stress for 6 hours induces the most marked changes in transcription.

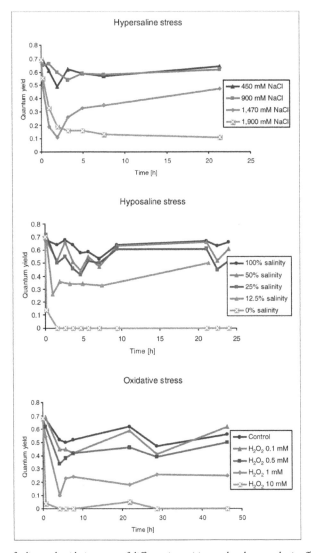

Figure 1. Effects of saline and oxidative stress of different intensities on the photosynthetic efficiency (quantum yield) of E. siliculosus. The conditions in red (1,470 mM NaCl, 12.5% seawater, and 1 mM H2O2) were the conditions chosen for the microarray analysis.

Initially, we had considered a fourth stress condition, 2 M sorbitol in artificial sea water (ASW), to imitate the osmotic pressure of the hypersaline treatment without the possible effects of the salts. However, this treatment was not included in the final experiment because cultures did not survive this treatment for 6 hours. For the other stresses, we observed 100% recovery of photosynthesis after about 6 days, even after 24 hours of stress.

Intracellular Osmolarity and Na+ Concentration

Apart from the photosynthetic activity, we also measured intracellular osmolarity and Na+ concentrations (Figure 2). After 6 hours of exposure to different salinities, the intracellular osmolarity was always about 500 mOsm higher than that of the extracellular medium. The intracellular Na+ concentration was about 500 mM lower than in the extracellular medium under hypersaline stress, 60 mM lower under control conditions, and the same under hyposaline stress. Oxidative stress had no detectable effect on the intracellular ion composition or osmolarity (data not shown).

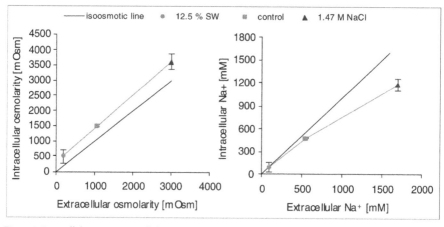

Figure 2. Intracellular versus extracellular osmolarity and Na+ concentration under saline stress. Oxidative stress samples are not shown as they did not differ significantly from the control sample. Every point represents the mean of five biological replicates ± standard deviation.

The E. Siliculosus Microarray Represents 17,119 Sequences

We designed a microarray based on 90,637 ESTs obtained by sequencing clones from 6 different cDNA libraries: immature sporophyte (normalized and non-normalized), mature sporophyte, immature gametophyte, mature gametophyte, and stress (sporophyte). Cleaning and assembly resulted in the generation of 8,165 contigs and 8,874 singletons. In addition, 21 genomic sequences and 231 E. siliculosus Virus 1 (EsV-1) genes were included. The array design file has been deposited under the accession number [ArrayExpress:A-MEXP-1445] and is also available on our Ectocarpus transcriptomics homepage [22].

Of the 17,119 genes represented on the array, 12,250 gave a significant signal over background in our experiments and were considered to be expressed under the conditions tested. The analysis focused on these 12,250 genes (see Materials and methods). A first comparison with the data obtained from a tiling experiment with E. siliculosus (MP Samanta and JM Cock, personal communication), where 12,600 genes were considered strongly expressed, demonstrates that our array offers a rather complete coverage of at least the highly transcribed parts of the E. siliculosus genome, suggesting that we are working at the whole genome scale.

cDNA Synthesis and Amplification Provided Consistent Results with Both mRNA and Total RNA Samples

For reasons as yet unknown, cDNA synthesis reactions with E. siliculosus are inhibited at high concentrations of RNA. Therefore, we decided to synthesize cDNAs from a small quantity of total RNA or mRNA, and to include a PCR amplification step in the protocol to obtain sufficient double-stranded cDNA (4 µg) for each hybridization. A comparison of the four four-fold replicates synthesized from 30 ng of mRNA and the single four-fold replicate synthesized from 100 ng total RNA showed that these two protocols yielded similar results. All total RNA replicates clustered with the mRNA replicates of the same stress (data not shown). Nevertheless, at a false discovery rate (FDR) of 5%, 163 transcripts gave significantly different results with the two types of sample. These transcripts represented mainly constituents of the ribosome, as revealed by a Kyoto Encyclopedia of Genes and Genomes (KEGG) Orthology Based Annotation System (KOBAS) analysis and by an analysis of overrepresented GO terms.

Validation of Microarray Results using Quantitative PCR

Nineteen genes that exhibited significant changes in their expression patterns in the microarray analysis were analyzed by real time quantitative PCR (RT-qPCR). Eighteen of these had similar expression profiles in both the microarray experiment and the RT-qPCR experiment (correlation coefficient r of between 0.57 and 0.99; Table 1). Only one gene, which codes for a microsomal glutathione S-transferase, displayed a different pattern in the two experiments (r = -0.48). Furthermore, the seven most stable 'housekeeping genes' as identified by qPCR in a previous report [28] showed only statistically non-significant relative changes of <1.5-fold (log2-ratio <0.58) in the microarray experiment (Table 2). This demonstrated that the protocol for cDNA amplification provided reliable measures of the relative transcript abundances. Although this method has been successfully applied in several small-scale expression studies [29-35], to our knowledge, the use of this technique has not been reported with commercial photolithographically synthesized arrays.

Table 1. Comparison of microarray and RT-qPCR results for genes changing expression

Comparison of microarray and RT-qPCR results for genes changing expression

ID	Genome ID	Name	r	Function
CL4038Contig1	[Esi0355_0025]	HSP70	0.87	HSP70
LQ0AAB7YD09FM1.SCF	[Esi0155_0065]	NADH	0.94	NADH dehydrogenase
CL7513Contig1	[Esi0269_0011]	ProDH	0.79	Proline dehydrogenase
CL3741Contig1	[Esi0024_0066]	TF	0.90	Putative transcription factor
LQ0AAB12YN05FM1.SCF	[Esi0399_0008]	WD_rep	0.66	WD repeat gene
CL1Contig3	[Esi0085_0055]	CLB1	0.95	Chlorophyll binding protein
CL43Contig1	[Esi0199_0054]	CLB2	0.98	Fucoxanthin binding protein
CL7742Contig1	[Esi0026_0055]	TagS	0.69	TAG synthase
CL2765Contig1	[Esi0526_0006]	NH4-Tr	0.96	Ammonium transporter
CL3832Contig1	[Esi0437_0012]	FOR	0.67	Phycoerythrobilin:ferredoxin oxidoreductase
LQ0AAA16YN10FM1.SCF	[Esi0153_0004]	Arg-MetTr	0.71	Arginine N-methyltransferase
CL7099Contig1	[Esi0018_0111]	HSD	0.83	Homoserine dehydrogenase
CL6576Contig1	[Esi0107_0059]	IGPS	0.97	Indole-3-glycerol-phosphate synthase
CL7231Contig1	[Esi0686_0001]	CDPK	0.85	cAMP-dependent protein kinase
CL4027Contig1	[Esi0122_0054]	mGST	-0.48	Microsomal glutathione S-transferase
CL4274Contig1	[Esi0023_0183]	SNR	0.57	SNR (vesicular transport)
CL5850Contig1	[Esi0109_0088]	mG	0.99	Glycin-rich protein
CL455Contig1	[Esi0159_0021]	G6PD	0.91	Glucose-6-phosphate 1-dehydrogenase
CL6746Contig1	[Esi0116_0065]	IF4E	0.91	Eukaryotic initiation factor 4E

R is the Pearson correlation coefficient between the microarray and the RT-qPCR expression profile. ID corresponds to the name of the sequence on the array.

Table 2. Comparison of microarray and RT-qPCR results for housekeeping or stable genes

Comparison of microarray and RT-qPCR results for housekeeping or stable genes

ID	Genome ID	Name	Maximum change ARRAY	Maximum change QPCR	Function
LQ0AAB30YA12FM1.SCF	[Esi0298_0008]	Dyn	0.23	0.77	Dynein
CL1914Contig1	[Esi0021_0024]	ARP2.1	0.22	0.44	Actin related protein
CL3Contig2	[Esi0387_0021]	EFIA	0.08	0.46	Elongation factor 1 alpha
CL8Contig12	[Esi0053_0059]	TUA	0.57	0.91	Alpha tubulin
CL1073Contig1	[Esi0054_0059]	UBCE	0.22	0.38	Ubiquitin-conjugating enzyme
CL29Contig4	[Esi0302_0019]	UBQ	0.18	0.82	Ubiquitin
CL461Contig1	[Esi0072_0068]	R26S	0.22	n/a	Ribosomal protein S26

The table displays the maximum log2-ratio between any stress and the control condition for both the microarray and the RT-qPCR analysis. No RT-qPCR value is available for R26S, as this gene was used for normalization of the RT-qPCR samples. ID corresponds to the name of the sequence on the array.

Ribosomal Protein Genes are among those whose Transcript Abundances are Least affected by Stress

The 100 most stably expressed genes in these microarray experiments included 51 genes with unknown functions. Nineteen genes code for ribosomal proteins, and

21 genes are known housekeeping genes with functions related to protein turnover (transcription, 4 genes; translation, 3 genes; degradation, 3 genes), energy production (6 genes), and the cytoskeleton (5 genes).

Classification of Stress Response Genes Using Automatic Annotations

Overall, 8,474 genes were identified as being differentially expressed in at least one of the conditions compared to the control, allowing a FDR of 10% (5,812 were labeled significant at an FDR of 5%). As can be seen in Figure 3, the relative change for these genes ranged from 1.2-fold (log2-ratio ≈0.3) to more than 32-fold (log2-ratio >5). Of these 8,474 genes, 2,569 (30%) could be automatically annotated with GO terms using the GO-term Prediction and Evaluation Tool (GOPET) [36] and 1,602 (19%) with KEGG orthology annotations using the KOBAS software [37]. These automatic annotations were analyzed for each stress condition individually, to identify GO categories and KEGG pathways that were significantly over-represented.

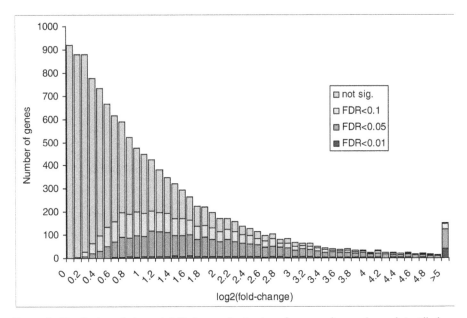

Figure 3. Distribution of observed fold-changes (log2-ratios of stress and control samples). All three comparisons between stress and control treatments were considered and the observed frequencies averaged. The color coding shows how many transcripts were labeled as differentially expressed at different FDRs. Not sig., not significant.

The KOBAS results (Figure 4) indicated that under hyposaline and hypersaline stresses most of the changes involved down-regulation of the synthesis and metabolism of amino acids. More precisely, genes involved in the synthesis of valine, leucine, and isoleucine, as well as that of the aromatic amino acids (phenylalanine, tyrosine, tryptophan), and arginine and proline metabolism were affected. This effect on amino acid synthesis was less marked for oxidative stress, where glutamate metabolism was the only amino acid metabolism affected. Under hypersaline conditions, there was also an increase in transcripts coding for enzymes that metabolize valine, leucine, and isoleucine. In addition, photosynthesis and vesicular transport seemed to be altered by both hyposaline and oxidative stress. Pathways that appeared to be specifically affected by one stress included the up-regulation of fatty acid metabolism and down-regulation of translation factors under hypersaline stress, the up-regulation of the proteasome and down-regulation of nitrogen metabolism under hyposaline stress, and an increase in glycerophospholipid metabolism under oxidative stress (Figure 4).

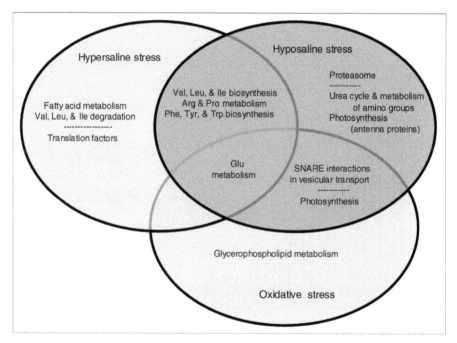

Figure 4. Venn diagram of KEGG pathways identified as over-represented among the transcripts significantly up- or down-regulated (FDR <0.1) in the different stress conditions. Only KEGG pathways with q-values < 0.1 in at least two conditions or for both datasets (FDR of 0.05 and FDR of 0.1) were considered. The general category 'other enzymes' was not included. Further 'SNARE interactions in vesicular transport' includes the category 'SNARE,' and 'photosynthesis' includes 'photosynthesis proteins' and 'porphyrin and chlorophyll metabolism.' No pathways were found to be common only to hyposaline and hypersaline stress. SNARE, soluble N-ethylmaleimide-sensitive factor attachment receptor.

The GOPET analysis (Table 3) was focused on the molecular function of the individual genes rather than their role in a specific pathway. Only three GO terms were identified as being over-represented among the up-regulated genes: arginase and agmatinase activity under hypersaline conditions, and microtubule motor activity under oxidative stress. Most GO terms were found to be significantly over-represented among the down-regulated genes. In agreement with the down-regulation of amino acid metabolism identified by the KOBAS analysis, we observed a decrease in the abundance of transcripts encoding aminoacyl-tRNA ligases in hypersaline and hyposaline conditions using the GOPET annotations.

Table 3. GO terms identified to be over-represented among the transcripts of significantly up- or down-regulated in the different stress conditions

GO terms identified to be over-represented among the transcripts of significantly up- or down-regulated in the different stress conditions

Condition	Change in expression	Category	Function	GO ID
Hyper	Down	Nucleic acid binding	RNA binding (mRNA, rRNA, snoRNA)	[GO:0003723]; [GO:0003729]; [GO:0019843]; [GO:0030515]
		Nucleic acid binding	Translation factor activity (elongation and initiation)	[GO:0008135]; [GO:0003746]; [GO:0003743]
		Lyase	UDP-glucuronate decarboxylase activity	[GO:0048040]
		Ligase	CTP synthase activity	[GO:0003883]
		Isomerase activity	Intramolecular oxidoreductase activity	[GO:0016860]
	Up	Hydrolase	Agmatinase activity	[GO:0008783]
		Hydrolase	Arginase activity	[GO:0004053]
Hyper Hypo	Down	Ligase	Aminoacyl-tRNA ligase activity (inlcuding Pro, Ser, Ile, Glu)	[GO:0004812]; [GO:0016876]; [GO:0004828]; [GO:0004829]; [GO:0004822]
Hypo	Down	Oxidoreductase (S, peroxide)	Antioxidant activity (glutathione-disulfide reductase and catalase, cytochrome-c peroxidase)	[GO:0016209]; [GO:0004362]; [GO:0004096]; [GO:0004130]
		Nucleic acid binding	Structure-specific DNA binding	[GO:0000404]; [GO:0032134]; [GO:0000403]; [GO:0032137]; [GO:0032138]; [GO:0032139]
		Tetrapyrrole binding	Chlorophyll binding	[GO:0016168]
		Lyase	Carbon-oxygen lyase activity	[GO:0016835]
		Transferase (N)	Transaminase activity (including TYR, ASP, histidinol-P, aromatic amino acids)	[GO:0008483]; [GO:0004838]; [GO:0004400]; [GO:0008793]; [GO:0004069]
		Transferase (C1)	Aspartate carbamoyltransferase activity	[GO:0004070]
		Transferase (glycosyl)	Transferase activity, transferring pentosyl groups	[GO:0016763]
		Oxidoreductase CH-CH	Biliverdin reductase activity	[GO:0004074]
		Oxidoreductase (CH-NH2)	Glutamate synthase activity	[GO:0015930]
		Isomerase	Isomerase activity	[GO:0016853]
		Transporter	NAD(P)+ transhydrogenase (B-specific) activity	[GO:0003957]
Hypo Oxi	Down	Oxidoreductase	Oxidoreductase activity	[GO:0016491]
			Oxidoreductase activity, acting on NADH or NADPH	[GO:0016651]; [GO:0016652]
			Oxidoreductase activity, acting on the CH-OH group of donors, NAD or NADP as acceptor (including L-iditol 2-dehydrogenase activity)	[GO:0016616]; [GO:0016614]; [GO:0003939]
Oxi	Down	Lyase	3-Isopropylmalate dehydratase activity	[GO:0003861]

Table 3. *(Continued)*

GO terms identified to be over-represented among the transcripts of significantly up- or down-regulated in the different stress conditions

	Transferase (P)	Amino acid kinase activity	[GO:0019202]
	Transporter	Nitrate transmembrane transporter activity	[GO:0015112]
	Transferase (C1)	S-adenosylmethionine-dependent methyltransferase activity (including nicotinate phosphoribosyltransferase)	[GO:0008757]
	Oxidoreductase (steroids)	Steroid dehydrogenase activity, acting on the CH-OH group of donors, NAD or NADP as acceptor	[GO:0033764]
	Transferase (glycosyl)	Transferase activity, transferring pentosyl groups	[GO:0016763]; [GO:0004853]
	Transferase (glycosyl)	Uracil phosphoribosyltransferase activity	[GO:0004845]
	Transferase (P)	Phosphoribulokinase activity	[GO:0008974]
Up	Motor activity	Microtubule motor activity	[GO:0003777]

The table shows only pathways that were labeled significant at an FDR <10% in both sets of significant genes (5% FDR and 10% FDR).

Also, under hypersaline stress, we observed down-regulation of genes associated with the GO terms RNA binding and translation factor activity, which corresponds to the KEGG category translation factors, and down-regulation of transcripts coding for proteins with a CTP synthase activity, which are involved in purine and pyrimidine metabolism. Under hyposaline stress, we observed that NAD(P)+ transhydrogenases, a number of transferases and oxidoreductases involved in amino acid metabolism, as well as genes with functions in nucleic acid and chlorophyll binding, were most affected, the latter matching well with the pathways 'photosynthesis-antenna proteins' identified by KOBAS. Under oxidative stress, using the GOPET annotations, we detected down-regulation of several different categories of transferases, nitrate transporters, oxidoreductases involved in steroid metabolism, and 3-isopropylmalate dehydratase-like enzymes that are involved in amino acid metabolism. Here, the KOBAS analysis did not identify any significantly up- or down-regulated pathways. Also in contrast to the KOBAS results, no GO terms were significantly over-represented among the genes identified as being up- or down-regulated in both oxidative and hypersaline stresses, or in all three stresses at the same time.

Manual Classification of Stress Response Genes with the Most Significant Changes in Expression

To identify the most important mechanisms involved in the stress response, we manually classified and examined in detail 966 genes that exhibited the most significant changes in one of the stress conditions compared to the control (that is, genes that meet both criteria: significance at an FDR <1% and a relative change in expression of more than two-fold).

We identified 519 genes (53.7%) with no homologues in either the National Center for Biotechnology Information (NCBI) databases or other heterokont genomes (e-value > 1e-10). An additional 122 genes (12.6%) code for conserved genes with unknown function. Of these 122 conserved genes, 23 (18.9%) are conserved only within the heterokont lineage. The remaining 325 genes (33.6%) were divided into 12 groups according to their putative functions in amino acid metabolism, DNA replication and protein synthesis, protein turnover, carbohydrate metabolism, photosynthesis-related processes, fatty acid metabolism, transporters, vesicular trafficking and cytoskeleton, classical stress response pathways, autophagy, signaling, and other processes. The following section gives a brief overview of the different groups of genes identified among the most significantly regulated genes.

Among genes involved in amino acid metabolism, we found a total of 32 down-regulated genes related to the metabolism of all 20 standard amino acids except aspartic acid. In contrast, nine genes were induced in at least one abiotic stress condition. These were involved in the metabolism of proline, arginine, cysteine, alanine, phenylalanine, tyrosine, tryptophan, leucine, isoleucine, and valine. Highly regulated genes involved in the different steps of DNA replication and protein synthesis coded for proteins, including helicases, DNA polymerases and related enzymes, proteins involved in purine and pyrimidine synthesis, DNA repair proteins, transcription factors, RNA processing enzymes, proteins involved in translation, ribosomal proteins, and proteins for tRNA synthesis and ligation. Most of these genes were down-regulated in all stress conditions, but some genes were up-regulated in response to abiotic stress. These genes include some helicases, transcription factors, and DNA repair proteins. We also found seven genes related to protein turnover to be down-regulated and six to be up-regulated in one or more of the stress conditions. Among the up-regulated genes, there were two ubiquitin conjugating enzymes, which play a potential role in targeting damaged proteins to the proteasome, or control the stability, function, or subcellular localization of proteins.

The situation was similar for genes involved in carbohydrate metabolism, where we found both glycolysis- and citric acid cycle-related genes to be strongly down-regulated under all the stresses tested (six and seven genes down-regulated, respectively). However, four genes, encoding a gluconolactonase, a xylulokinase, a phosphoglycerate kinase, and an isocitrate lyase, were up-regulated. In particular, an isocitrate lyase gene was 19- to 212-fold up-regulated under the different stress conditions. Photosynthesis-related genes that were regulated in response to abiotic stress included eight chlorophyll a/c binding proteins as well as genes responsible for the assembly of photosystem 2, electron transport, light sensing, and carotenoid synthesis. Many of these genes were strongly affected in the hypersaline condition, with the majority being down-regulated (17 versus 11 that were up-regulated). There was at least one gene that was up-regulated under one or more stress condition in every group. Genes with roles in fatty acid metabolism altered their expression patterns in a similar way under all stress conditions. We were able to distinguish between two groups: three genes involved in the synthesis of fatty acids, which were down-regulated; and genes functioning in the degradation of fatty acids, among which five of six genes were up-regulated. We further observed that three genes involved in lipid synthesis were up-regulated, and genes involved in inositol metabolism were also affected.

With respect to transporters, we identified five genes encoding nitrogen transporters (all down-regulated) as well as three genes encoding sugar transporters (all up-regulated). Genes coding for ion transporters were also mainly down-regulated under hypersaline and hyposaline conditions, although two potassium and magnesium transporter genes were up-regulated under hypersaline stress. Among genes responsible for the transport of solutes and proteins to the mitochondrion, we observed an up-regulation mainly in the hyposaline stress condition. Regarding genes related to vesicular trafficking and the cytoskeleton, we identified 13 up- and 6 down-regulated genes, many of these genes containing an ankyrin repeat domain and showing strongest changes in transcription under hyposaline and oxidative stress conditions.

We further found several classical stress response genes to be up-regulated. Four genes coding for heat shock proteins (HSPs) were up-regulated mainly under hyposaline and oxidative stress, but there were also two genes coding for a chaperonin cpn60 and a prefoldin, each of which was down-regulated. In addition, we found genes involved in protection against oxidative stress to be induced. These include a glutaredoxin (oxidative stress), a methionine sulfoxide reductase (hyposaline stress), and three glutathione peroxidases (mainly hypersaline stress). At the same time, however, a catalase-coding gene was down-regulated in all stress conditions, most strongly under hyposaline stress.

Two genes involved in autophagy, one of which is represented by two sequences on the microarray, were up-regulated in all stress conditions and several genes with putative signaling functions were affected. Six protein kinases were among the most significantly up-regulated genes: three equally under all stress conditions, and one each specifically under hyposaline, hypersaline and oxidative stress. Furthermore, one protein kinase and one WD-40 domain containing gene were down-regulated under hyper- and hyposaline stress, respectively.

Several other genes are not mentioned here, either because only a very vague prediction of their function was possible, or because they are difficult to put into categories with other genes.

Stress Response Genes with Unknown Functions

All unknown and conserved unknown genes present among the most significantly regulated genes were sorted into groups according to their sequence similarity. Among the groups with three or more members, there were three (I to III) that had no known homologs in species other than E. siliculosus, and three (IV to VI) for which we were able to find homologs in other lineages for most of the sequences.

Known Brown Algal Stress Genes

Many of the brown alga-specific stress response genes identified in L. digitata by Roeder et al. [21] were not among the most regulated genes identified in this study. Nevertheless, we decided to examine their expression patterns in more detail. The array used in this study contained probes for one vanadium-dependent bromoperoxidase (CL83Contig2), but this gene was not strongly regulated under the different stress conditions (1.06-fold to 1.4-fold induced, P = 0.75). Twenty-four C5-epimerases were represented, but none of these genes were among the most significantly regulated loci, although several of them were either induced or repressed under the different stress conditions. Finally, we decided to consider genes involved in the synthesis of mannitol, a well-known osmolyte in brown algae [16,17]. Only one enzyme specific to the synthesis of this polyol could be clearly identified based on sequence homology: mannitol 1-phosphate dehydrogenase (see [38] for a description of the mannitol synthesis pathway in brown algae). Our array contains probes for two genes identified as potential mannitol 1-phosphate dehydrogenases: one (CL200Contig2 corresponding to Esi0017_0062 in the Ectocarpus genome), which was among the most significantly regulated genes and six-fold down-regulated in hyposaline condition, and one (CL2843Contig corresponding to Esi0020_0181), which was generally expressed at a very low level but was up-regulated approximately five-fold under hypersaline stress (P = 0.066).

Clusters of Genes with Similar Expression Patterns

Based on a figure of merit (FOM) graph, we decided to divide the set of expressed genes into seven different clusters (A to G). These clusters, along with the GO terms and KEGG pathways that are over-represented among each of them, are shown in Figure 5. We identified one cluster (A) representing the stably expressed genes, three clusters included mainly up-regulated genes (B-D), and the remaining three clusters included mainly down-regulated genes (E-G). Among both the up- and down-regulated clusters, we found one cluster each that was equally affected by all stress conditions (B and E), one each where gene expression was affected only by hyposaline and oxidative stress conditions (C and G), and one cluster each where gene expression was affected mainly by hypersaline stress (D and F). Most of the principal functions identified for each cluster by GOPET and KOBAS fit well with the results from our earlier analysis of the up- and down-regulated genes.

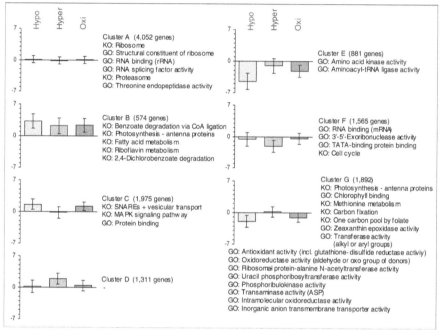

Figure 5. Expression graphs of clusters identified by the k-means algorithm. The graphs display the log2-ratio of all stress conditions (hypo = green, hyper = red, oxi = blue) with the control condition. GO terms and KEGG pathways (KO) identified as over-represented in these clusters (FDR = 10%) are shown next to the graph.

Discussion

This study presents the first global gene expression analysis of a brown alga. Our goal was to determine the transcriptomic changes in response to short-term hypersaline, hyposaline and oxidative stress—three stresses that play an important role in the natural habitat of many brown algae, the intertidal zone [20,26]. Our results show that almost 70% of the expressed genes had a modified expression pattern in at least one of the examined stress conditions. This is in contrast to what has been observed in flowering plants, where the proportion of significantly regulated genes generally ranges from 1% to 30%, depending on types of abiotic stress examined, their number, and the statistical treatment applied (see [27,39,40] for some examples). Our findings demonstrate that, rather than relying on a few specific stress response proteins, E. siliculosus responds to abiotic stress by extensive reprogramming of its transcriptome.

A more detailed analysis of the manual annotation of the 966 most significantly regulated genes and the results for the GOPET and KOBAS analysis for all three stress conditions, reveals two major themes concerning the short-term stress response of E. siliculosus: down-regulation of primary metabolism and growth processes; and activation of energy stores and of genes and pathways involved in 'stress management.' These findings are summarized in Figure 6. In the following sections we will first discuss the differences observed between the different stress conditions that were tested, and then highlight some of the general trends that emerged from our data.

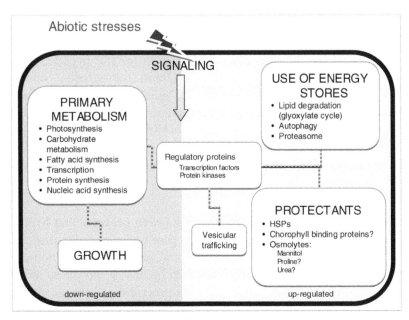

Figure 6. Major transcriptomic changes in E. siliculosus under short-term oxidative and saline stress. This schema summarizes the most important transcriptomic changes discussed in the text. Processes on the left (blue) were repressed, while processes on the right (yellow) were activated. Please note that this graph displays only the general trends; some of the pathways are not regulated in all stress conditions and not all genes of one pathway are always regulated in the same way (see text for details).

Comparison of Stress Conditions

We have compared each stress condition to the control condition, and analyzed these results using KOBAS and GOPET. Due to the necessity to control the FDR with multiple testing, the chance of beta-errors (that is, the chance of falsely labeling a gene or a pathway as not significantly regulated) greatly increased, making a direct comparison of the genes and groups of genes identified as being up- or down-regulated under the different stress conditions prone to false conclusions. Therefore, we based our comparison of the different stresses on the cluster analysis and the results from the manually analyzed 966 most significantly changing genes.

The first and most apparent observation from the clustering was that the changes in gene expression induced by hyposaline and oxidative stress were more similar to each other than to those observed under hypersaline stress. One explanation for this might be that hypersaline stress, although it is a common stressor in the natural habitat of brown algae, is not likely to occur at the same intensity in the field as that used for our laboratory experiments (about three times the concentration of normal seawater). Even though we did not observe a strong difference in the efficiency of photosynthesis under the different stress conditions, it is possible that hypersaline stress, at the intensities applied in our experiments, represents a condition the alga is less able to adjust to. This hypothesis is supported by the fact that in cluster F (down-regulated in hypersaline conditions) cell cycle-related genes were over-represented, indicating that growth was most strongly affected under hypersaline stress conditions. Furthermore, other classical stress responses, such as the up-regulation of SNARE (soluble N-ethylmaleimide-sensitive factor attachment receptor)-related genes, which are important for cellular transport of vesicles and their fusion with membranes and play a role in plant development and abiotic stress response [41], were observed mainly under oxidative and hyposaline stress.

There are also several smaller differences between oxidative and hyposaline stress, such as the induction of a glutaredoxin gene under oxidative stress conditions. Glutaredoxins are known to play a central role in the protection against oxidative damage [42], as they can be oxidized by diverse substrates, including ROSs, and are reduced by glutathione. Other genes did not change expression under oxidative stress, but were specifically regulated under saline stress. Such examples are given in the following paragraphs.

Down-Regulation of Growth and Primary Metabolism

Many genes involved in several pathways related to growth and primary metabolism were identified to be down-regulated by more than one of our analyses (GO-PET, KOBAS, k-means clustering, and manual analysis). We observed a decrease in the abundance of transcripts of genes that are important in the synthesis of purine and pyrimidine nucleotides and, correspondingly, of several genes responsible for the replication of DNA. This function is essential for cell division, a process affected in all of the stress conditions examined.

A reduction in growth implies a reduction in the requirements for primary metabolites necessary to fuel this growth. Within our dataset, we found widespread evidence for down-regulation of processes involved in primary metabolism. This was most pronounced in the case of protein synthesis. Several genes encoding enzymes involved in this and related processes (the synthesis of amino acids, their ligation with the appropriate tRNAs, the production of mRNA (that is, transcription), and the actual assembly of polypeptide chains (that is, translation) were down-regulated under stress conditions. Furthermore, genes responsible for the uptake of nitrogen, which is used mainly for the synthesis of amino acids, were also down-regulated. Together, this provides a strong indication that the overall rate of protein synthesis was reduced under the stress conditions examined.

However, there were additional primary metabolic processes that were at least partially affected by the stress treatments. These included the synthesis of fatty acids, photosynthesis and pigment synthesis, and carbohydrate metabolism. Again, these changes probably reflected a decreased need for metabolites for growth.

One possible explanation for the observed down-regulation of genes involved in growth and primary metabolism can be found in the results of our pulse amplitude modulation fluorometer experiment: we observed that the efficiency of photosynthesis decreased almost immediately after the application of the stress treatments. Photosynthesis is strongly affected by environmental stress [43]. A decrease in photosynthetic efficiency is synonymous with a decrease in energy production, and we can assume that reducing all of these aspects of primary metabolism might represent a means of conserving energy. This phenomenon is known in higher plants, where Kovtun et al. [44] have reported cross-talk between oxidative stress and auxin signaling, auxin being a major growth hormone in higher plants. Most likely, this cross-talk allows plants to shift their energy from growth-related processes to stress protection and survival. This might also be true for E. siliculosus, where the observed down-regulation of growth- and primary metabolism-related genes might represent a way of compensating for reduced energy production under stress conditions and redirecting energy to specific stress response processes.

Activation of Protein Degradation, Energy Stores, and Nutrient Recycling

We observed an up-regulation of two genes involved in autophagy in all three of the stresses examined. Under nutrient-limited and under stress conditions, this process has been shown to play a role in the re-allocation of sugar and nutrients to essential biological processes in several organisms [45]. The activation of autophagy-related genes might, therefore, just like the down-regulation of growth-related processes, represent a mechanism of compensating for reduced energy production under stress conditions and provide sugars and nutrients for both core biological processes and synthesis of stress proteins. This coincides with the strong up-regulation of an isocitrate lyase gene under all stress conditions. Isocitrate lyases are enzymes located in the glyoxysome and catalyze a rate-controlling step in the glyoxylate cycle (reviewed in [46]), one function of which is the conversion of lipids to carbohydrates when non-lipid-derived storage reserves are depleted. Corresponding to this and the up-regulation of autophagy-related genes, we observed an up-regulation of three genes coding for sugar transporters in all stress conditions and of genes coding for mitochondrial exchange proteins under the hyposaline and oxidative stress conditions. These transporters may direct recycled sugars and nutrients to the mitochondrion, where they can be used for energy production.

Recently, additional roles of autophagy have been discovered in higher plants, including the degradation and removal of oxidized or damaged proteins during stress [47]. To a certain degree, these roles overlap with the role of the proteasome. Although we found some genes involved in protein turnover to be down-regulated, the KEGG pathway 'proteasome' was identified as up-regulated in the hyposaline stress condition and we identified two genes involved in ubiquitination among the most significantly up-regulated genes under all stress conditions. Ubiquitination is a process in which proteins are labeled with a small polypeptide (ubiquitin) [48], thereby modifying their stability, function, or subcellular localization. This could serve regulatory purposes (for example, by targeting transcription factors or other regulatory proteins for degradation by the proteasome), accelerate translation of the large-scale transcriptomic changes into changes in protein abundance, and/or, just like autophagy, facilitate nitrogen recycling [49].

Activation of Signaling Pathways

Large-scale transcriptomic reprogramming as observed in our dataset most certainly requires a large number of signals for coordination. We have already discussed a possible regulatory role of ubiquitination in the previous paragraph. This, however, is not the only regulatory mechanism highlighted by the transcriptomic changes in our dataset. We found several other genes that might play roles in orchestrating the abiotic stress response of E. siliculosus. For example, three protein kinases with a potential role in cell signaling were strongly up-regulated under all stress conditions, while three other members of this large family appeared each to be specific to one particular stress. Furthermore, several potential transcription factors were strongly regulated under different stress conditions. Since there is still very little known about the molecules and proteins involved in the stress sensing signaling cascades of brown algae, these genes provide particularly interesting candidates for more targeted experiments such as targeted mutagenesis and chromatin immunoprecipitation in the case of the putative transcription factors.

In addition to transcription factors and protein kinases, we detected an up-regulation of genes involved in fatty acid metabolism, and more specifically fatty acid catabolism. Fatty acid derivates such as oxylipins have been shown to function in signaling in both terrestrial plants [50] and marine algae [51]. Thus, we can conclude that our data show an up-regulation of genes putatively involved in several different signaling pathways.

Synthesis of 'Classical' Stress Response Proteins

The only medium throughput transcriptomic analysis of the abiotic stress response in brown algae so far [21], conducted with protoplasts of L. digitata, reported the transcriptional activation of vanadium-dependent bromoperoxidases and C5-epimerases. Neither of these genes was regulated in our study. While in L. digitata vanadium-dependent bromoperoxidases (enzymes implicated in the synthesis of halogenated organic compounds associated with defense of seaweeds against biotic stressors [52]) comprise a multigenic family [21], the E. siliculosus genome contains only a single copy of a vanadium-dependent bromoperoxidase, indicating a possibly different or subordinate role in this organism. Regarding C5-epimerases, which are enzymes responsible for the modification of brown algal cell walls [10] highly represented in the Ectocarpus genome, we observed differences in regulation between the study of Roeder et al. [21] and our study. This can be explained by the nature of the stress (generation of protoplasts (that is, removal of the cell wall) in [21] versus milder saline or oxidative stress in our study).

Generally, only a few 'classical' stress response genes changed expression in our experiments. In most organisms, up-regulation of genes coding for HSPs and other chaperones has been observed under abiotic stress conditions. These molecules stabilize proteins and membranes, and have been shown to play a vital role in protecting against stress by re-establishing normal protein conformation and, thus, cellular homeostasis [53]. In E. siliculosus, we observed four HSPs or chaperones to be among the most significantly up-regulated genes in hyposaline and oxidative stress conditions. However, this was not the case under hypersaline stress conditions. Moreover, two chaperone-like proteins were down-regulated under all stress conditions. The situation was similar for genes coding for proteins with antioxidant activity. Three genes coding for glutathione peroxidases were up-regulated under hypersaline and oxidative stress, and one glutaredoxin under hyposaline stress. At the same time, two genes encoding a glutathione S-transferase and a catalase were among the most significantly down-regulated in all stress conditions. Consequently, the GO term 'antioxidant activity' was significantly over-represented among the most down-regulated genes in the hyposaline condition. In flowering plants all of these proteins are known to carry out important functions in the protection against reactive oxygen species [25]. Our finding that these genes were not induced was, at first, surprising, but is in accordance with a transcriptomic analysis of the abiotic stress response of the intertidal red alga C. crispus. Collén et al. [20] have reported that the average expression of HSP-coding genes was only moderately elevated (approximately 1.3-fold) under hypersaline and hyposaline conditions. Furthermore, in E. siliculosus, the average expression of genes coding for proteins with antioxidant activity was slightly repressed (1.17-fold) under hyposaline stress conditions and slightly induced under hypersaline stress conditions (1.15-fold). One possible explanation for this might lie in the fact that transcriptional regulation is not the most important mechanism regulating the activity of these enzymes. This hypothesis would be compatible with an earlier study by Collén and Davison [54], who found that the cellular activity of ROS scavenging enzymes correlated with vertical zonation of different species of the brown algal order Fucales in the intertidal zone. To our knowledge, it is currently not known whether the activity of the ROS scavengers that were examined also changes upon exposure to abiotic stress. Such studies could greatly aid our understanding of the role of these enzymes in the brown algal stress response.

An alternative or additional explanation to that of non-transcriptional mechanisms regulating the activity of ROS scavenging enzymes could be that other, yet unknown proteins and mechanisms play more important roles in the defense against ROS in brown algae. One candidate for this kind of protein could be the chlorophyll a/c binding proteins. Thirty chlorophyll a/c binding proteins were represented on our microarray, most of them being down-regulated mainly under hyposaline and oxidative stress conditions. As chlorophyll a/c binding proteins serve as light-harvesting antennae, this down-regulation is likely to represent a response to the reduced photosynthesis efficiency (quantum yield) under stress conditions. Reducing the amount of energy that reaches the photosynthetic reaction center would also reduce the need for non-photochemical quenching and decrease the risk of the formation of ROS. However, there were also three genes coding for chlorophyll a/c binding proteins among the most significantly up-regulated genes under hyposaline and hypersaline stresses. A similar observation was made by Hwang et al. [55] in the Antarctic diatom Chaetoceros neogracile, where heat stress induced the up-regulation of five and the down-regulation of ten genes coding for chlorophyll a/c binding proteins. It is possible that these up-regulated chlorophyll a/c binding proteins, in spite of their high sequence similarity with the other transitionally down-regulated ones, have evolved or are evolving to serve different functions within the heterokont lineage.

Ions and Potential Osmolytes

In parallel to the induction of 'classical' stress response genes, we observed the transcriptional regulation of several genes involved in the synthesis or degradation of potential organic osmolytes and the transport of ions under saline stress. Under hypersaline stress, changes in the extracellular salt concentration as well as cell volume are likely to cause imbalances in ion concentrations, explaining the need for transporters to maintain homeostasis. Two genes coding for a magnesium and a potassium transporter were up-regulated specifically under hypersaline stress. Interestingly, we did not observe transcriptional activation of sodium transporter genes under salt stress, although many glycophytes (non- or moderately salt toler-ant terrestrial plants) use these transporters to exclude NaCl from their cytosol, al-lowing a certain degree of salt tolerance [56]. The latter observation suggests that the large quantities of NaCl accumulated upon exposure to saline stress (Figure 2) are stored within the cytoplasm rather than in the vacuole. A similar observa-tion was made by Miyama and Tada [57] in the Burma mangrove. In this tree, exposure to sub-lethal NaCl stress did not cause an activation of Na transporters but led to a slow increase of the NaCl concentration in the leaves. One possible explanation for this, as proposed by Miyama and Tada [57], is that NaCl itself could serve as an osmolyte within the cells of the Burma mangrove. This may also be true for brown algae—a hypothesis that is strengthened by the observation that, in Ectocarpus as well as in Lamninaria [58] and the Burma mangrove [57], sorbitol, added at the same osmolarity as NaCl, had irreversible toxic effects.

While there is no evidence of the synthesis of additional compatible osmolytes in the Burma mangrove, we observed an increasing difference between intracel-lular osmolarity and intracellular Na+ concentration with rising salinity in E. si-liculosus, demonstrating the accumulation of other osmotically active substances in the cell. Mannitol has frequently been suggested to be a compatible osmolyte in brown alga [16,17], and indeed, one of the two mannitol 1-phosphate-dehy-dogenases in E. siliculosus was down-regulated in hyposaline stress, and the other up-regulated in hypersaline stress (though with a weak P-value of 0.066). In ad-dition, under hyposaline conditions, we observed a strong up-regulation of a pro-line dehydrogenase gene, an enzyme responsible for the degradation of proline, which is known as a compatible osmolyte in higher plants [59] and diatoms [60]. Degrading proline under conditions of low salinity might help E. siliculosus to reduce the osmotic pressure between the intracellular and extracellular medium. Finally, a possible role of urea as a compatible osmolyte was suggested in diatoms

[61]. E. siliculosus possesses the complete urea cycle and genes encoding arginases were up-regulated in hypersaline conditions. As arginases catalyze the last reaction of this cycle—that is, the degradation of arginine to ornithine and urea—their up-regulation supports the hypothesis of urea as a compatible osmolyte in heterokonts. There are, however, other or additional possible roles of arginases: in higher plants, for example, arginases have been shown to play a regulatory role in nitric oxide metabolism, increasing both the synthesis of proline and polyamines [62], both of which, in turn, are part of their osmotic stress response [59]. Additional experiments addressing the question of the possible compatible osmolytes in brown algae—for example, metabolite profiling—are now required to further test these hypotheses.

Stress Response Genes with Unknown Functions

Although our study has revealed several major themes underlying the abiotic stress response of E. siliculosus, we should not forget that this analysis was based on only a subset of the genes that actually changed expression. Our manual analysis of the 966 of the most significantly regulated genes has shown that 53.7% of these genes, to date, have no known homologs in current databases, including diatoms, and for another 12.6% there is no indication of their function, even though homologs exist in other organisms. This demonstrates both the discovery potential working with E. siliculosus and the amount of work that lies ahead for the phycological community.

Conclusions

In this study, which presents the first large-scale transcriptomic study within the brown algal lineage, we have developed and compiled the tools and protocols necessary to perform microarray experiments in the emerging model brown alga E. siliculosus, and used these tools to study the transcriptional response to three different stresses. Our results show that E. siliculosus undergoes large-scale transcriptomic reprogramming during the short-term acclimation to abiotic stress. The observed changes include several modifications to transcription, translation, amino acid metabolism, protein turnover, and photosynthesis, and indicate a shift from primary production to protein and nutrient recycling.

Although E. siliculosus shares many stress responses with flowering plants, for example, the induction of genes involved in vesicular trafficking, some classical stress responses, such as the induction of several ROS scavengers, could not be observed. On the other hand, our data highlighted many novel reactions such as the up-regulation of several genes coding for chlorophyll a/c binding proteins or the regulation of a large percentage of unknown genes, many of which are unique to E. siliculosus. In particular, the latter result, that is, that the functions of two-thirds of the regulated genes are unknown, underlines the fact that many of the molecular mechanisms underlying the acclimation to environmental stresses in brown algae are still entirely unknown. Understanding these mechanisms is a challenge that will still require much research, and our study provides a valuable starting point to approach this task.

Materials and Methods

Growth Conditions, Stress Treatments, and Measurements of Osmolarity and Na+ Concentration

E. siliculosus (Dillwyn) Lyngbye (Ectocarpales, Phaeophyceae) unialgal strain 32 (accession CCAP 1310/4, origin San Juan de Marcona, Peru) was cultivated in 10-liter plastic flasks in a culture room at 14°C using filtered and autoclaved natural seawater enriched in Provasoli nutrients [63]. Light was provided by Philips daylight fluorescence tubes at a photon flux density of 40 μmol m-2 s-1 for 14 h per day. Cultures were bubbled with filtered (0.22 μm) compressed air to avoid CO2 depletion. Ten days before the stress experiments, tissues were transferred to ASW with the following ion composition: 450 mM Na+, 532 mM Cl-, 10 mM K+, 6 mM Ca2+, 46 mM Mg2+, 16 mM $SO_4$2-.

Three different stress media were prepared based on ASW. For hyposaline stress, ASW was diluted to 12.5% of its original concentration with distilled water, resulting in a final NaCl concentration of 56 mM. For hypersaline stress, ASW with a final concentration of 1.47 M NaCl was used. For oxidative stress, H2O2 (30% w/w; Sigma-Aldrich, St. Louis, MO, USA) was added to the ASW immediately before beginning the stress experiment at a final concentration of 1 mM. Identical quantities of Provasoli nutrients were added in each of these media.

In order to monitor the intensity of a stress, we measured the quantum yield, a fluorometric marker for the photosynthetic efficiency, using a Walz Phyto-pulse amplitude modulation fluorometer (Waltz, Effeltrich, Germany) and default parameters (actinic light intensity 3, approximately 90 μE m-2 s-1; saturation pulse intensity 10, approximately 2,000 μE m-2 s-1, 200 ms) before harvesting the cultures.

The stress experiment was started by filtering 20 liters of ASW-acclimated E. siliculosus and transferring approximately 4 g of tissue each to 20 flasks (5 replicates per condition) containing 1 liter of one of the three stress media or the control medium (ASW with Provasoli nutrients). After 6 h the content of each flask was harvested by filtration, dried with a paper towel, and immediately frozen in liquid nitrogen.

Immediately after harvesting, about 300 mg (wet weight) of sample were thoroughly ground at room temperature (RT) and centrifuged for 1 minute at 16,000 g. Both the supernatant and a sample of the culture medium were then used to measure the concentration of osmolytes employing a freezing point depression osmometer (Osmometer Type 15, Löser Messtechnik, Berlin, Germany), and to determine the intracellular concentration of Na+ with a FLM3 flame photometer (Radiometer, Copenhagen, Denmark).

Sequence Preparation and Array Design

The 90,637 EST sequences used for the microarray design were cleaned using Phred [64] (trim-cutoff 0.05) and SeqClean and assembled using TGICL [65] and default parameters. Forty-one sequences that had been removed by Phred were re-included in the dataset, because they had significant BLAST hits with known eukaryotic proteins. In addition, 231 E. siliculosus virus 1 (Es-V1) sequences and 21 genomic intron sequences were included in the design. All assembled sequences are available directly from our homepage [22] and the ESTs have been deposited in public databases [EMBL: FP245546-FP312611].

Four 60-mere probes were designed for 17,119 of the 17,332 sequences (132 genes are not represented on the array) by Roche NimbleGen (Madison, WI, USA) and synthesized on 4-plex arrays with 72,000 features per hybridization zone. Roche NimbleGen also carried out cDNA labeling and hybridization as part of their gene expression array service.

Automatic Annotation and Correspondence Table

All sequences were automatically annotated with KEGG orthology (KO) numbers using KOBAS [37] and with GO terms [66] using GOPET [36]. Protein sequences corresponding to the assembled EST sequences were then predicted using ORF predictor [67]. The automatic annotation of these sequences yielded 2,383 and 3,148 annotated sequences, respectively. As 37.5% of the cDNA sequences that were represented on the array contained mainly, or exclusively, 3' untranslated region sequence, their function could not be assessed directly. In these cases the corresponding genome sequence was annotated, yielding an additional 1,047 KO and 2,743 GO annotations. The correspondence table used to relate the assembled ESTs to supercontigs was generated by blasting all of the assembled EST sequences against the full E. siliculosus genome (coding and non-coding sequences) and selecting the best hit (best identity, longest alignments). Wherever this best hit was part of a predicted coding sequence (CDS), the corresponding CDS was chosen. In cases where the hit region was upstream of only one CDS, this CDS was chosen. In some cases the best hit was located upstream of two CDSs on opposite strands. Here the closest CDS was selected, if the distance to the closer CDS was half as long as or shorter than that to the other CDS. Otherwise no corresponding genome sequence was selected. In total, 3,430 (20%) of all represented genes were annotated with KEGG and 5,891 (34%) were annotated with GO terms.

Sample Preparation, Hybridization and Verification

RNA was extracted from approximately 100 mg (wet weight) of tissue following Apt et al. [68] with modifications as described by Le Bail et al. [28], using a cetyltrimethylammonium bromide (CTAB)-based buffer and subsequent phenol-chloroform purification, LiCl-precipitation, and DNAse (Turbo DNAse, Ambion, Austin, TX, USA) steps. RNA quality and quantity was then verified on 1.5% agarose gel stained with ethidium bromide and a NanoDrop ND-1000 spectrophotometer (NanoDrop products, Wilmington, DE, USA). For four of the five flasks, mRNA was isolated from the total RNA using the PolyATtract® mRNA Isolation System III (Promega, Madison, WI, USA). These samples were concentrated in a SpeedVac concentrator (Savant, Ramsy, MN, USA) and again quantified using the NanoDrop. One replicate was used to verify if our procedure would directly work from total RNA.

Double-stranded cDNA was synthesized and amplified with the SMART cDNA synthesis kit (Clontech, Mountain View, CA, USA) and the M-MuLV reverse transcriptase (Finnzymes, Espoo, Finland) starting from 30 ng of mRNA or 100 ng of total RNA. In this kit, the first strand of cDNA is synthesized using an oligo(dT) primer with an attached SMART priming site. The terminal C-transferase activity of the reverse transcriptase will create an oligo(dC) tail at the 5' end of each mRNA, which is used to add a second SMART priming site. The two priming sites are then used to produce the second strand of the cDNA and to amplify it by PCR according to the Clontech protocol, using the Advantage2 polymerase (Clontech). The optimal number of amplification cycles was determined by semi-quantitative PCR, and ranged between 20 and 25 for our samples.

The PCR reactions were purified by first vortexing with one volume of phenol:chloroform:isoamyl alcohol (25:24:1), then by precipitating the aqueous phase with 0.5 volume of 7.5 M NH4OH, 6 µg of nuclease-free Glycogen (Ambion), and 2.4 volumes of ethanol. After centrifugation at RT (20 minutes at 14,000 g), the pellet was washed with 70% EtOH, centrifuged (10 minutes at 14,000 g and RT) and resuspended in 14 µl of H2O. To finish, cDNAs were once more quantified and checked on an agarose gel to ensure that they met the requirements of Roche NimbleGen for hybridization (concentration >250 μg l-1, $A_{260/280} \geq 1.7$, $A_{260/230} \geq 1.5$, median size ≥ 400 bp).

RT-qPCR validation of the microarray was performed as described by Le Bail et al. [28].

Statistical Analysis

Expression values were generated by Roche NimbleGen using quantile normalization [69], and the Robust Multichip Average algorithm [70,71].

To increase the power of subsequent statistical tests, non-expressed genes were removed from the dataset. This was done by comparing the raw expression values of each gene with those of the random probes included in the array design by Roche NimbleGen. Most of the random probes (>99.66%) had raw expression values inferior to 450, so all genes that had raw expression values over 450 in one of the replicates of the different experimental conditions were considered expressed. The probability p' of labeling a non-expressed gene as being expressed can be calculated according to the laws of a binomial distribution. Since 16 hybridizations were considered, for our dataset p' equals 5.3%, meaning that we have theoretically removed 94.7% of all genes without detectable expression.

The most stable genes were defined by ranking the genes according to the sum of squares of the log2-ratios of all stress conditions with the control. Differentially expressed genes were identified in the reduced dataset by t-test using TigrMEV 4.1 [72] with subsequent calculation of the FDR according to Benjamini and Hochberg [73]. Clustering was performed on the log2-ratios of all expressed genes with the control, considering all mRNA replicates using TigrMEV 4.1. A k-means algorithm [74], 'Euclidian distance', and 'average linkage' were selected. The ideal number of clusters for our dataset (k = 7) was determined using a figure of merit graph [75]. After clustering, the different replicates were averaged to generate the expression graphs. Both the clusters and the over-expressed and repressed genes identified by the t-test (FDR 10%) were analyzed to identify over-represented groups of genes. Over-expressed KEGG categories were identified using the KOBAS web-site [37] and a binomial test. Over-represented GO terms were identified using the GO Local Exploration Map (GOLEM) software (version 2.1) [76] and the Benjamini and Yekutieli algorithm to determine the FDR [77]. Additionally, the 966 genes that showed the most significant changes in expression (that is, genes that meet both criteria: significance at an FDR <1% and a relative change in expression compared to the control of more than twofold) were annotated and grouped manually.

Stress Response Genes with Unknown Functions

Genes were considered unknown stress response genes if they significantly changed expression in at least one stress treatment compared to the control (FDR <1%, more than twofold change compared to the control) and when, for both the assembled EST sequence itself and for the corresponding genome sequence, no BLAST hits were found in either the NCBI databases or among the known heterokont genomes (Phytophthora sojae, P. ramosum, Thalassiosira pseudonana, and Phaeodactylum tricornutum). Where homologs (e-value < 1e-10) were found, but these homologs had no functional annotations, genes were considered conserved unknown stress response genes. To group these genes, all unknown stress response genes and conserved unknown stress response genes were blasted against themselves using the NCBI BLAST program (version 2.2.18) [78] and a cut-off e-value of 1e-10. If there were several alignments between two genes, only the alignment with the lowest e-value was considered. Self-hits were removed. All groups of genes with homologs among the (conserved) unknown stress response genes in E. siliculosus were then visualized using Cytoscape 2.6.0 [79]. Their subcellular localizations were identified using HECTAR [80] and transmembrane domains were searched for using TMAP [81].

Data Deposition

Microarray data have been deposited in a public database [ArrayExpress:E-TABM-578].

Abbreviations

ASW: artificial sea water; CDS: coding sequence; EST: expressed sequence tag; FDR: false discovery rate; GO: Gene Ontology; GOPET: GO-term Prediction and Evaluation Tool; HSP: heat shock protein; KEGG: Kyoto Encyclopedia of Genes and Genomes; KOBAS: KEGG Orthology-Based Annotation System; NCBI: National Center for Biotechnology Information; ROS: reactive oxygen species; RT: room temperature; RT-qPCR: real time quantitative PCR; SNARE: soluble N-ethylmaleimide-sensitive factor attachment receptor.

Authors' Contributions

SMD, together with and under supervision of TT, performed the experiments, analyzed the data and wrote the manuscript, with help from CB and JMC; DS,

BS and JLP have generated cDNA libraries (normalized and non-normalized); CDS was involved in EST data treatment and data formatting for sequence submissions; EC helped assemble the ESTs and set up the internet site; RK and KHG provided the GOPET annotations; MD modified ArrayLIMS and EMMA to work with the Roche NimbleGen arrays; JMC, PR, YVDP, and LS were involved in genome-wide gene annotation; all authors read and approved the manuscript.

Acknowledgements

We would like to thank Declan Schroeder for providing the E. siliculosus virus 1 sequences, Maela Kloareg for her advice on the analysis of transcriptomic data, Gurvan Michael for his help identifying the E. siliculosus C5-epimerases, Gildas Le Corguillé for his help installing KOBAS and setting up the website, and Jonas Collén, as well as the anonymous reviewers, for critical reading of the manuscript. Part of this work was performed within the framework of the Network of Excellence 'Marine Genomics Europe' (European Commission contract no. GOCE-CT-2004-505403). SD has received funding from the European community's Sixth Framework Programme (ESTeam contract no. MESTCT 2005-020737).

References

1. Boyen C, Oudot MP, Loiseaux-De Goër S: Origin and evolution of plastids and mitochondria: the phylogenetic diversity of algae. Cah Biol Mar 2001, 42:11–24.

2. Baldauf SL: The deep roots of eukaryotes. Science 2003, 300:1703–1706.

3. Graham MH, Kinlan BP, Druehl LD, Garske LE, Banks S: Deep-water kelp refugia as potential hotspots of tropical marine diversity and productivity. Proc Natl Acad Sci USA 2007, 104:16576–16580.

4. Santelices B: The discovery of kelp forests in deep-water habitats of tropical regions. Proc Natl Acad Sci USA 2007, 104:19163–19164.

5. Küpper FC, Carpenter LJ, McFiggans GB, Palmer CJ, Waite TJ, Boneberg E-M, Woitsch S, Weiller M, Abela R, Grolimund D, Potin P, Butler A, Luther GW, Kroneck PMH, Meyer-Klaucke W, Feiters MC: Iodide accumulation provides kelp with an inorganic antioxidant impacting atmospheric chemistry. Proc Natl Acad Sci USA 2008, 105:6954–6958.

6. Ritter A, Goulitquer S, Salaün JP, Tonon T, Correa JA, Potin P: Copper stress induces biosynthesis of octadecanoid and eicosanoid oxygenated derivatives in the brown algal kelp Laminaria digitata. New Phytol 2008, 180:809–821.

7. Bartsch I, Wiencke C, Bischof K, Buchholz CM, Buck BH, Eggert A, Feuerpfeil P, Hanelt D, Jacobsen S, Karez R, Karsten U, Molis M, Roleda MY, Schubert H, Schumann R, Valentin K, Weinberger F, Wiese J: The genus Laminaria sensu lato: recent insights and developments. Eur J Phycol 2008, 43:1–86.

8. Kloareg B, Quatrano RS: Structure of the cell-walls of marine-algae and ecophysiological functions of the matrix polysaccharides. Oceanogr Mar Biol 1988, 26:259–315.

9. Tonon T, Rousvoal S, Roeder V, Boyen C: Expression profiling of the mannuronan C5-epimerase multigenic family in the brown alga Laminaria digitata (Phaeophyceae) under biotic stress conditions. J Phycol 2008, 44:1250–1256.

10. Nyvall P, Corre E, Boisset C, Barbeyron T, Rousvoal S, Scornet D, Kloareg B, Boyen C: Characterization of mannuronan C-5-epimerase genes from the brown alga Laminaria digitata. Plant Physiol 2003, 133:726–735.

11. Charrier B, Coelho SM, Le Bail A, Tonon T, Michel G, Potin P, Kloareg B, Boyen C, Peters AF, Cock JM: Development and physiology of the brown alga Ectocarpus siliculosus: two centuries of research. New Phytol 2008, 177:319–332.

12. Peters AF, Marie D, Scornet D, Kloareg B, Cock JM: Proposal of Ectocarpus siliculosus (Ectocarpales, Phaeophyceae) as a model organism for brown algal genetics and genomics. J Phycol 2004, 40:1079–1088.

13. Coelho SM, Peters AF, Charrier B, Roze D, Destombe C, Valero M, Cock JM: Complex life cycles of multicellular eukaryotes: new approaches based on the use of model organisms. Gene 2007, 406:152–170.

14. Peters AF, Scornet D, Ratin M, Charrier B, Monnier A, Merrien Y, Corre E, Coelho SM, Cock JM: Life-cycle-generation-specific developmental processes are modified in the immediate upright mutant of the brown alga Ectocarpus siliculosus. Development 2008, 135:1503–1512.

15. Soeder C, Stengel E: Physico-chemical factors affecting metabolism and growth rate. In Algal Physiology and Biochemistry. Volume 10. Edited by: Stewart WDP. Oxford, London, Edinburgh, Melbourne: Blackwell Scientific Publications; 1974. [Brunett JH, Baker HG, Beevers H, Whatley FR (Series Editors): Botanical Monographs].

16. Davison IR, Reed RH: The physiological significance of mannitol accumulation in brown algae: the role of mannitol as a compatible solute. Phycologia 1985, 24:449–457.

17. Reed RH, Davison IR, Chudek JA, Foster R: The osmotic role of mannitol in the phaeophyta–an appraisal. Phycologia 1985, 24:35–47.

18. Davison IR, Pearson GA: Stress tolerance in intertidal seaweeds. J Phycol 1996, 32:197–211.

19. Collén J, Hervé C, Guisle-Marsollier I, Léger JJ, Boyen C: Expression profiling of Chondrus crispus (Rhodophyta) after exposure to methyl jasmonate. J Exp Bot 2006, 57:3869–3881.

20. Collén J, Guisle-Marsollier I, Léger JJ, Boyen C: Response of the transcriptome of the intertidal red seaweed Chondrus crispus to controlled and natural stresses. New Phytol 2007, 176:45–55.

21. Roeder V, Collén J, Rousvoal S, Corre E, Leblanc C, Boyen C: Identification of stress gene transcripts in Laminaria digitata (Phaeophyceae) protoplast cultures by expressed sequence tag analysis. J Phycol 2005, 41:1227–1235.

22. Ectocarpus Transcriptomics Homepage [http://www.sb-roscoff.fr/UMR7139/ectocarpus/transcriptomics/].

23. Geissler U: Die salzbelastete Flusstrecke der Werra–ein Binnenlandstandort für Ectocarpus confervoides (Roth) Kjellmann. Nova Hedwigia 1983, 37:193–217.

24. West J, Kraft G: Ectocarpus siliculosus (Dillwyn) Lyngb. from Hopkins River Falls, Victoria–the first record of a freshwater brown alga in Australia. Muelleria 1996, 9:29–33.

25. Mittler R, Vanderauwera S, Gollery M, Van Breusegem F: Reactive oxygen gene network of plants. Trends Plant Sci 2004, 9:490–498.

26. Cosse A, Leblanc C, Potin P: Dynamic defense of marine macroalgae against pathogens: from early activated to gene-regulated responses. Adv Bot Res 2007, 46:221–266.

27. Seki M, Narusaka M, Ishida J, Nanjo T, Fujita M, Oono Y, Kamiya A, Nakajima M, Enju A, Sakurai T, Satou M, Akiyama K, Taji T, Yamaguchi-Shinozaki K, Carninci P, Kawai J, Hayashizaki Y, Shinozaki K: Monitoring the expression profiles of 7000 Arabidopsis genes under drought, cold and high-salinity stresses using a full-length cDNA microarray. Plant J 2002, 31:279–292.

28. Le Bail A, Dittami SM, de Franco PO, Rousvoal S, Cock M, Tonon T, Charrier B: Normalisation genes for expression analyses in the brown alga model Ectocarpus siliculosus. Bmc Mol Biol 2008, 9:75.

29. Spirin KS, Ljubimov AV, Castellon R, Wiedoeft O, Marano M, Sheppard D, Kenney MC, Brown DJ: Analysis of gene expression in human bullous ker-

atopathy corneas containing limiting amounts of RNA. Invest Ophthalmol Vis Sci 1999, 40:3108–3115.

30. Livesey FJ, Furukawa T, Steffen MA, Church GM, Cepko CL: Microarray analysis of the transcriptional network controlled by the photoreceptor homeobox gene Crx. Curr Biol 2000, 10:301–310.

31. Wang E, Miller LD, Ohnmacht GA, Liu ET, Marincola FM: High-fidelity mRNA amplification for gene profiling. Nat Biotechnol 2000, 18:457–459.

32. Zhumabayeva B, Diatchenko L, Chenchik A, Siebert PD: Use of SMART (TM)-generated cDNA for gene expression studies in multiple human tumors. Biotechniques 2001, 30:158–163.

33. Puskas LG, Zvara A, Hackler L, Van Hummelen P: RNA amplification results in reproducible microarray data with slight ratio bias. Biotechniques 2002, 32:1330–1340.

34. Stirewalt DL, Pogosova-Agadjanyan EL, Khalid N, Hare DR, Ladne PA, Sala-Torra O, Zhao LP, Radich JP: Single-stranded linear amplification protocol results in reproducible and reliable microarray data from nanogram amounts of starting RNA. Genomics 2004, 83:321–331.

35. Katsuta H, Koyanagi-Katsuta R, Shiiba M, Anzai K, Irie T, Aida T, Akehi Y, Nakano M, Yasunami Y, Harada M, Nagafuchi S, Ono J, Tachikawa T: cDNA microarray analysis after laser microdissection in proliferating islets of partially pancreatectomized mice. Med Mol Morphol 2005, 38:30–35.

36. Vinayagam A, del Val C, Schubert F, Eils R, Glatting KH, Suhai S, Konig R: GOPET: A tool for automated predictions of Gene Ontology terms. BMC Bioinformatics 2006, 7.

37. Wu JM, Mao XZ, Cai T, Luo JC, Wei LP: KOBAS server: a web-based platform for automated annotation and pathway identification. Nucleic Acids Res 2006, 34:W720-W724.

38. Yamaguchi T, Ikawa T, Nisizawa K: Pathway of mannitol formation during photosynthesis in brown algae. Plant Cell Physiol 1969, 10:425–440.

39. Kreps JA, Wu YJ, Chang HS, Zhu T, Wang X, Harper JF: Transcriptome changes for Arabidopsis in response to salt, osmotic, and cold stress. Plant Physiol 2002, 130:2129–2141.

40. Kimura M, Yamamoto YY, Seki M, Sakurai T, Sato M, Abe T, Yoshida S, Manabe K, Shinozaki K, Matsui M: Identification of Arabidopsis genes regulated by high light-stress using cDNA microarray. Photochem Photobiol 2003, 77:226–233.

41. Lipka V, Kwon C, Panstruga R: SNARE-Ware: The role of SNARE-domain proteins in plant biology. Annu Rev Cell Dev Biol 2007, 23:147–174.

42. Rodriguez-Manzaneque MT, Ros J, Cabiscol E, Sorribas A, Herrero E: Grx5 glutaredoxin plays a central role in protection against protein oxidative damage in Saccharomyces cerevisiae. Mol Cell Biol 1999, 19:8180–8190.

43. Foyer CH, Lelandais M, Kunert KJ: Photooxidative stress in plants. Physiol Plant 1994, 92:696–717.

44. Kovtun Y, Chiu WL, Tena G, Sheen J: Functional analysis of oxidative stress-activated mitogen-activated protein kinase cascade in plants. Proc Natl Acad Sci USA 2000, 97:2940–2945.

45. Kourtis N, Tavernarakis N: Autophagy and cell death in model organisms. Cell Death Differ 2009, 16:21–30.

46. Eastmond PJ, Graham IA: Re-examining the role of the glyoxylate cycle in oilseeds. Trends Plant Sci 2001, 6:72–78.

47. Bassham DC: Plant autophagy-more than a starvation response. Curr Opin Plant Biol 2007, 10:587–593.

48. von Kampen J, Wettern M, Schulz M: The ubiquitin system in plants. Physiol Plant 1996, 97:618–624.

49. Belknap WR, Garbarino JE: The role of ubiquitin in plant senescence and stress responses. Trends Plant Sci 1996, 1:331–335.

50. Fujita M, Fujita Y, Noutoshi Y, Takahashi F, Narusaka Y, Yamaguchi-Shinozaki K, Shinozaki K: Crosstalk between abiotic and biotic stress responses: a current view from the points of convergence in the stress signaling networks. Curr Opin Plant Biol 2006, 9:436–442.

51. Gerwick WH, Roberts MA, Vulpanovici A, Ballantine DL: Biogenesis and biological function of marine algal oxylipins. Adv Exp Med Biol 1999, 447:211–218.

52. Colin C, Leblanc C, Wagner E, Delage L, Leize-Wagner E, Van Dorsselaer A, Kloareg B, Potin P: The brown algal kelp Laminaria digitata features distinct bromoperoxidase and iodoperoxidase activities. J Biol Chem 2003, 278:23545–23552.

53. Wang WX, Vinocur B, Shoseyov O, Altman A: Role of plant heat-shock proteins and molecular chaperones in the abiotic stress response. Trends Plant Sci 2004, 9:244–252.

54. Collén J, Davison IR: Reactive oxygen metabolism in intertidal Fucus spp. (Phaeophyceae). J Phycol 1999, 35:62–69.

55. Hwang YS, Jung G, Jin E: Transcriptome analysis of acclimatory responses to thermal stress in Antarctic algae. Biochem Biophys Res Commun 2008, 367:635–641.

56. Blumwald E: Sodium transport and salt tolerance in plants. Curr Opin Cell Biol 2000, 12:431–434.

57. Miyama M, Tada Y: Transcriptional and physiological study of the response of Burma mangrove (Bruguiera gymnorhiza) to salt and osmotic stress. Plant Mol Biol 2008, 68:119–129.

58. Butler DM, Ostgaard K, Boyen C, Evans LV, Jensen A, Kloareg B: Isolation conditions for high yields of protoplasts from Laminaria saccharina and Laminaria digitata (Phaeophyceae). J Exp Bot 1989, 40:1237–1246.

59. Bouchereau A, Aziz A, Larher F, Martin-Tanguy J: Polyamines and environmental challenges: recent development. Plant Sci 1999, 140:103–125.

60. Krell A, Funck D, Plettner I, John U, Dieckmann G: Regulation of proline metabolism under salt stress in the psychrophilic diatom Fragilariopsis cylindrus (Bacillariophyceae). J Phycol 2007, 43:753–762.

61. Armbrust EV, Berges JA, Bowler C, Green BR, Martinez D, Putnam NH, Zhou S, Allen AE, Apt KE, Bechner M, Brzezinski MA, Chaal BK, Chiovitti A, Davis AK, Demarest MS, Detter JC, Glavina T, Goodstein D, Hadi MZ, Hellsten U, Hildebrand M, Jenkins BD, Jurka J, Kapitonov VV, Kroger N, Lau WW, Lane TW, Larimer FW, Lippmeier JC, Lucas S, et al.: The genome of the diatom Thalassiosira pseudonana: ecology, evolution, and metabolism. Science 2004, 306:79–86.

62. Morris SM Jr: Recent advances in arginine metabolism. Curr Opin Clin Nutr Metab Care 2004, 7:45–51.

63. Starr RC, Zeikus JA: Utex–the Culture Collection of Algae at the University-of-Texas at Austin 1993 List of Cultures. J Phycol 1993, 29:1–106.

64. Ewing B, Hillier L, Wendl MC, Green P: Base-calling of automated sequencer traces using phred. I. Accuracy assessment. Genome Res 1998, 8:175–185.

65. Pertea G, Huang X, Liang F, Antonescu V, Sultana R, Karamycheva S, Lee Y, White J, Cheung F, Parvizi B, Tsai J, Quackenbush J: TIGR Gene Indices clustering tools (TGICL): a software system for fast clustering of large EST datasets. Bioinformatics 2003, 19:651–652.

66. Ashburner M, Ball CA, Blake JA, Botstein D, Butler H, Cherry JM, Davis AP, Dolinski K, Dwight SS, Eppig JT, Harris MA, Hill DP, Issel-Tarver L, Kasarskis A, Lewis S, Matese JC, Richardson JE, Ringwald M, Rubin GM, Sherlock

G, Consortium GO: Gene Ontology: tool for the unification of biology. Nat Genet 2000, 25:25–29.

67. Min XJ, Butler G, Storms R, Tsang A: OrfPredictor: predicting protein-coding regions in EST-derived sequences. Nucleic Acids Res 2005, 33:W677-680.

68. Apt KE, Clendennen SK, Powers DA, Grossman AR: The gene family encoding the fucoxanthin chlorophyll proteins from the brown alga Macrocystis Pyrifera. Mol Gen Genet 1995, 246:455–464.

69. Bolstad BM, Irizarry RA, Astrand M, Speed TP: A comparison of normalization methods for high density oligonucleotide array data based on variance and bias. Bioinformatics 2003, 19:185–193.

70. Irizarry RA, Bolstad BM, Collin F, Cope LM, Hobbs B, Speed TP: Summaries of Affymetrix GeneChip probe level data. Nucleic Acids Res 2003, 31:e15.

71. Irizarry RA, Hobbs B, Collin F, Beazer-Barclay YD, Antonellis KJ, Scherf U, Speed TP: Exploration, normalization, and summaries of high density oligonucleotide array probe level data. Biostatistics 2003, 4:249–264.

72. Saeed AI, Sharov V, White J, Li J, Liang W, Bhagabati N, Braisted J, Klapa M, Currier T, Thiagarajan M, Sturn A, Snuffin M, Rezantsev A, Popov D, Ryltsov A, Kostukovich E, Borisovsky I, Liu Z, Vinsavich A, Trush V, Quackenbush J: TM4: A free, open-source system for microarray data management and analysis. Biotechniques 2003, 34:374–377.

73. Benjamini Y, Hochberg Y: Controlling the false discovery rate–a practical and powerful approach to multiple testing. J R Stat Soc Ser B–Methodol 1995, 57:289–300.

74. Soukas A, Cohen P, Socci ND, Friedman JM: Leptin-specific patterns of gene expression in white adipose tissue. Genes Dev 2000, 14:963–980.

75. Yeung KY, Haynor DR, Ruzzo WL: Validating clustering for gene expression data. Bioinformatics 2001, 17:309–318.

76. Sealfon RS, Hibbs MA, Huttenhower C, Myers CL, Troyanskaya OG: GOLEM: an interactive graph-based gene-ontology navigation and analysis tool. BMC Bioinformatics 2006, 7:443.

77. Benjamini Y, Yekutieli D: The control of the false discovery rate in multiple testing under dependency. Ann Stat 2001, 29:1165–1188.

78. Altschul SF, Madden TL, Schaffer AA, Zhang J, Zhang Z, Miller W, Lipman DJ: Gapped BLAST and PSI-BLAST: a new generation of protein database search programs. Nucleic Acids Res 1997, 25:3389–3402.

79. Shannon P, Markiel A, Ozier O, Baliga NS, Wang JT, Ramage D, Amin N, Schwikowski B, Ideker T: Cytoscape: a software environment for integrated

models of biomolecular interaction networks. Genome Res 2003, 13:2498–2504.

80. Gschlössl B, Guermeur Y, Cock JM: HECTAR: a method to predict subcellular targeting in heterokonts. BMC Bioinformatics 2008, 9:393.

81. Milpetz F, Argos P, Persson B: TMAP: a new email and WWW service for membrane-protein structural predictions. Trends Biochem Sci 1995, 20:204–205.

CITATION

Originally published under the Creative Commons Attribution License. Dittami SM1, Scornet D, Petit JL, Ségurens B, Da Silva C, Corre E, Dondrup M, Glatting KH, König R, Sterck L, Rouzé P, Van de Peer Y, Cock JM, Boyen C, Tonon T. Global expression analysis of the brown alga Ectocarpus siliculosus (Phaeophyceae) reveals large-scale reprogramming of the transcriptome in response to abiotic stress. Genome Biol. 2009;10(6):R66. doi:10.1186/gb-2009-10-6-r66.

Chloroplast Genome Sequence of the Moss *Tortula ruralis*: Gene Content, Polymorphism, and Structural Arrangement Relative to Other Green Plant Chloroplast Genomes

Melvin J. Oliver, Andrew G. Murdock, Brent D. Mishler,
Jennifer V. Kuehl, Jeffrey L. Boore, Dina F. Mandoli,
Karin D. E. Everett, Paul G. Wolf, Aaron M. Duffy
and Kenneth G. Karol

ABSTRACT

Background

*Tortula ruralis, a widely distributed species in the moss family Pottiaceae, is
increasingly used as a model organism for the study of desiccation tolerance*

and mechanisms of cellular repair. In this paper, we present the chloroplast genome sequence of T. ruralis, only the second published chloroplast genome for a moss, and the first for a vegetatively desiccation-tolerant plant.

Results

The Tortula chloroplast genome is ~123,500 bp, and differs in a number of ways from that of Physcomitrella patens, the first published moss chloroplast genome. For example, Tortula lacks the ~71 kb inversion found in the large single copy region of the Physcomitrella genome and other members of the Funariales. Also, the Tortula chloroplast genome lacks petN, a gene found in all known land plant plastid genomes. In addition, an unusual case of nucleotide polymorphism was discovered.

Conclusions

Although the chloroplast genome of Tortula ruralis differs from that of the only other sequenced moss, Physcomitrella patens, we have yet to determine the biological significance of the differences. The polymorphisms we have uncovered in the sequencing of the genome offer a rare possibility (for mosses) of the generation of DNA markers for fine-level phylogenetic studies, or to investigate individual variation within populations.

Background

Tortula ruralis (Hedw.) Gaertn., also known as Syntrichia ruralis (Hedw.) F. Weber & D. Mohr (Pottiaceae) is a moss with a cosmopolitan distribution in relatively dry habitats. In North America the species is widespread in northern latitudes but is more common in the Western U.S., south into Mexico [1]. Tortula ruralis has received considerable attention over the last forty years as a model for the study of vegetative desiccation tolerance, i.e., the ability to equilibrate to the water potential of dry air and survive, regaining growth and development upon rehydration. Tortula ruralis offers much as an experimental model for the study of environmental impacts on plants: it grows easily in culture, has a limited number of cell types, and, because of its morphology, experimental treatments act directly at the cellular level [2,3]. It is the latter property that also makes it an ideal choice for an indicator species in air pollution studies [4,5].

Tortula ruralis is among the most desiccation-tolerant of land plants and it can recover from desiccation even after at least three years in the dried state [6,7]. Physcomitrella patens is relatively tolerant of dehydration but cannot tolerate the levels of drying that T. ruralis can survive [8]. It is well established that the chloroplast plays a central role in the recovery of vegetative plants cells from desiccation

[9] and it is possible that differences between the chloroplast genomes of T. ruralis and P. patens may relate to this fundamental difference between the two mosses.

The rapid recovery of photosynthesis is critical in order to recover and re-establish growth when water is available, thus maximizing the time available to the moss for carbon fixation and productivity [10]. Following slow drying to -100 Mpa, photosystem II (PSII) activity in T. ruralis recovers within minutes after rewetting [11], with normal rates of carbon fixation returning within an hour [2]. Photosynthesis is essential for the production of the energy required for repair and protein synthesis following the desiccation event. Obviously, the integrity and metabolic capacity of the chloroplast is central to the speed of recovery of photosynthesis. It is clear from electron microscopic observation of freeze-fracture preparations that chloroplast membranes, both the envelope and thylakoid membranes, in T. ruralis are unaltered by desiccation [12], which supports the idea that desiccation does not damage the photosynthetic apparatus. Such protection of chloroplast structure has also been demonstrated for gametophytes of Polytrichum formosum, which also appear to be unaltered by the imposition of desiccation and the rigors of rehydration [9]. Thus it is clear that the chloroplast holds a central role in the response of T. ruralis to desiccation and rehydration and it is important to study the nature of its genome in this plant, the first vegetatively desiccation-tolerant plant to have its chloroplast genome sequenced. The genome sequence of T. ruralis, and its comparison to other chloroplast genomes, is critical if we want to understand how the interplay between the nuclear and chloroplast genomes plays a role in desiccation tolerance.

In addition to the relevance of the T. ruralis chloroplast genome to the important trait of desiccation tolerance, the genome sequence has considerable relevance to our current understanding of evolutionary history of the land plants. Current evidence suggests that mosses are the sister group of hornworts plus tracheophytes, diverging at least 450 million years ago [13,14]. As an early diverging lineage, mosses hold a place in the phylogeny of land plants that is important for comparative purposes to seed plants [15], although comparisons are currently hampered because only one published chloroplast genome is available for mosses, while hundreds are available for its sister group. Several chloroplast genome sequences will be required to estimate the ancestral genome sequence for mosses, which will in turn allow comparisons with tracheophyte genomes.

The interest in Tortula ruralis as a model desiccation-tolerant organism has increased as our need to understand how plants survive dehydration stress grows and the global impact of climate change becomes more critical. This impetus and the need for increasing sampling within the mosses for phylogenetic comparative purposes led to the choice of T. ruralis for chloroplast genome sequencing. The assembly and annotation of the T. ruralis chloroplast genome sequence is presented here, only the second chloroplast genome sequenced for a moss and the first for a desiccation tolerant plant. The first chloroplast genome for a moss, Physcomitrella patens, was completed in 2003 [16]. Because P. patens was found to have a major rearrangement in the chloroplast compared to what PCR-based methods show for most other moss lineages [17], the T. ruralis chloroplast genome will serve as an important point of comparison and will assist in ongoing efforts to utilize whole-genomic sequences and structural characters in a comparative phylogenetic framework [13,18].

Results

Whole Chloroplast Genome Description

The chloroplast genome sequence comprises 122,530 bp; a gap of ~750 bp (an estimate based on comparison to Physcomitrella patens) remains undetermined within the coding region of ycf2 despite repeated attempts to sequence this region using long-distance PCR and gene walking. While the precise length of the chloroplast genome remains unknown, it is estimated to be 123.5 kb. Figure 1 summarizes the structure of the T. ruralis chloroplast genome. Structurally, T. ruralis lacks the large ~71 kb inversion in the LSC region of the genome that is found in P. patens. The gene list reveals the absence of petN, a gene that is found in all other known land plant chloroplast genomes. The trnP^{GGG} gene, although containing an altered anticodon region similar to that seen in Physcomitrella, also contains significant mismatches in the stem regions of the predicted tRNA structure. The inference is that this gene has become a pseudo gene in both lineages (Figure 2.). The gene content of the inverted repeat (IR) regions is conserved between T. ruralis and P. patens.

Polymorphisms

One surprising discovery was the identification of apparent polymorphism in the T. ruralis chloroplast DNA. Overall, we observed 29 clearly polymorphic sites, each of which appears to have two possible states (Table 1). Two of the polymorphisms appear in the non-coding region of the IR and so only 28 of the polymorphic sites are unique. Eleven of the sites were situated within protein coding sequences: five result in synonymous amino acid codons and six are non-synonymous. The remaining polymorphic sites are situated in non-coding regions: six in introns and ten in the intergenic regions.

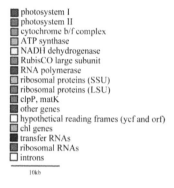

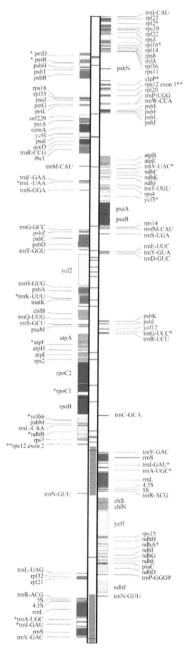

Figure 1. Tortula ruralis chloroplast genome structure. Genes (colored boxes) on the right side of the map are transcribed in the top down direction, whereas those on the left side are transcribed bottom up. The tRNA genes are indicated by the one-letter amino acid code (fM = initiator methionine) followed by the anticodon. Introns are show with an asterisk (*), the trans-spliced gene rps12 is shown with two asterisks (**) and the pseudogene trnPGGG is shown with a Psi (Ψ). Horizontal red lines along the genome indicate polymorphic nucleotides.

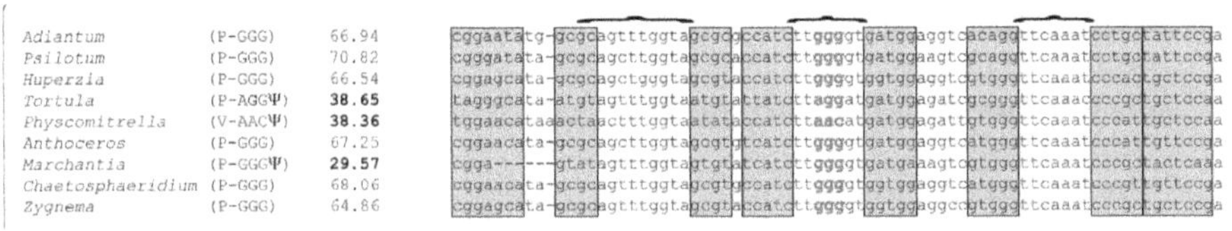

Figure 2. Alignment of trnPGGG coding regions from Tortula and eight other green plant chloroplast genomes. Columns from left to right are: Species name, tRNA anticodon, and Cove score for tRNA structure derived from tRNAscan-SE [23]. The stem regions of the tRNA molecule are indicated by grey boxes with corresponding internal stem segments indicated by brackets: the tRNA anticodon region is shown in red. Alteration of the anticodon region is found in both moss lineages Tortula ruralis and Physcomitrella patens.

Table 1. List of all sequence polymorphisms located on the Tortula ruralis chloroplast genome sequence.

position	location	state	strand	codon
7810	petD intron	Y (C/T)	-	
7821	petD intron	M (A/C)	-	
8076	petD intron	M (A/C)	-	
9525	petB intron	Y (C/T)	-	
19020	psbJ/petA IGR	K (G/T)	na	
23480	psaI/ycf4 IGR	K (G/T)	na	
23623	psaI/ycf4 IGR	K (G/T)	na	
29394	trnM/trnV IGR	M (A/C)	na	
41006	psaB/rps14 IGR	R (A/G)	na	
45811	trnT/trnE IGR	W (A/T)	na	
47128	ycf2	R (A/G)	-	GTA or GTG - synonymous
47698	ycf2	M (A/C)	-	TTA or TTC - nonsynonymous
49026	ycf2	K (G/T)	-	GTA or TTA - nonsynonymous
49946	ycf2	K (G/T)	-	AGT or ATT - nonsynonymous
50907	ycf2	R (A/G)	-	TAA or TAG - synonymous
53115	trnH/psbA IGR	R (A/G)	na	
55502	matK	K (G/T)	-	TTG or TTT - nonsynonymous
56736	trnK intron	Y (C/T)	-	
65765	atpI/atpH IGR	K (G/T)	na	
67161	rps12	R (A/G)	-	CGA or CGG - synonymous
72206	rpoC1-exon2	K (G/T)	-	TGT or TTT - nonsynonymous
75462	rpoB	Y (C/T)	-	CGC or CGT - synonymous
81152	ndhB intron	R (A/G)	-	
82247	ndhB exon1	Y (C/T)	-	CCA or TCA nonsynonymous
82463	rps7/ndhB IGR	W (A/T)	na	
83334	rps12 exon2/rps7 IGR	W (A/T)	na	
84909	Noncoding IR	K (G/T)	na	
105822	ndhG	M (A/C)	+	GTA or GTC - synonymous
121800	Noncoding IR	M (A/C)	na	

Discussion

Whole Chloroplast Genome Description

Compared to other published chloroplast genomes, the chloroplast genome of Tortula ruralis is most similar to that of Physcomitrella patens. However, T. ruralis lacks the large ~71 kb inversion in the LSC region of the genome that is found in P. patens. This large inversion is found only in the moss order Funariales, whereas other moss lineages have a more plesiomorphic gene order similar to the liverwort Marchantia polymorpha [16,17].

It is interesting that the Tortula ruralis chloroplast genome lacks the petN gene because the protein encoded by this gene plays an important role in photosynthesis electron transport. The PetN protein is a subunit of the cytochrome b6f (Cyt b6f) complex of cyanobacteria and plants. The Cytb6f complex is a plastoquinol-plastocyanin oxidoreductase within thylakoid membranes that functions as the photosynthetic redox control of energy distribution between the two photosystems, PSII and PSI, and gene expression [19]. With such a critical role in chloroplast function it is clear that T. ruralis cannot be devoid of the PetN polypeptide and it is very likely, given the conservation of gene order in plant chloroplasts, that the petN gene does not reside in the region of the genome for which we do not yet have a sequence (e.g., within ycf2). What remains is the probability that in T. ruralis the petN gene has moved into the nuclear genome or there is a nuclear-encoded gene product that serves the same function as the PetN protein. This hypothesis remains to be tested but it offers an exciting speculation that the translocation of the petN gene from the chloroplast to nuclear genome may be related to the stability of the chloroplast in desiccation-tolerant mosses. Obviously, a much more detailed survey of moss chloroplast genomes needs to be accomplished and further experimental evidence needs to be collected before such speculation can be tested.

Polymorphism

Few studies have detected nucleotide polymorphism for plastid DNA within plant populations [20]. Our results suggest a considerable level of population polymorphism in the Tortula ruralis chloroplast DNA. Mosses have monoplastidic cell division [21,22], thus it is unlikely that the polymorphism occurs within individual plants. Because the sequenced material was grown from wild-collected spores from multiple parents, the polymorphism is more likely to be due to genotypic races or otherwise cryptic lineages within the source population. One alternative source of polymorphism that is difficult to discount completely is that the variation is caused by inclusion of nuclear DNA which can occasionally contain fragments of chloroplast DNA [23]. We consider this unlikely because our source of plastid DNA was from isolated chloroplasts, so nuclear DNA levels would be very low and the chances of capturing rare inserts would be negligible. Also, the 29 polymorphisms were distributed across the genome rather than clustered in one region which is what would be expected if one or a few nuclear regions had been captured. Many recently sequenced plastid genomes have used shotgun or PCR-based approaches, and completely eliminating nuclear DNA is impossible.

The polymorphisms that give rise to non-synonymous changes in amino acid at a defined site in a protein coding region are possible targets for post transcriptional editing should the non-synonymous change result in a loss or detrimental change in protein function. For example, the polymorphic position 744 in the matK gene (A or C) could result in either leucine (UUA) or phenylalanine (UUC) in the resultant peptide. Given the importance of this group II intron maturase gene to chloroplast function [24] this change either does not result in a significant change in the normal activity of the polypeptide or it would be corrected by the RNA editing activity within the chloroplast prior to translation. Testing this hypothesis, and estimating the extent of RNA editing will require sequencing of cDNA from Tortula chloroplast genes.

The shotgun sequencing method used for assembling this genome makes a determination of the number of haplotypes in the sequenced sample impossible. However, there are some indications that only two haplotypes may be present: (1) all variable sites were bimodal (e.g., either an A or G); and (2) in 5 instances, multiple polymorphic sites were located within a single DNA fragment and in all 5 cases only two versions of the sequence were present. Interestingly, one third of the variable sites occurred in coding regions of the chloroplast, including two sites in rbcL and three in ndhF. Polymorphic sites that can be easily sequenced from chloroplast DNA have potential for future use in population-level studies in T. ruralis. Future studies that include cloning and sequencing chloroplasts genomic regions from single individuals, gathered from across the range of this species, would be very productive. It would also be of interest to determine if the polymorphisms within the coding regions of the chloroplast genes generate functional allelic variation in the gene products involved.

trnP^{GGG}

Among the known chloroplast genomes of land plants, the trnP^{GGG} gene shows an unusual evolutionary pattern. In Physcomitrella patens and Tortula ruralis the sequence data indicate that this gene has become a pseudogene in these lineages (Figure 2); further substantiated by the low Cove scores for the tRNA structure prediction from tRNAscan-SE [25]. In the liverwort Marchantia polymorpha, the anticodon is intact but there are stem mismatches and a deletion of 5 bases, indicating that this gene may also be pseudogenized in M. polymorpha. However, in the hornwort Anthoceros formosae and in vascular plants trnP^{GGG} appears to be functional. It is of course possible that the change in the anticodon sequence seen in T. ruralis and P. patens are modified post-transcriptionally to allow non-wobble interactions and thus rendering the tRNA functional. However, if these alterations in the trnP^{GGG} seen in T. ruralis and P. patens do indicate a loss of function then given the current understanding of land plant phylogeny, this would imply at least two independent deactivation events for the trnP^{GGG} gene in the chloroplast. Functional copies of trnP^{GGG} occur in the chloroplast genomes of the green algae Chaetosphaeridium and Zygnema, otherwise this gene appears to have been lost in Chara and is absent in other examined green-algal chloroplast genomes [26]. The GGG codon is used in coding regions of the chloroplast with roughly equal frequency in all of these lineages.

Relationships to Desiccation-Tolerance

Protection of chloroplast structure appears to be a major aspect of the mechanism for desiccation tolerance in bryophytes (for review see Oliver [27]). It is unclear at the moment as to how much the chloroplast genome, and the genes it encodes, influences the stability of the chloroplast during dehydration, or how much it contributes to the rapid resumption of photosynthetic electron transport. However, the importance of the chloroplast in the phenotype of desiccation tolerance cannot be underestimated. This is evident in the percentage of transcripts encoding chloroplast-directed proteins that are present in the transcriptome of recovering rapid-dried gametophytes of Tortula ruralis [28]. Of the transcripts that can be annotated in the rehydration transcriptome, 12.5% are classified as chloroplastic in the Gene Ontology (GO) functional classification scheme. One of the more prominent being the transcript that encode the Early Light Inducible Protein (ELIP) which is described as being an important protein for chloroplast structural protection during desiccation and upon rehydration of T. ruralis gametophytes [29]. Unfortunately, because of the nature of plastid transcripts (lack of a polyA tail) they are generally not present in the cDNA preparations used for transcriptome sequencing projects and thus we know little of how their synthesis responds to desiccation and rehydration or what is the role of the chloroplast transcriptome in the mechanism of desiccation tolerance we see in T. ruralis. Sequencing the chloroplast genome is an important first step in answering these questions about the interplay between the two genomes, nuclear and plastid, in this important trait. The role of the chloroplast genome and the expression of its genes remains a fertile area for study.

Conclusions

The Tortula ruralis genome differs from that of the only other published chloroplast genome sequence, that of Physcomitrella patens. The differences in chloroplast genome structure and gene content offer some tantalizing hypotheses in relation to one of the fundamental differences between the two mosses in their ability to tolerate the stresses associated with extreme water loss. The most intriguing observation is the loss of the petN gene, presumably to the nuclear genome, given its important role in photosynthetic electron transport and the significance of this with regards to desiccation tolerance.

The polymorphisms we have uncovered in the sequencing of the genome offer the possibility of the generating chloroplast DNA markers for future fine-level phylogenetic studies, or for future population genetic studies examining individual variation within populations.

Methods

DNA Isolation and Sequencing

Material of Tortula ruralis for DNA extraction was grown in sterile culture from spores collected from a wild population located under a pine canopy along the southern bank of the Bow River west of Calgary, Alberta, Canada, approximately 51° 06' 04° N, 114° 17' 10° W (voucher specimen deposited in the University and Jepson Herbaria, University of California, Berkeley, CA). Isolated sporophytes with intact capsules were carefully removed from the parent gametophyte, washed twice in 5% bleach solution each followed by a sterile water wash, and placed on sterile minimal media and the spores removed by breaking open the capsules. The moss cultures used in the preparation of chloroplasts for genome sequencing were derived from five different capsules. Isolation of intact chloroplasts was achieved using fluorescence-activated cell sorting (FACS) wherein fluorescently stained organelles can be visualized, separated, and collected using flow cytometry. Total chloroplast DNA was amplified using rolling circle amplification (RCA) with random hexamer primers [30]. Amplified products were sheared in approximately equal-sized fragments of ~3000 bp by repeated passage through a narrow aperture using a Hydroshear (Genemachines, San Carlos, CA). The resulting fragments were subcloned and sequenced by the Department of Energy Joint Genome Institute (JGI). Detailed methods for the complete process can be found in Wolf et al., [31].

Genome Assembly, Finishing & Annotation

Because the sequences covered random sections of the chloroplast, sufficient numbers of DNA fragments were sequenced to provide an average of eight times coverage for the majority of the genome. Sequence fragments were assembled using the Phrap software package, and Consed was used to visualize assembly of contigs (see http://www.phrap.org). Remaining gaps, low-coverage regions, and areas of questionable assembly were manually resequenced from specific clones or from genomic DNA. The finishing sequences were obtained by cloning, and sequencing PCR-generated fragments obtained using flanking primers from known sequence or by use of a GenomeWalker™ kit (Clontech, Mountain View, CA) using either purified chloroplast genomic DNA or total genomic DNA extractions that contain significant levels of chloroplast DNA.

The resulting genome was annotated using the on-line Dual Organellar GenoMe Annotator (DOGMA) [32] and tRNAscan-SE [25], and gene content was compared to published annotated chloroplast genomes and in particular to the chloroplast genome of Physcomitrella patens (NC_005087). Location of introns, pseudogenes, and beginning and endpoints of many genes was aided by the libraries of plastid DNA and annotated genomes available through NCBI GenBank, http://www.ncbi.nlm.nih.gov [33].

The inverted repeat (IR) characteristic of the chloroplast genome in embryophytes causes problems for automatic alignment programs such as Phrap, because the two IR copies cannot be differentiated by the program. However, this apparent problem can be turned into an advantage and can be used to determine the ends of the IR. Consed (an assembly viewing program) visually indicates when paired sequences from the same clone are placed too far apart in a contig, making the general area of the IR visually apparent if sufficient reads are presented. A consensus sequence for this region with an additional ~2 kb on either side was excised, and all reads were then aligned to this sequence. Using the "color means match" option in Consed, the ends of the IR were visually determined by finding the points at which there was a transition from all reads agreeing to the presence of two distinct sequence motifs.

The sequence of the chloroplast genome for Tortula ruralis (=Syntrichia ruralis) was deposited into the NCBI GenBank http://www.ncbi.nlm.nih.gov, [33] and given the accession number FJ546412.

Authors' Contributions

MJO cultured and maintained the experimental material, conducted the finishing sequencing, participated in assembly and sequence analysis, and drafted the manuscript, AGM and BDM participated in the genome assembly, data analysis and comparative analyses, JVK and JLB sequenced the isolated chloroplast DNA, ensured quality sequence data, and participated in assembly of the genome, DFM and KDE were involved in the conception of the project, isolated (FACS) and purified the chloroplast DAN for sequencing, PGW, APD and KGK were involved in the conception of the project, participated in the assembly of the genome, annotated the genome, and participated in the comparative analyses. All authors have read and approved the final manuscript.

Acknowledgements

This research was supported in part by a collaborative grant from the US National Science Foundation: ATOL: Collaborative Research: Deep Green Plant Phylogenetics: Novel Analytical Methods for Scaling from Genomics to Morphology http://ucjeps.berkeley.edu/TreeofLife/, NSF grant numbers 0228729 to BDM and JLB, 0228660 to DFM, and 0228432 to PGW. Additional funding was provided by the US National Institutes of Health Interdisciplinary Training in Genomic Sciences Grant T32-HG00035 to KGK and USDA-CSREES NRI grant 2007-02007 to MJO. The authors would like to thank Dr Aru K. Arumuganathan, Director of the Flow Cytometry Core, Benaroya Research Institute at Virginia Mason for his excellent technical assistance in the isolation of intact chloroplasts using fluorescence-activated cell sorting (FACS). We would also like to acknowledge the excellent technical assistance of Dean Kelch, Jeremy Hudgeons, and Jim Elder. Work presented here was also supported in part by USDA-ARS CRIS project 3622-21000-027-00 (MJO). This work was also partly performed under the auspices of the US Department of Energy's Office of Science, Biological and Environmental Research Program, and by the University of California, Lawrence Berkeley National Laboratory under Contract No. DE-AC02-05CH11231. Mention of a trademark or proprietary product does not constitute a guarantee or warranty of the product by the United States Department of Agriculture, and does not imply its approval to the exclusion of other products that may also be suitable.

References

1. Mishler BD: Syntrichia. In Flora of North America North of Mexico. Volume 27. the Flora of North America Editorial Committee: Oxford University Press, USA; 2007:618–627.

2. Bewley JD: Physiological aspects of desiccation-tolerance. Ann Rev Plant Physiol 1979, 30:195–238.

3. Oliver MJ, Bewley JD: Desiccation-tolerance of plant tissues: A mechanistic overview. Hort Reviews 1997, 18:171–214.

4. Winner WE, Bewley JD: Terrestrial mosses as bioindicators of SO2 pollution stress. Oecologia 1978, 35:221–230.

5. Naszradi T, Badacsonyi A, Németh N, Tuba Z, Bati F: Zinc, Lead, and Cadmium content in meadow plants and mosses along the M3 motorway (Hungary). J Atmos Chem 2004, 49:593–603.

6. Alpert P, Oliver MJ: Drying without Dying. In Desiccation and survival in plants: Drying without Dying. Edited by: Black M, Pritchard HW. CABI Publishing, Wallingford, Oxon; 2002:3–43.

7. Oliver MJ, Mishler BD, Quisenberry JE: Comparative measures of desiccation-tolerance in the Tortula ruralis complex. I. Variation in damage control and repair. Am J Bot 1993, 80:127–136.

8. Wood AJ: The nature and distribution of vegetative desiccation-tolerance in hornworts, liverworts and mosses. The Bryologist 2007, 110:163–177.

9. Proctor MCF, Oliver MJ, Wood AJ, Alpert P, Stark LR, Cleavitt N, Mishler BD: Desiccation Tolerance in Bryophytes. The Bryologist 2007, 110:595–621.

10. Oliver MJ, Velten J, Mishler BD: Desiccation tolerance in bryophytes: a reflection of the primitive strategy for plant survival in dehydrating habitats? Integ Comp Biol 2005, 45:788–799.

11. Tuba Z, Csintalan Z, Proctor MCF: Photosynthetic responses of a moss, Tortula ruralis, ssp. ruralis, and the lichens Cladonia convoluta and C. furcata to water deficit and short periods of desiccation, and their ecophysiological significance: a baseline study at present-day CO2 concentration. New Phytol 1996, 133:353–361.

12. Platt KA, Oliver MJ, Thomson WW: Membranes and organelles of dehydrated Selaginella and Tortula retain their normal configuration and structural integrity: freeze fracture evidence. Protoplasma 1994, 178:57–65.

13. Kelch DG, Driskell A, Mishler BD: Inferring phylogeny using genomic characters: a case study using land plant plastomes. In Molecular Systematics of Bryophytes [Monographs in Systematic Botany 98]. Edited by: Goffinet B, Hollowell V, Magill R. Missouri Botanical Garden Press, St. Louis; 2004:3–12.

14. Qiu Y-L, Li L, Wang B, Chen Z, Knoop V, Groth-Malonek M, Dombrovska O, Lee J, Kent L, Rest J, Estabrook GF, Hendry TA, Taylor DW, Testa CM, AMbros M, Crandall-Stotler B, Duff RJ, Stech M, Frey W, Quandt D, Davis

CC: The deepest divergences in land plants inferred from phylogenomic evidence. PNAS 2006, 103:15511–1551.

15. Mishler BD, Oliver MJ: Putting Physcomitrella patens on the Tree of Life: The evolution and ecology of mosses. In The Moss Physcomitrella patens: Annual Plant Reviews. Volume 36. Edited by: Knight C, Perroud P-F, Cove D. Wiley-Blackwell, New Jersey; 2009:1–15.

16. Sugiura C, Kobayashi Y, Aoki S, Sugita C, Sugita M: Complete chloroplast DNA sequence of the moss Physcomitrella patens : evidence for the loss and relocation of rpoA from the chloroplast to the nucleus. Nuc Acids Res 2003, 31:5324–5331.

17. Goffinet B, Wickett N, Werner O, Ros RM, Shaw AJ, Cox CJ: Distribution and phylogenetic significance of a 71 kb inversion in the chloroplast genome of the Funariidae (Bryophyta). Annals of Botany 2007, 99:747–753.

18. Mishler BD, Kelch DG: Phylogenomics and early land plant evolution. In Bryophyte Biology. Second edition. Edited by: Shaw AJ, Goffinet B. Cambridge University Press, New York; 2009:173–197.

19. Allen JF: Cytochrome b6f: structure for signalling and vectorial metabolism. Trends Plant Sci 2004, 3:130–137.

20. Mccauley DE, Stevens JE, Peroni PA, Raveill JA: The spatial distribution of chloroplast DNA and allozyme polymorphisms within a population of Silene alba (Caryophyllaceae). Am J Bot 1996, 83:727–731.

21. Brown RC, Lemmon BL: Monoplastidic cell division in lower land plants. Am J Bot 1990, 77:559–571.

22. Renzaglia KS, Brown RC, Lemmon BE, Duckett JG, Ligrone R: The occurrence and phylogenetic significance of monoplastidic meiosis in liverworts. Can J Bot 1994, 72:65–7.

23. Guo XY, Ruan SL, Hu WM, Ca DG, Fan LJ: Chloroplast DNA insertions into the nuclear genome of rice: the genes, sites and ages of insertion involved. Functional & Integrative Genomics 2008, 8:101–108.

24. Barthet MM, Hilu KW: Expression of matK : functional and evolutionary implications. Am J Bot 2007, 94:1402–1412.

25. Lowe TM, Eddy SR: tRNAscan-SE: a program for improved detection of transfer RNA genes in genomic sequence. Nucleic Acids Res 1997, 25:955–964.

26. Turmel M, Otis C, Lemieux C: The chloroplast genome sequence of Chara vulgaris sheds new light into the closest green algal relatives of land plants. Mol Biol Evol 2006, 23:1324–1338.

27. Oliver MJ: Lessons on dehydration tolerance from desiccation tolerant plants. In Plant Desiccation Tolerance. Edited by: Jenks MA, Wood AJ. Wiley-Blackwell, New Jersey; 2007:11–50.

28. Oliver MJ, Dowd SE, Zaragoza J, Mauget SA, Payton PR: The rehydration transcriptome of the desiccation-tolerant bryophyte Tortula ruralis : Transcript classification and analysis. BMC Genomics 2004, 5(89):1–19.

29. Zeng Q, Chen X, Wood AJ: Two early light-inducible protein (ELIP) cDNAs from the resurrection plant Tortula ruralis are differentially expressed in response to desiccation, rehydration, salinity, and high light. J Exp Bot 2002, 53:1197–1205.

30. Dean FB, Nelson JR, Giesler TL, Lasken RS: Rapid amplification of plasmid and phage DNA using Phi 29 DNA polymerase and multiply-primed rolling circle. Genome Res 2001, 11:1095–1099.

31. Wolf PG, Karol KG, Mandoli DF, Kuehl J, Arumuganathan K, Ellis MW, Mishler BD, Kelch DG, Olmstead RG, Boore JL: The first complete chloroplast genome sequence of a lycophyte, Huperzia lucidula (Lycopodiaceae). Gene 2005, 350:117–128.

32. Wyman SK, Boore JL, Jansen RK: Automatic annotation of organellar genomes with DOGMA. Bioinformatics 2004, 20:3252–3255.

33. National Center for Biotechnology Information's Genbank Database [http://www.ncbi.nlm.nih.gov].

CITATION

Originally published under the Creative Commons Attribution License. Oliver MJ, Murdock AG, Mishler BD, Kuehl JV, Boore JL, Mandoli DF, Everett KDE, Wolf PG, Duffy AM, Karol KG. Chloroplast genome sequence of the moss Tortula ruralis: gene content, polymorphism, and structural arrangement relative to other green plant chloroplast genomes. BMC Genomics 2010, 11:143. doi:10.1186/1471-2164-11-143.

Erwinia Carotovora Elicitors and Botrytis Cinerea Activate Defense Responses in Physcomitrella patens

Inés Ponce de León, Juan Pablo Oliver, Alexandra Castro, Carina Gaggero, Marcel Bentancor and Sabina Vidal

ABSTRACT

Background

Vascular plants respond to pathogens by activating a diverse array of defense mechanisms. Studies with these plants have provided a wealth of information on pathogen recognition, signal transduction and the activation of defense responses. However, very little is known about the infection and defense responses of the bryophyte, Physcomitrella patens, to well-studied phytopathogens. The purpose of this study was to determine: i) whether two representative broad host range pathogens, Erwinia carotovora ssp. carotovora

(E.c. carotovora) and Botrytis cinerea (B. cinerea), could infect Physcomitrella, and ii) whether B. cinerea, elicitors of a harpin (HrpN) producing
E.c. carotovora strain (SCC1) or a HrpN-negative strain (SCC3193), could
cause disease symptoms and induce defense responses in Physcomitrella.

Results

B. cinerea and E.c. carotovora were found to readily infect Physcomitrella
gametophytic tissues and cause disease symptoms. Treatments with B. cinerea
spores or cell-free culture filtrates from E.c. carotovoraSCC1 (CF(SCC1)),
resulted in disease development with severe maceration of Physcomitrella tissues, while CF(SCC3193) produced only mild maceration. Although
increased cell death was observed with either the CFs or B. cinerea, the occurrence of cytoplasmic shrinkage was only visible in Evans blue stained protonemal cells treated with CF(SCC1) or inoculated with B. cinerea. Most cells
showing cytoplasmic shrinkage accumulated autofluorescent compounds and
brown chloroplasts were evident in a high proportion of these cells. CF treatments and B. cinerea inoculation induced the expression of the defense-related genes: PR-1, PAL, CHS and LOX.

Conclusion

B. cinerea and E.c. carotovora elicitors induce a defense response in Physcomitrella, as evidenced by enhanced expression of conserved plant defense-related
genes. Since cytoplasmic shrinkage is the most common morphological change
observed in plant PCD, and that harpins and B. cinerea induce this type of
cell death in vascular plants, our results suggest that E.c. carotovora CFSCC1
containing HrpN and B. cinerea could also induce this type of cell death in
Physcomitrella. Our studies thus establish Physcomitrella as an experimental host for investigation of plant-pathogen interactions and B. cinerea and
elicitors of E.c. carotovora as promising tools for understanding the mechanisms involved in defense responses and in pathogen-mediated cell death in
this simple land plant.

Background

Plants are continuously subjected to pathogen attack and respond by activating
a range of defense mechanisms. Recognition of the pathogen or elicitors derived
either from the pathogen or released from the plant cell wall is accompanied with
the production of molecular signals including salicylic acid [1], jasmonic acid [2]
and ethylene [3] that lead to the induction of defense gene expression. This in turn
results in the accumulation of functionally diverse pathogenesis-related (PR) proteins and metabolites (e.g., phenylpropanoids) [4,5]. Recognition of the pathogen

or elicitors is usually accompanied by the rapid death of the infected cells, known as the hypersensitive response (HR), which limits the access of the pathogen to water and nutrients thereby restricting its growth [6,7]. HR can be triggered either by non-specific elicitors recognized by plant receptors, or by specific elicitors (encoded by pathogen avirulence (avr) genes) recognized by corresponding encoded products of plant resistance (R) genes [8,9]. Several studies have suggested that plant cell death resulting from the HR is a type of programmed cell death (PCD). Plant cells undergoing PCD share a number of characteristic morphological and biochemical features in common with animal cell apoptosis [7,10,11]. Moreover, cell death with apoptotic features has also been observed in plants susceptible to virulent pathogens [12,13].

Although bryophytes are non-vascular plants and are considered to be primitive among the embryophyta, mosses have been shown to respond to a variety of environmental stimuli and to several common plant growth factors much like vascular plants. Thus, in spite of having diverged from vascular plants approximately 700 million years ago [14], mosses are well-suited for the study of fundamental processes in plant biology. Furthermore, mosses have a simple developmental program and a life cycle with a predominant haploid phase which greatly facilitates genetic analysis [15].

Physcomitrella patens, a relatively small moss, has recently become a model plant to study plant gene function in that it exhibits high-frequency homologous recombination comparable with that of Saccharomyces cerevisiae, enabling the construction of gene knock-outs [16,17]. The assembled Physcomitrella genome has recently been released and full-length cDNAs in addition to 80,000 ESTs are available in the databases [18-20]. These advantages together with the presence of a great number of Physcomitrella ESTs with high sequence identity to defense-related genes of vascular plants, many of them with unknown functions, makes this plant a very useful model to study plant-pathogen interactions. The susceptibility of distinct tissues to pathogens can also be studied, since Physcomitrella can be maintained as a haploid gametophyte with distinct developmental stages. These consist of the protonema which is a filamentous network of cells, and the radially symmetric gametophore which is a leafy shoot composed of a non-vascular stem with leaves as well as rhizoids [21]. Disease development can be visualized microscopically in that the leaves and protonemal filaments are formed of a monolayer of cells.

There have been very few reports on either pathogen infection or the activation of defense responses in mosses. In silico analysis of the Physcomitrella genome, however, indicates the presence of several encoded proteins with high similarity to R gene products found in flowering plants [22]. Regarding natural infection, the fungus Scleroconidioma sphagnicola (S. sphagnicola) can infect and cause disease symptoms in the moss Sphagnum fuscum (S. fuscum) [23] and viruses were detected in Antarctic mosses [24]. S. sphagnicola hyphae can grow inside the cell wall of S. fuscum, digesting wall components, penetrating into cells of leaves and causing chlorosis of the tissue. In more advanced stages of disease development, necrosis of infected leaf and stem cells, as well as host death can be observed [23].

In this study we aimed to identify plant pathogens capable of infecting and triggering a defense response in Physcomitrella, with the goal of establishing a model system to conduct molecular, cellular and genetic studies on Physcomitrella-pathogen interactions. We used two pathogens with a broad host range, the bacterium Erwinia carotovora ssp carotovora (E.c. carotovora) and the fungus Botrytis cinerea (B. cinerea). E.c. carotovora is a soft-rot Erwinia which causes disease on many vascular plants [25,26]. The main virulence factors of E.c. carotovora are the plant cell wall-degrading enzymes including cellulases, proteases and pectinases [26]. These enzymes cause maceration of the infected tissues and the released cell wall fragments can act as elicitors of the plant defense response [27-31]. Previous studies have shown that cell-free culture filtrate (CF) containing plant cell wall-degrading enzymes from E.c. carotovora$_{SCC3193}$ produces similar symptoms and defense gene expression as those caused by E.c. carotovora$_{SCC3193}$ infection and enhanced disease resistance in CF-treated plants [28,30-32]. Some E.c. carotovora strains produce harpins, which are small, acidic, glycine-rich, heat-stable proteins, that elicit HR and induction of plant defense responses [33,34]. In the present study we have used two strains of E.c. carotovora; i) E.c. carotovora$_{SCC1}$ which is a harpin (HrpN) producing strain [35], and ii) E.c. carotovora$_{SCC3193}$ which is a HrpN-negative strain [36]. B. cinerea is a necrotrophic fungal pathogen that attacks over 200 different plant species [37], by producing multiple proteins and metabolites that kill the host cells [38]. The main virulence factors of B. cinerea vary depending on the isolate, and include toxins and cell wall degrading enzymes such as endopolygalacturonases and xylanases [39-41]. In this study, we demonstrate that E.c. carotovora-derived elicitors and B. cinerea cause disease symptoms and induce a defense response in the moss Physcomitrella.

Results

E.c. carotovora and B. cinerea Infect Physcomitrella Patens

In order to determine whether E.c. carotovora infects Physcomitrella tissues, a gfp-labelled E.c. carotovora$_{SCC3193}$ strain was inoculated onto Physcomitrella leaves as described in methods. Two days after inoculation tissue examined by confocal microscopy indicated that the labelled cells of E.c. carotovora had occupied the apoplast between leaf cells, as well as the cellular space of some plant cells in this same tissue (Figure 1A).

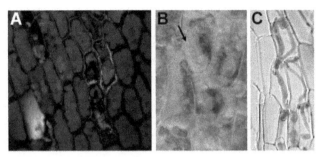

Figure 1. E.c. carotovora and B. cinerea inoculation of Physcomitrella leaves. A. Leaves of Physcomitrella gametophores inoculated with E.c. carotovora$_{SCC3193}$ carrying a GFP-expressing plasmid at 2 dpi. B. B. cinerea inoculated leaves at 2 dpi. C. Trypan blue stained B. cinerea hyphae in inoculated leaves at 2 dpi. Arrow indicates hyphae growing in Physcomitrella.

When B. cinerea inoculated leaf tissue was examined, outlines of fungal hyphae were apparent inside the cell cavity displacing the cytoplasmic contents (Figure 1B). Infection of Physcomitrella tissues by B. cinerea was examined in more detail by staining fungal hyphae with trypan blue. Two days post infection (dpi) hyphae appeared to be within the limits of the cell walls in Physcomitrella leaves (Figure 1C). Our observation that B. cinerea hyphae appeared within plant cells is likely in that Physcomitrella leaves are composed of a contiguous monolayer of adjacent cells.

E.c. carotovora, CFs and B. cinerea Cause Disease Symptoms in Physcomitrella

Development of disease symptoms by E.c. carotovora was initiated by inoculating the harpin HrpN-producing E.c. carotovora$_{SCC1}$ strain and the HrpN-negative E.c. carotovora$_{SCC3193}$ strain onto Physcomitrella leaves. Inoculation with both strains caused visible symptoms around the wounded tissue within 2 days when observed with a magnifying glass, while mock inoculated tissues did not (Figure 2).

Physcomitrella infection by E.c. carotovora and subsequent development of symptoms required that we first wound the plant mechanically (Figures 1 and 2). Growth or colonization of E.c. carotovora in planta as determined by enumeration of bacteria in a given tissue could not be done due to the difficulty of consistently wounding the tissue sufficiently without causing excessive damage and dessication. Instead of continuing our studies with E.c. carotovora bacterial inoculation, cell-free culture filtrate (CF) was used to elicit a defense response since: i) in vascular plants CF incites the same disease symptoms and induces defense gene expression in the same way as does inoculation with E.c. carotovora [30-32], ii) CF is sprayed onto the colonies allowing for direct and homogeneous contact, and iii) it overcomes the technical difficulty of introducing a sufficient number of small wounds on the moss tissue to allow inoculation by E.c. carotovora for a comprehensive evaluation of the plant defense response.

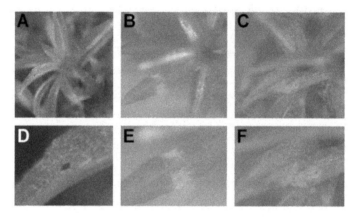

Figure 2. Symptom development in response to E.c. carotovora inoculation. Leaves of Physcomitrella gametophores were wounded and inoculated with 0.9% NaCl (A, D), E.c. carotovoraSCC3193 (B, E) or E.c. carotovoraSCC1 (C, F). Pictures of representative colonies were taken at 2 dpi.

Disease symptom development was observed upon treating Physcomitrella colonies with: (i) CF of E.c. carotovoraSCC1 (CF(SCC1)), (ii) CF of E.c. carotovoraSCC3193 (CF(SCC3193)), or (iii) B. cinerea spores. Disease symptoms such as tissue maceration developed in colonies two days after treatment with either CF(SCC1), CF(SCC3193) or after inoculation with B. cinerea spores, as shown in Figure 3. In control experiments, moss colonies treated with either Luria-Bertani (LB) (Figure 3A), potato dextrose broth (PBD) growth media (not shown) or with spores of a non-pathogen, Aspergillus nidulans (not shown), did not develop disease symptoms. Protonemal filaments treated with either CF(SCC1) or CF(SCC3193) developed maceration symptoms, with CF(SCC3193) causing less tissue damage than CF(SCC1) (Figures 3C and 3F). Additionally, CF(SCC1)-treated colonies acquired a brownish aspect as seen in the protonemal filaments shown in Figures 3C and 3E. CF-treated gametophores, also known as leafy shoots, did not show maceration symptoms, although brownish stems were observed in CF(SCC1)-treated colonies (Figure 3D).

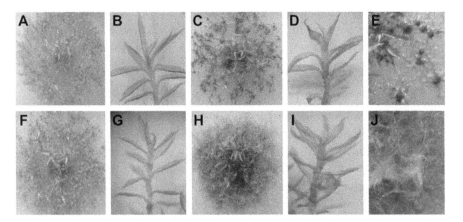

Figure 3. Symptom development in response to CF treatment and B. cinerea inoculation. Moss colonies and gametophores treated with LB (A, B), CF(SCC1) (C, D), CF(SCC3193) (F, G) or with B. cinerea spores (H, I). A closer view of colonies treated with CF(SCC1) (E), or inoculated with B. cinerea spores (J) is shown. Pictures of representative colonies were taken 2 days after treatment.

Physcomitrella showed a clear susceptibility to B. cinerea, involving a characteristic proliferation of mycelium and the appearance of necrotic protonemal tissue in addition to the browning of stems (Figures 3H, 3J and 3I). Inoculated tissues were soft, macerated and were easily separated from the rest of the moss colony. Four dpi B. cinerea-infected moss tissues were completely macerated (data not shown). Protonemal filaments were more susceptible than leaves to CF treatments and B. cinerea inoculation. Taken together, these results show that CF(SCC1), CF(SCC3193) and B. cinerea are capable of causing disease symptoms in Physcomitrella.

CF(SCC1) and B. cinerea Trigger Cytoplasmic Shrinkage, Accumulation of Autofluorescent Compounds and Chloroplast Browning

Pathogen infection or elicitor treatment can induce plant cell death with characteristic changes in cells, including cytoplasmic shrinkage, alteration of chloroplast organization and accumulation of autofluorescent compounds [13,42-44]. In the present study, we examined cellular changes occurring in Physcomitrella tissue showing macroscopic disease symptoms after exposure to B. cinerea spores or CF of the E.c. carotovora strains. CF(SCC1)-treated (Figures 4C and 4D) and B. cinerea-inoculated (Figure 4E) protonemal cells showed cytoplasmic shrinkage after 2 days. In contrast, no cytoplasmic shrinkage was evident in cells treated with LB (control for CF treatments, Figure 4A), CF(SCC3193) (Figure 4B) or

PDB (control for B. cinerea inoculation, data not shown). Other morphological changes were also observed in Physcomitrella CF(SCC1)-treated and B. cinerea-inoculated cells. After 2 days, both treatments caused browning of the chloroplasts in a high proportion of cells (Figures 4D and 4E). Chloroplast browning was evident only in cells showing cytoplasmic shrinkage (Figures 4D and 4E). Additionally, CF(SCC1)-treated (Figures 4F and 4G) and B. cinerea-inoculated cells (data not shown) with brownish chloroplasts showed lack of red autofluorescence of chlorophyll. CF(SCC1)-treated protonemal cells having cytoplasmic shrinkage and brown chloroplasts were more abundant after 4 days. Within this treated protonemal tissue, cells with fewer brown chloroplasts were also observed suggesting that they were being brokendown (Figure 4H). In contrast, CF(SCC3193)- or LB-treated protonemal filaments did not show such changes and green chloroplasts were evident at least 6 days after treatment (data not shown).

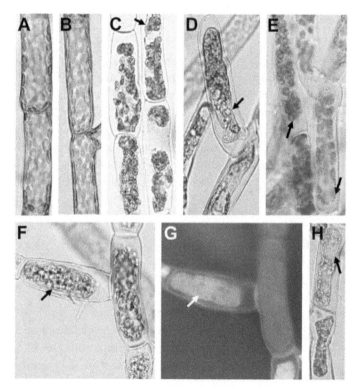

Figure 4. Analysis of protonemal filament changes in response to treatments with CF and B. cinerea spores. Protonemal filaments examined under transmitted light after 2 days of treatment with LB (A), CF(SCC3193) (B), CF(SCC1) (C, D), and B. cinerea spores (E). Cytoplasmic shrinkage observed with CF(SCC1) and B. cinerea spores are indicated with arrows. CF(SCC1)-treated protonemal cells showing browning of chloroplasts and loss of red chlorophyll autofluorescence after UV-excitation are indicated with arrows (F, G). A cell with collapsed cytoplasm and fewer chloroplasts is shown 4 days after treatment (indicated with an arrow) (H).

After an apparent initial contact of B. cinerea hyphae with individual cells within the leaf, plant cells were observed to respond by accumulating autofluorescent compounds (Figure 5A). Cells in B. cinerea-inoculated leaves developing light blue to yellow autofluorescence (AF) were also observed. This AF appeared confined to the cytoplasm now separated from the cell wall (Figure 5C). No AF, however, was observed in CF(SCC1)- or CF(SCC3193)-treated leaves (data not shown). AF was clearly evident in protonemal filaments of B. cinerea-inoculated and CF(SCC1)-treated colonies (Figures 5E and 5G). In contrast, no AF was seen in PDB-treated leaves (Figure 5B) or PDB- or LB-treated protonemal filaments (Figures 5D and 5F), and only a few CF(SCC3193)-treated cells accumulated autofluorescent compounds (data not shown). Also, accumulation of autofluorescent compounds was generally observed once cytoplasmic shrinkage occurred (Figures 5H and 5I). In summary, our results show that CF(SCC1) and B. cinerea induce cellular changes in Physcomitrella protonemal cells, including cytoplasmic shrinkage, browning of chloroplasts and accumulation of autofluorescent compounds, suggesting a cell death process.

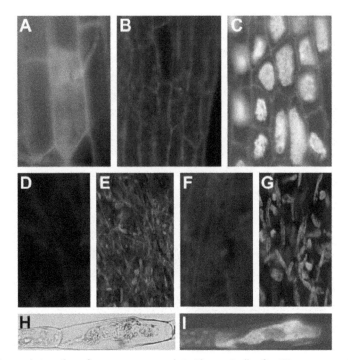

Figure 5. Accumulation of autofluorescent compounds in Physcomitrella after CF treatment and B. cinerea inoculation. Examination of UV-stimulated autofluorescence of B. cinerea-inoculated leaf (A, C), PDB-treated leaf (B), PDB- (D), B. cinerea spores- (E), LB- (F) and CF(SCC1)-treated protonemal filaments (G). A closer view of a CF(SCC1)-treated protonemal cell with cytoplasmic shrinkage and UV-stimulated autofluorescence is shown (H, I). Observations were made 2 days after treatments.

E.c. carotovora Elicitors and B. cinerea Trigger Cell Death in Physcomitrella

To assess whether CF and B. cinerea caused cell death of Physcomitrella tissues, we stained moss colonies with Evans blue, a dye that is excluded by membranes of living cells but diffuses into dead cells [45]. Figure 6 shows pictures of representative tissues. Two days after treatments, an increase in stained cells was observed with either CF(SCC3193), CF(SCC1) or B. cinerea spores, compared with control treatments. Although, while in CF(SCC3193)-treated tissue a low proportion of stained protonemal cells was observed (Figure 6B), CF(SCC1)-treated or B. cinerea-inoculated tissues showed a high proportion of stained protonemal cells (Figures 6C and 6E). In control treatments, almost no stained cells were visible (Figures 6A, 6D).

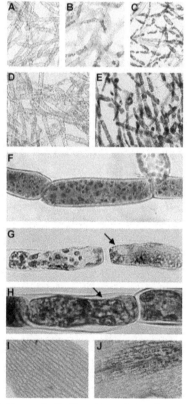

Figure 6. Analysis of cell death in Physcomitrella. Evans blue staining of protonemal tissues after treatments with LB (A), CF(SCC3193) (B), CF(SCC1) (C), PDB (D) and B. cinerea spores (E). A closer view of CF(SCC3193)-(F), CF(SCC1)- (G) and B. cinerea inoculated (H) protonemal cells is shown. Arrows indicate cytoplasmic shrinkage. Leaves treated with CF(SCC1) (I) and B. cinerea spores (J) were also stained with Evans blue. Pictures of representative tissues were taken 2 days after treatment.

Cytoplasmic shrinkage was evident in most Evans blue stained protonemal cells treated with CF(SCC1) or B. cinerea spores (Figures 6G and 6H). Whenever cytoplasmic shrinkage occurred, cells were stained with Evans blue, indicating that these cells were dying or dead. In contrast, most CF(SCC3193)-treated protonemal cells did not show cytoplasmic shrinkage and the dye was distributed homogeneously in the cells (Figure 6F). Three days after treatment, stained CF(SCC3193)-treated protonemal cells did not exhibit cytoplasmic shrinkage suggesting that this response does not develop at a later stage (data not shown). In gametophores, Evans blue stained cells could be detected in leaves inoculated with B. cinerea (Figure 6J), whereas stained cells were not seen after CF(SCC1) (Figure 6I), CF(SCC3193) or control treatments (not shown).

B. cinerea and E.c. carotovora Elicitors Mediate Activation of Defense-Related Genes

To analyze whether CF treatment or B. cinerea inoculation trigger Physcomitrella defense gene expression, we characterized the expression of a number of defense-related gene homologues including; (i) PR-1, (ii) LOX, (iii) PAL, and (iv) CHS. LOX (lipoxygenase) is a key enzyme in the synthesis of defense-related compounds including JA [46], PAL (phenylalanine ammonia-lyase) mediates the biosynthesis of phenylpropanoids and SA [5,47] and CHS (chalcone synthase) is the first enzyme in the synthesis of flavonoids [5]. The results in Figure 7 show that expression of the Physcomitrella homologues increased after CF treatment or B. cinerea inoculation. Clearly, three types of expression patterns were observed. The level of PR-1 expression peaked at 24 h in CF(SCC3193)-treated moss colonies, while in CF(SCC1)-treated and B. cinerea-inoculated tissues expression of PR-1 peaked at 4 h. Expression of PAL and CHS peaked at 4 h in tissues treated with either of the CFs or with B. cinerea spores, although among the treatments, higher expression levels were observed with CF(SCC1). In the case of CHS, two transcripts with an identical expression pattern were detected. LOX expression was moderately induced in CF(SCC1)- and CF(SCC3193)-treated moss colonies at 4 and 24 h, while transcript levels increased significantly in B. cinerea inoculated Physcomitrella tissues, reaching the highest expression level at 24 h. The results obtained in this study show that several conserved defense-related gene homologues of Physcomitrella were induced in response to treatment with E.c. carotovora elicitors or B. cinerea spores.

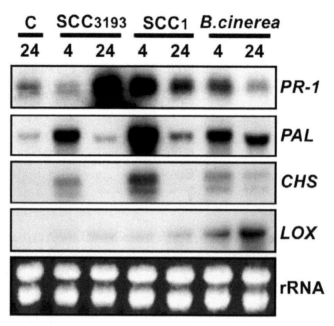

Figure 7. CF and B. cinerea-induced expression of defense-related genes in Physcomitrella. Expression of PR-1, PAL, CHS and LOX genes was characterized by RNA-gel blot hybridization after the following treatments: moss colonies sprayed with LB (control, C), CF(SCC1), CF(SCC3193) or inoculated with B. cinerea spores (2 × 105 spores/ml). Plant samples were harvested at the indicated times (hours) after treatment. 10 μg of RNA was separated on formaldehyde-agarose gels, transferred to nylon membranes and hybridized to 32P-labeled DNA probes. Ethidium bromide staining of rRNA was used to ensure equal loading of RNA samples. Similar results were obtained from two independent experiments.

Discussion

The developmental simplicity, ease of genetic analysis and the evolutionarily relationship between Physcomitrella and other plants has prompted us to study the interaction between this moss and two broad host range pathogens, the bacterium E.c. carotovora and the fungus B. cinerea. Our results indicate that both B. cinerea and E.c. carotovora can infect Physcomitrella tissues and cause disease symptoms. Since Physcomitrella gametophytes do not have stomata, E.c. carotovora entered plant tissues through wounds, while B. cinerea hyphae probably entered by hydrolyzing the plant cell wall using hydrolytic enzymes or by secreting cell wall permeable toxins that kill plant cells. We have observed B. cinerea hyphae within plant cells, including cells in which hyphae were apparently inside the cell cavity displacing the cytoplasm. Hyphae of other necrotrophic fungi, including S. sphagnicola, Tephrocybe palustris and Nectria mnii, are capable of penetrating live cells of moss leaves, resulting in cell death a posteriori [23,48]. In case of

Nectria mnii, it was shown that intracellular hyphae could displace the host cell contents [48].

E.c. carotovoraSCC1, but not E.c. carotovoraSCC3193, was previously shown to harbour the harpin-encoding gene hrpN [35,36]. Harpins are bacterial effector proteins released into the host cells, through a type III secretion system encoded by the hypersensitive reaction and pathogenicity (hrp) gene cluster. When present in plant tissue, harpins cause HR and induction of defense mechanisms [49,50]. In Erwinia spp. hrp genes have been shown to contribute to virulence and to the ability of the pathogen to grow in the plant [35,51]. The higher maceration rate of the protonemal tissues observed with CF(SCC1) compared with CF(SCC3193) is consistent with previous studies showing that polygalacturonase, together with harpin HrpN from E.c. carotovoraSCC1 greatly enhanced lesion formation in Arabidopsis [52].

Cytoplasmic shrinkage is the most common morphological change occurring in plant PCD and has been observed in cells undergoing HR, as well as in tissues of plants susceptible to virulent pathogens [53-55]. Cytoplasmic shrinkage was observed only in Evans blue stained protonemal cells treated with CF(SCC1) but not with CF(SCC3193), probably suggesting that a different mechanism leading to cell death had occurred. This finding is consistent with the induction of HR by harpins and with previous results showing that Pseudomonas syringae pv phaseolicola induced cytoplasmic shrinkage in plant cells, while a hrpD mutant did not [43].

Breakdown of chloroplast membranes and chlorophyll has been observed in cells undergoing PCD, including those treated with elicitors, infected with pathogens or those undergoing senescence [56-59]. Our results showed that treatments with CF(SCC1) and B. cinerea spores induce browning of chloroplasts, which is likely followed by the breakdown of these organelles. This is consistent with previous results showing that S. sphagnicola hyphae are capable of causing degeneration of chloroplasts in the moss S. fuscum [23]. Chloroplasts remained green in protonemal CF(SCC3193)-treated cells, while boiled CF(SCC1), containing the heat-stable HrpN, still induced browning of the chloroplasts in cells also showing cytoplasmic shrinkage (data not shown). These findings suggest that HrpN might trigger cell death associated with cytoplasmic shrinkage and chloroplasts browning. In addition, browning of chloroplasts was associated with chlorophyll breakdown in CF(SCC1)-treated and B. cinerea-inoculated cells, since no red chlorophyll autofluorescence was observed (although quenching by other compounds cannot be excluded). These results are supported by findings showing that E.c. carotovoraSCC1 induced the expression of chlorophyllase 1 in Arabidopsis, which could be involved in the degradation of photoactive chlorophylls to avoid higher levels of reactive oxygen species (ROS) production and cellular damage during pathogen infection [60]. It is also interesting to note that whenever browning of chloroplasts was observed in CF(SCC1)-treated or B. cinerea-inoculated cells, cytoplasmic shrinkage was also present. Since CF(SCC1)-treated protonemal cells with cytoplasmic shrinkage but green chloroplasts were also observed, browning of chloroplasts could be a process occurring later in dying cells after collapse of the cytoplasm. Browning of chloroplasts could be indicative of oxidative processes due to excessive accumulation of ROS in the chloroplasts at late stages of CF(SCC1) and B. cinerea treatments, finally leading to chloroplasts breakdown. To our knowledge, this is the first report in which browning of chloroplasts was observed after pathogen and elicitor treatment. The ability to observe changes in the coloration of the chloroplasts was facilitated in that the leaf tissue, like the protonemal filaments, is composed of a single monolayer of cells.

Accumulation of autofluorescent compounds has been associated with the occurrence of HR in vascular plants [6,44]. CF(SCC1)-treated or B. cinerea-inoculated Physcomitrella tissues developed AF. A previous report demonstrated localized deposition of phenolic compounds at the sites of fungal penetration and also as a second major response that appeared to follow cell death [44]. These findings are consistent with our results, in showing AF confined to the collapsed cytoplasm of dead cells in B. cinerea-inoculated leaves and protonemal filaments.

B. cinerea induce PCD to enable rapid colonization of vascular plants, and Erwinia harpins have been shown to elicit cell death [49,50,61,38]. Cytoplasmic shrinkage, an indicator of plant PCD, correlated with accumulation of autofluorescent compounds and chloroplast browning after inoculation with B. cinerea or treatment with CF of HrpN producing E.c. carotovoraSCC1 suggesting that either treatment results in PCD in Physcomitrella.

Our results also showed that E.c. carotovora elicitors and B. cinerea induced defense-related gene expression in Physcomitrella. Earlier induction of the PR-1--like gene expression and the higher levels of PAL and CHS mRNA accumulation triggered by CF(SCC1) compared with CF(SCC3193), corresponded well with the higher levels of tissue maceration observed with CF(SCC1). CHSs are encoded by multiple genes in vascular plants and Physcomitrella [5,62], and in our study two CHS transcripts with an identical expression pattern were detected. Recently, a new enzymatic activity was described for the same Physcomitrella LOX gene product induced by B. cinerea in this study [63]. Novel oxylipins were generated by this enzyme suggesting a possible involvement in defense responses. In vascular plants PR-1, PAL and LOX are induced by inoculation with E.c. carotovora or by CF treatments [52,64,65] and PR-1 transcript accumulation is increased after B. cinerea infection [66,67]. The results obtained in this study suggest that E.c. carotovora elicitors and B. cinerea similarly induce expression of Physcomitrella defense gene homologues of those studied in vascular plants, and thus validate the use of non-specific plant pathogens or elicitors derived from them to study moss-pathogen interactions.

Conclusion

In the present study, we demonstrate that E.c. carotovora elicitor treatment and B. cinerea inoculation cause disease symptoms and induce defense responses in Physcomitrella. CF(SCC1), CF(SCC3193) and B. cinerea induced the expression of defense-related genes, including PR-1, LOX, PAL and CHS homologues. Compounds produced by LOX, PAL and CHS are involved in the synthesis of JA, phenylpropanoids and SA and flavonoids, respectively, in vascular plants. These compounds could play a role in the defense response of Physcomitrella as has been shown in vascular plants. As such our results further establish E.c. carotovora elicitors, as well as B. cinerea as promising systems to analyze induction of defense responses in Physcomitrella.

Since cytoplasmic shrinkage is the most common morphological change observed in plant PCD, and that harpins and B. cinerea induce this type of cell death in vascular plants, our results suggest that E.c. carotovora CFSCC1 containing HrpN and B. cinerea could also induce this type of cell death in Physcomitrella. Finally, the occurrence of distinct cellular responses leading to cell death by CF(SCC1) and CF(SCC3193) provides a useful system to analyze pathogen-induced cell death and to characterize the key elements involved in its regulation by targeted gene disruption in Physcomitrella.

Methods

Plant Material and Growth Conditions

Physcomitrella patens Gransden WT isolate [68] was grown on cellophane overlaid BCDAT agar medium consisting of 1.6 g l -1 Hoagland's, 1 mM $MgSO_4$, 1.8 mM KH_2PO_4 pH 6.5, 10 mM KNO_3, 45 μM $FeSO_4$, 1 mM $CaCl_2$, 5 mM ammonium tartrate and 10 g l-1 agar [69]. Protonemal cultures and moss colonies were grown as described previously [70]. Plants were grown at 22°C under a photoperiod of 16 h light and three-week-old colonies were used for all the experiments.

Pathogen Inoculation and Culture Filtrate Treatments

Erwinia carotovora ssp carotovora strains SCC3193 [71] and SCC1 [72] were propagated on LB medium [73] at 28°C. Cell-free culture filtrates were prepared by growing bacteria in LB broth overnight, removing bacterial cells by centrifuga-tion (10 min at 4000 g) and filter sterilizing the supernatant (0.2 μm pore size). This filter-sterilized supernatant (CF) was applied by spraying the moss colonies (3 ml per Petri dish containing 16 moss colonies). E.c. carotovoraSCC3193 and E.c. carotovoraSCC1 were grown on LB, and E.c. carotovoraSCC3193 trans-formed with plasmid pUC18 containing the GFP sequence as reporter gene under control of the lac promoter was grown on LB containing 100 μg/ml ampicillin. After 16 h bacterial cells were centrifuged and suspended in 0.9% NaCl to a final concentration of 5×10^8 cfu/ml. These suspensions were used for inoculation of Physcomitrella leaves previously wounded with a needle to create small lesions. An isolate of B. cinerea from a lemon plant was cultivated on 39 g/L potato dex-trose agar (DIFCO) at room temperature. B. cinerea was inoculated by spraying a 2×10^5 spores/ml suspension in half-strength PDB (DIFCO). Symptom de-velopment of CF-treated and B. cinerea-inoculated Physcomitrella colonies was analyzed in three independent experiments using two Petri dishes containing16 colonies each. The experiments involving leaves inoculated with E.c. coratovora strains SCC1, SCC3193 and SCC3193 carrying the GFP marker were performed at least three times.

Evans Blue and Trypan Blue Staining, Autofluorescence Detection and Microscopy

For detection of cell death, moss colonies were incubated for 2 hours with 0.05% Evans blue and washed 4 times with deionized water to remove excess and un-bound dye. Growth and development of B. cinerea mycelium inside leaf tissues was monitored by staining with lactophenol-trypan blue and destaining in satu-rated chloral hydrate as described previously [74]. For autofluorescent compound detection, leaves were boiled in alcoholic lactophenol and rinsed in ethanol and water [75]. Material was then mounted on a slide in 50% glycerol and examined for Evans blue or trypan blue staining or using ultraviolet epifluorescence for de-tection of autofluorescent compounds (Microscope Olympus BX61). The infec-tion of Physcomitrella leaves by GFP-tagged E.c. carotovora was visualized with a laser scanning confocal microscope FV 300 (Olympus).

RNA Gel Blot Analysis

Total RNA was isolated from control and treated plant tissue corresponding to 64 moss colonies, using standard procedures based on phenol/chloroform extraction followed by LiCl precipitation. Ten micrograms of total RNA separated by dena-turing agarose-formaldehyde gels was transferred to a nylon membrane (Hybond

N) following standard procedures [76]. Membranes were prehybridized at 65°C in 6 × SCC, 0.5% SDS, 0.125 mg milk powder and 0.5 mg ml-1 denatured salmon sperm DNA. Hybridizations were performed at 65°C overnight. The DNA fragments to be used as probes were obtained by PCR using the plasmid harbouring the corresponding cDNA as template and the primers M13 forward and reverse. The cDNA clones used were: [DDBJ:BJ182301 (PR-1), DDBJ:BJ201257 (PAL), DDBJ:BJ192161 (CHS) and DDBJ:BJ159508 (LOX)]. PCR fragments were purified using Qiaquick columns (Qiagen), and were labelled with [α32P]-dCTP using Rediprime II Random Prime labelling system (Amersham Biosciences). After hybridization, membranes were washed twice for 30 min at 65°C with 5 × SCC, 0.1% SDS and twice 30 min with 2 × SCC, 0.1% SDS. Subsequently, membranes were exposed on autoradiography film. The amount of RNA loaded was verified by addition of ethidium bromide to the samples and photography under UV light after electrophoresis.

Authors' Contributions

IPDL participated in the Northern blot analysis and the microscopic studies, designed this study, drafted and edited the manuscript. JPO carried out the analysis of symptom development and all the microscopic studies. AC and MB participated in the cell death analysis by Evans blue staining. CG participated in the Northern blot analysis. SV helped to draft the manuscript. All authors read and approved the final manuscript.

Acknowledgements

We gratefully acknowledge E. Tapio Palva, Tarja Kariola and Anne Tuikkala for their generous gift of GFP-tagged E.c. carotovora strain. We thank E. Tapio Palva for the E.c. carotovora strains and Luiz Diaz for the B. cinerea isolate. We would also thank Tomas Cascón for excellent technical assistance, José Roberto Sotelo-Silveira and Anabel Fernández for confocal microscopy assistance. We are grateful to Marcos Montesano and Paul Gill for critical reading of the manuscript and to Carmen Castresana for helpful discussions. This work was supported by Fondo Clemente Estable (Project 9008) DINACYT. The Physcomitrella ESTs were obtained from the RIKEN Biological Research Center.

References

1. Lee HI, Leon J, Raskin I: Biosynthesis and metabolism of salicylic acid. Proc Natl Acad Sci USA 1995, 92(10):4076–4079.

2. Creelman RA, Mullet JE: Biosynthesis and action of jasmonates in plants. Annu Rev Plant Physiol Plant Mol Biol 1997, 48:355–381.

3. Enyedi AJ, Yalpani N, Silverman P, Raskin I: Signal molecules in systemic plant resistance to pathogens and pests. Cell 1992, 70(6):879–886.

4. Linthorst HJM: Pathogenesis-related proteins of plants. Crit Rev Plant Sci 1991, 10:123–150.

5. Dixon RA, Paiva NL: Stress-induced phenylpropanoid metabolism. Plant Cell 1995, 7(7):1085–1097.

6. Goodman RN, Novacky AJ: The Hypersensitive Reaction in Plants to Pathogens: A Resistance Phenomenon. St. Paul: American Phytopathological Society Press; 1994.

7. Dangl JL, Dietrich RA, Richberg MH: Death don't have no mercy: Cell death programs in plant-microbe interactions. Plant Cell 1996, 8(10):1793-1807.

8. Dangl JL, Jones JD: Plant pathogens and integrated defence responses to infection. Nature 2001, 411(6839):826–833.

9. Nimchuk Z, Eulgem T, Holt IB, Dangl JL: Recognition and response in the plant immune system. Annu Rev Genet 2003, 37:579–609.

10. Greenberg JT: Programmed cell death in plant-pathogen interactions. Annu Rev Plant Physiol Plant Mol Biol 1997, 48:525–545.

11. Lam E, Kato N, Lawton M: Programmed cell death, mitochondria and the plant hypersensitive response. Nature 2001, 411:848–853.

12. Wang H, Li J, Bostock RM, Gilchrist DG: Apoptosis: A functional paradigm for programmed plant cell death induced by a host-selective phytotoxin and invoked during development. Plant Cell 1996, 8(3):375–391.

13. Greenberg JT, Yao N: The role and regulation of programmed cell death in plant-pathogen interactions. Cell Microbiol 2004, 6(3):201–211.

14. Heckman DS, Geiser DM, Eidell BR, Stauffer RL, Kardos NL, Hedges SB: Molecular evidence for the early colonization of land by fungi and plants. Science 2001, 293(5532):1129–1133.

15. Cove D, Benzanilla M, Harries P, Quatrano R: Mosses as model systems for the study of metabolism and development. Annu Rev Plant Biol 2006, 57:497–520.

16. Schaefer DG, Zrÿd JP: The moss Physcomitrella patens, now and then. Physcomitrella patens 2001, 127(4):1430–1438.

17. Schaefer DG: A new moss genetics: targeted mutagenesis in Physcomitrella patens. Annu Rev Plant Biol 2002, 53:477–501.

18. Quatrano RS, McDaniel SF, Khandelwal A, Perroud PF, Cove DJ: Physcomitrella patens: mosses enter the genomic age. Current Opinion in Plant Biology 2007, 10:182–189.

19. Nishiyama T, Fujita T, Shin-I T, Seki M, Nishide H, Uchiyama I, Kamiya A, Carninci P, Hayashizaki Y, Shinozaki K, Kohara Y, Hasebe M: Comparative genomics of Physcomitrella patens gametophytic transcriptome and Arabidopsis thaliana: implication for land plant evolution. Proc Natl Acad Sci USA 2003, 100(13):8007–8012.

20. Lang D, Eisinger J, Reski R, Rensing SA: Representation and high-quality annotation of the Physcomitrella patens transcriptome demonstrates a high proportion of proteins involved in metabolism in mosses. Plant Biology 2005, 7(3):238–250.

21. Reski R: Physcomitrella and Arabidopsis: the David and Goliath of reverse genetics. Trends Plant Sci 1998, 3:209–210.

22. Akita M, Valkonen JP: A novel gene family in moss (Physcomitrella patens) shows sequence homology and a phylogenetic relationship with the TIR-NBS class of plant disease resistance genes. J Mol Evol 2002, 55(5):595–605.

23. Tsuneda A, Chen MH, Currah RS: Characteristics of a disease of Spagnum fuscum caused by Scleroconidioma sphagnicola. Can J Bot 2001, 79:1217–1224.

24. Polischuk V, Budzanivska I, Shevchenko T, Oliynik S: Evidence for plant viruses in the region of Argentina islands, Antartica. FEMS Microbiol Ecol 2007, 59:409–417.

25. Pérombelon MCM, Kelman A: Ecology of the soft-rot Erwinia. Annu Rev Phytopathol 1980, 12:361–387.

26. Toth IK, Bell KS, Holeva MC, Birch PRJ: Soft rot erwiniae: From genes to genomes. Mol Plant Pathol 2003, 4(1):17–30.

27. Collmer A, Keen NT: The role of pectic enzymes in plant pathogenesis. Annu Rev Phytopathol 1986, 24:383–409.

28. Palva TK, Holmström KO, Heino P, Palva ET: Induction of plant defense response by exoenzymes of Erwinia carotovora ssp. carotovora. Mol Plant-Microbe Interact 1993, 6(2):190–196.

29. Vidal S, Ponce de León I, Denecke J, Palva ET: Salicylic acid and the plant pathogen Erwinia carotovora induce defense genes via antagonistic pathways. Plant J 1997, 11(1):115–123.

30. Vidal S, Eriksson ARB, Montesano M, Denecke J, Palva ET: Cell wall-degrading enzymes from Erwinia carotovora cooperate in the salicylic acid-indepen-

dent induction of a plant defense response. Mol Plant-Microbe Interact 1998, 11(1):23–32.

31. Montesano M, Brader G, Ponce de Leon I, Palva ET: Multiple defense signals induced by Erwinia carotovora ssp. carotovora in potato. Molecular Plant Pathol 2005, 6(5):541–549.

32. Norman C, Vidal S, Palva ET: Interacting signal pathways control defense gene expression in Arabidopsis in response to the plant pathogen Erwinia carotovora. Mol Plant-Microbe Interact 2000, 13(4):430–438.

33. Desikan R, Reynolds A, Hancock JT, Neill SJ: Harpin and hydrogen peroxide both initiate programmed cell death but have differential effects on defense gene expression in Arabidopsis suspension cultures. Biochem J 1998, 330:115–120.

34. Krause M, Durner J: Harpin inactivates mitochondria in Arabidopsis suspension cells. Mol Plant-Microbe Interact 2004, 17(2):131–139.

35. Rantakari A, Virtaharju O, Vähämiko S, Taira S, Palva ET, Saarilahti HT, Romantschuk M: Type III secretion contributes to the pathogenesis of the soft-rot pathogen Erwinia carotovora Partial characterization of the hrp gene cluster. Mol Plant-Microbe Interact 2001, 14(8):962–968.

36. Mattinen L, Tshuikina M, Mäe A, Pirhonen M: Identification and characterization of Nip, necrosis-inducing virulence protein of Erwinia carotovora subsp. carotovora. Mol Plant-Microbe Interact 2004, 17(12):1366–1375.

37. Elad Y, Williamson B, Tudzynski P, Delen N: Botrytis: Biology, Pathology and Control. Dordrecht Kluwer Academic Publishers; 2004.

38. van Kan JAL: Licensed to kill: the lifestyle of a necrotrophic plant pathogen. Trends in Plant Sci 2006, 11(5):247–253.

39. Reino JL, Hernández-Galán R, Durán-Patrón R, Collado IG: Virulence-toxin production relationship in isolates of the plant pathogenic fungus Botrytis cinerea. J Phytopathol 2004, 152:563–566.

40. Kars I, Geja H, Krooshof GH, Wagemakers L, Joosten R, Benen JAE, van Kan JAL: Necrotising activity of five Botrytis cinerea endopolygalacturonases produced in Pichia pastoris. Plant J 2005, 43(2):213–225.

41. Brito N, Espino JJ, Gonzalez C: The endo-beta-1,4-xylanase xyn11A is required for virulence in Botrytis cinerea. Mol Plant-Microbe Interact 2006, 19(1):25–32.

42. Mittler R, Simon L, Lam E: Pathogen-induced programmed cell death in tobacco. J Cell Science 1997, 110:1333–1344.

43. Bestwick CS, Bennett MH, Mansfield JW: Hrp mutant of Pseudomonas syringae pv phaseolicola induces cell wall alterations but not membrane damage leading to the hypersensitive reaction in lettuce (Lactuca sativa). Plant Physiol 1995, 108(2):503–516.

44. Bennett M, Gallagher M, Fagg J, Bestwick C, Paule T, Beale M, Mansfield J: The hypersensitive reaction, membrane damage, and accumulation of autofluorescent phenolics in lettuce cells challenged by Bremia lactucae. Plant J 1996, 9(6):851–865.

45. Gaff DF, Okong'o-Ogola O: The use of nonpermeating pigments for testing the survival of cells. J Exp Bot 1971, 22:756–758.

46. Feussner I, Wasternack C: The lipoxygenase pathway. Annu Rev Plant Biol 2002, 53:275–297.

47. Hahlbrock K, Scheel D: Physiology and molecular biology of phenylpropenoid metabolism. Annu Rev Plant Physiol Plant Mol Biol 1989, 40:347–369.

48. Davey ML, Currah RS: Interactions between mosses (Bryophyta) and fungi. Can J Bot 2006, 84:1509–1519.

49. He SY, Huang HC, Collmer A: Pseudomonas syringae pv. syringae harpinPss: a protein that is secreted via the Hrp pathway and elicits the hypersensitive response in plants. Cell 1993, 2(7):1255–66.

50. Wei ZM, Laby RJ, Zumoff CH, Bauer DW, He SY, Collmer A, Beer SV: Harpin, elicitor of the hypersensitive response produced by the plant pathogen Erwinia amylovora. Science 1992, 257(5066):85–88.

51. Yang CH, Gavilanes-Ruiz M, Okinaka Y, Vedel R, Berthuy L, Boccara M, Chen JW, Perna NT, Keen NT: hrp genes of erwinia chrysanthemi 3937 are important virulence factors. Mol Plant-Microbe Interact 2002, 15(5):472–480.

52. Kariola T, Palomäki TA, Brader G, Palva ET: Erwinia carotovora subsp. carotovora and Erwinia -derived elicitors HrpN and PehA trigger distinct but interacting defense responses and cell death in Arabidopsis. Mol Plant-Microbe Interact 2003, 16(3):179–187.

53. Levine A, Pennell RI, Alvarez ME, Palmer R, Lamb C: Calcium-mediated apoptosis in a plant hypersensitive disease resistance response. Curr Biol 1996, 6:427–437.

54. Morel J.-B, Dangl JL: The hypersensitive response and the induction of cell death in plants. Cell Death Differ 1997, 4:671–683.

55. de Pinto MC, Paradiso A, Leonetti P, De Gara L: Hydrogen peroxide, nitric oxide and cytosolic ascorbate peroxidase at the crossroad between defence and cell death. Plant J 2006, 48(5):784–95.

56. Yoshinaga K, Arimura SI, Hirata A, Niwa Y, Yun DJ, Tsutsumi N, Uchimiya H, Kawai-Yamada M: Mammalian Bax initiates plant cell death through organelle destruction. Plant Cell Rep 2005, 24(7):408–17.

57. Yao N, Tada Y, Park P, Nakayashiki H, Tosa Y, Mayama S: Novel evidence for apoptotic cell response and differential signals in chromatin condensation and DNA cleavage in victorin-treated oats. Plant J 2001, 28(1):13–26.

58. Navarre DA, Wolpert TJ: Victorin induction of an apoptotic, senescence-like response in oats. Plant Cell 1999, 11(2):237–250.

59. Keates SE, Kostman TA, Anderson JD, Bailey BA: Altered gene expression in three plant species in response to treatment with Nep1, a fungal protein that cause necrosis. Plant Physiol 2003, 132(3):1610–1622.

60. Kariola T, Brader G, Li J, Palva ET: Chlorophyllase 1, a damage control enzyme, affects the balance between defense pathways in plants. Plant Cell 2005, 17(1):282–294.

61. Govrin EM, Levine A: The hypersensitive response facilitates plant infection by the necrotrophic pathogen Botrytis cinerea . Curr Biol 2000, 10(13):751–757.

62. Jiang C, Schommer CK, Kim SY, Suh D-Y: Cloning and characterization of chalcone synthase from the moss, Physcomitrella patens . Phytochem 2000, 67(23):2531–2540.

63. Senger T, Wichard T, Kunze S, Gobel C, Lerchl J, Pohnert G, Feussner I: A multifunctional lipoxygenase with fatty acid hydroperoxide cleaving activity from the moss Physcomitrella patens. J Biol Chem 2005, 280(9):7588–7596.

64. Aguilar I, Poza-Carrin C, Gui A, Rodrguez-Palenzuela P: Erwinia chrysanthemi genes specifically induced during infection in chicory leaves. Mol Plant Pathol 2002, 3(4):271–275.

65. Vidal S: Molecular Defense responses against the plant pathogen Erwinia carotovora. Signal pathways in the regulation of pathogen-induced gene expression in plants. PhD thesis. Swedish University of Agricultural Sciences, Department of Plant Biology; 1998.

66. Govrin EM, Levine A: Infection of Arabidopsis with a necrotrophic pathogen, Botrytis cinerea, elicits various defense responses but does not induce systemic acquired resistance (SAR). Plant Mol Biol 2002, 48:267–276.

67. Ferrari S, Plotnikova JM, De Lorenzo G, Ausubel FM: Arabidopsis local resistance to Botrytis cinerea involves salicylic acid and camalexin and requires EDS4 and PAD2, but not SID2, EDS5 or PAD4. Plant J 2003, 35(2):193–205.

68. Schaefer D, Zryd JP, Knight CD, Cove DJ: Stable transformation of the moss Physcomitrella patens . Mol Gen Genet 1991, 226:418–424.

69. Ashton NW, Cove DJ: The isolation and preliminary characterization of auxotrophic and analogue resistant mutants in the moss Physcomitrella patens . Mol Gen Genet 1977, 154:87–95.

70. Saavedra L, Svensson J, Carballo V, Izmendi D, Welin B, Vidal S: A dehydrin gene in Physcomitrella patens is required for salt and osmotic stress tolerance. Plant J 2006, 45(2):237–249.

71. Pirhonen M, Heino P, Helander I, Harju P, Palva ET: Bacteriophage T4-resistant mutants of the plant pathogen Erwinia carotovora . Microbe Pathog 1988, 4:359–367.

72. Saarilahti HT, Palva ET: Major outer membrane proteins in the phytopathogenic bacteria Erwinia carotovora subsp. carotovora and subsp. atroseptica . FEMS Microbiol Lett 1986, 35:267–270.

73. Miller JH: Experiments in Molecular genetics. Cold Spring Harbor, Cold Spring Harbor Press; 1972.

74. Koch E, Slusarenko A: Arabidopsis is susceptible to infection by downy mildew fungus. Plant Cell 1990, 2:437–455.

75. Dietrich RA, Delaney TP, Uknes SJ, Ward ER, Ryals JA, et al.: Arabidopsis mutants simulating disease resistance response. Cell 1994, 77:565–577.

76. Sambrook J, Fitsch EF, Maniatis T: Molecular Cloning: A Laboratory Manual. Cold Spring Harbor, Cold Spring Harbor Press; 1989.

CITATION

Originally published under the Creative Commons Attribution License. de León IP, Oliver JP, Castro A, Gaggero C, Bentancor M, Vidal S. Erwinia carotovora elicitors and Botrytis cinerea activate defense responses in Physcomitrella patens. BMC Plant Biology 2007, 7:52. doi:10.1186/1471-2229-7-52.

Copyright

Index

cAMP cascade regulates pathogenicity
of, 177
maize tissue, 188

V

vanadium-dependent bromoperoxidases,
243, 255
transcriptional activation of, 261
Venn diagram, 226, 227
victorin, 16
VIGS silencing assays, 48
viral proteins, 45
viral RNAs, 31
virulence
of environmental pathogens, 14
of many Gibberella species, 16
in plant pathogens, 12
virulence factors, 14, 16, 17, 110
dual-use, 12
of human pathogens, 12
of plant pathogens, 14
in Streptomyces, 17
virus- and transgene-induced RNA silencing, 34
virus induced gene silencing (VIGS),
35–36

W

Walz Phyto-pulse amplitude modulation
fluorometer, 265

wheat, 171
leaves, gene expression of, 174
stem rust, 187
stripe rust, 170
Wheat GeneChip® technique, 145
wheat-Pst interactions, 144, 159, 178
whole-plant specific resistance (WPSR),
126
wide spectrum efflux pumps, genes coding
for, 15
efflux pump BcAtrB of Botrytis
cinerea, 15
transporter ABC1 from Magnaporthe
grisea, 15
Ws-0 ecotype, 62, 76
wt ecotypes, 70
wt-like HR and PR-1 gene expression, 65
wt-like phenotypes, 77
wt plants, 67

Y

yeast two-hybrid system, 33
Yekutieli algorithm, 268
yellow rust, 144
Yr8 resistance gene, 178

Z

zoonoses, 12

Milton Keynes UK
Ingram Content Group UK Ltd.
UKHW031143141024
449569UK00024B/1105